T0136494

Ecology and Management of Sitka Spruce, Emphasizing Its Natural Range in British Columbia

SOUTH
MORESBY
FOREST
REPLACEMENT
ACCOUNT

Funding for this publication
was provided by the
Canada - British Columbia South Moresby
Forest Replacement Account (SMFRA).

Canada

E.B. Peterson, N.M. Peterson,
G.F. Weetman, and P.J. Martin

Ecology and Management of Sitka Spruce, Emphasizing Its Natural Range in British Columbia

UBCPress / Vancouver

© UBC Press 1997

All rights reserved. No part of this publication may be reproduced, stored in a retrieval system, or transmitted, in any form or by any means, without prior written permission of the publisher, or, in Canada, in the case of photocopying or other reprographic copying, a licence from CANCOPY (Canadian Copyright Licensing Agency), 900 – 6 Adelaide Street East, Toronto, ON M5C 1H6.

Printed in Canada on acid-free paper ∞

ISBN 0-7748-0561-7

Canadian Cataloguing in Publication Data

Main entry under title:
Ecology and management of sitka spruce,
emphasizing its natural range in British Columbia

Includes bibliographical references and index.
ISBN 0-7748-0561-7

1. Sitka spruce – Ecophysiology – British Columbia. 2. Sitka spruce – British Columbia. 3. Forest management – British Columbia. 4. Forest ecology – British Columbia. I. Peterson, E. B. (Everett B.)
SD 397.S56E26 1997 634.9′752 C97-910472-6

UBC Press gratefully acknowledges the ongoing support to its publishing program from the Canada Council for the Arts, the British Columbia Arts Council, and the Department of Canadian Heritage of the Government of Canada.

Set in Stone by Irma Rodriguez, Artegraphica Design
Printed and bound in Canada by Friesens
Copy-editor: Francis J. Chow
Indexer: Annette Lorek

UBC Press
University of British Columbia
6344 Memorial Road
Vancouver, BC V6T 1Z2
(604) 822-5959
Fax: 1-800-668-0821
E-mail: orders@ubcpress.ubc.ca
http://www.ubcpress.ubc.ca

Contents

Acknowledgments

The assembly of information for this book began in the late 1980s at the initiative of the Pacific Forestry Centre, Canadian Forest Service, which enabled W. Stanek to focus on Sitka spruce. His initial compilation was supplemented by additional support from 1989 to 1996 through a contract to Western Ecological Services Ltd., Victoria, from the South Moresby Forest Replacement Account. The publication of this work would not have been possible without the interest, scientific support, and editorial suggestions of R.B. Smith, first as the Pacific Forestry Centre's Scientific Authority for the project and, in 1996, as technical and scientific advisor. In 1994 R. De Jong of the Pacific Forestry Centre also served as Scientific Authority. We also thank F. Nuszdorfer, K. Weaver, and G. Wiggins of the British Columbia Ministry of Forests, who were contract administrators. The suggestions of all these foresters are much appreciated and are now a part of this book.

This book incorporates some of the information originally compiled by W. Stanek. We are grateful to him for beginning the assembly of 1980s information about Sitka spruce, at a time when not many foresters in British Columbia were giving attention to this species. This book has also benefited from the input of many other persons. Discussions between W. Stanek and the following persons provided early guidance and information: V. Korelus, Forester and Nursery Superintendent, Canadian International Paper Inc., Saanichton; C. Ying, Ministry of Forests, Victoria; B. Voth, Western Forest Products Limited, Holberg; B. Wilson, Ministry of Forests, Terrace; D. Ashbee, Ministry of Forests, Victoria. W. Stanek's late-1980s Sitka spruce research was assisted by the contract work of J.C. Bartlett and P.J. McFie. We also wish to thank librarians S. Barker of the Ministry of Forests and A. Solyma of the Pacific Forestry Centre, and their staffs, who assisted in locating literature sources. In 1989 R.B. Smith and E.B. Peterson had the opportunity to view various aspects of Sitka spruce management on northern Vancouver Island, thanks to a field tour provided by S. Joyce of Port McNeill and K. McGourlick of Holberg on behalf of Western Forest Products Limited.

This book benefited from the information provided to W. Stanek during the International Union of Forest Research Organizations' Sitka Spruce Working Group meeting in Edinburgh in 1986. In particular, the assistance of A.M. Fletcher and R. Lines, both of the Forestry Commission, Edinburgh, and J. O'Driscoll, Forest and Wildlife Service, Dublin, was much appreciated. We also extend thanks to the staff and specialists at the Northern Research Centre, Roslin, Scotland, for their patience in answering questions during two days of interviews by G.F. Weetman in 1993.

If unidentified, photographs are those of the authors. We wish to acknowledge photographic contributions from the following: A. Dorst for Plate 13; British Columbia Archives and Records Services for Figures 3, 4, and 49; Image Library, Ministry of Forests, for Plates 3 and 8; R.B. Smith, formerly of the Canadian Forest Service, for Plates 1, 2, 4, 5, and 9-11, and Figures 7, 8, 10, and 23-26; F. Pendl, formerly of the B.C. Ministry of Forests, for Figures 30, 38, and 51; and J. Barker of Western Forest Products Limited for Figures 36 and 50. From the British Columbia Archives and Records Services, we thank Kelly Nolin for assistance in searching for Sitka spruce photographs, and Brian Young for waiving the permission fee for republication of three photographs from the archives collection.

We thank the following organizations for permission to republish certain figures: Canadian Forest Service (Figures 5, 6, 31, 33, 34, and 48); Research Branch, Ministry of Forests (Plate 7, and Figures 2, 18-22, 27-29, and 44); Institute of Terrestrial Ecology, Edinburgh (Figure 40); Forestry Commission, Edinburgh (Figures 37, 39, and 45); Faculty of Forestry, University of British Columbia (Figure 16); and Botanical Garden, University of British Columbia (Figure 9). Figures previously published by researchers of the United States Forest Service are acknowledged in the figure legends by author and date; permission to republish these figures was not specifically requested as they are in the public domain.

The portions of this book dealing with crop planning and growth and yield of Sitka spruce stands benefited from advice on various topics from the following persons:

C. Bartram, B.C. Ministry of Forests, Inventory Branch (Variable Density Yield Prediction – VDYP); M. DiLucca, B.C. Ministry of Forests, Research Branch (Research Branch Sitka spruce [Ss] data set and Tree and Stand Simulator – TASS); P. Kofoed, MacMillan Bloedel Ltd. (MB), Woodland Services (regeneration model for Ss in MB timber supply analysis); S. Northway, MacMillan Bloedel Ltd., Woodland Services (MB growth and yield modelling for Ss); N. Smith, MacMillan Bloedel Ltd., Woodland Services (MB growth and yield program and site quality evaluation); B. Wilson, MacMillan Bloedel Ltd., Woodland Services (MB PSP data set for Ss); B. Wilson, B.C. Ministry of Forests, Inventory Branch (Forest Productivity Councils' Ss database).

Mensurational advice was provided by K. Mitchell and G. Nigh, Forest Productivity and Decision Support Section, Ministry of Forests. J. Barker of Western Forest Products Limited provided the net merchantable volume data shown in Figure 43 and Table 33. A. Nussbaum and G. Nigh produced the revised Sitka spruce site index table and curves shown in Figure 44 and Table 36. C. Farnden kindly made available the stand density management diagrams (Figures 41 and 42) in advance of their publication elsewhere.

The manuscript benefited greatly from technical review by the following persons who provided suggestions for improvement of specific sections or subsections: from the Pacific Forestry Centre – R. Alfaro, J. Arnott, M. Bonnor, B. Callan, R. De Jong, and R.B. Smith; from the B.C. Ministry of Forests – K. Klinka, F. Nuszdorfer, F. Pendl, and K. Weaver; from the Northern Research Station, Forestry Commission, Scotland – C. Booth, A. Fletcher, and D. Redfern; and from Western Forest Products Limited – J. Barker.

For final editing of the manuscript, we are also very appreciative of the support and advice of J. Wilson and H. Keller-Brohman of UBC Press, and F. Chow, copy-editor. We also thank A. Lorek for preparation of the subject index. Our last note of thanks, the most important of all, is to the Sitka spruce researchers whose cited reports are the basis of this text.

Ecology and Management of Sitka Spruce, Emphasizing Its Natural Range in British Columbia

1
Introduction

Sitka spruce is an ecologically important component of North America's north temperate coastal rain forest. This tree species, *Picea sitchensis* (Bong.) Carr., reaches the largest size of any of earth's spruces. Aboriginal people were inescapably involved with this component of the west coast rain forest, as were later harvesters of the west coast's valuable timber resource. Today, too, Sitka spruce is part of the forest management challenge faced by resource managers who seek to maintain and renew these productive forest ecosystems in a sustainable way.

Sitka spruce's ecological and economic role is not dominant on a provincial scale in British Columbia. However, in certain regions of the province, and often in very productive ecosystems, this species has an importance for forestry far beyond its provincial ranking among British Columbia's major commercial tree species. This is especially true for Sitka spruce on the Queen Charlotte Islands. Because of this, Forestry Canada and the British Columbia Ministry of Forests, through a joint federal-provincial program known as the South Moresby Forest Replacement Account (SMFRA), decided to update several earlier publications that had focused on Sitka spruce (Day 1957; Ruth 1958; Harris 1964; Phelps 1973; Ruth and Harris 1979; Harris 1984).

This book is the result. It reiterates some of the information from the earlier reports on this species, but the emphasis is on recently acquired knowledge. Of the 834 references cited in this text, 41% were published in the 1990s and 11% are very recent (1995 and the first half of 1996). Clearly, Sitka spruce remains an active research subject. A forest manager or researcher who wishes to assemble a basic library of publications on Sitka spruce should supplement this text with the 1957 to 1984 reports cited above.

There is broad public interest in the north temperate rain forest where Sitka spruce occurs, but a specific practical purpose of this publication was to serve SMFRA's goal to increase wood production on the Queen Charlotte Islands to offset the loss of harvestable forest land resulting from the creation of Gwaii Haanas/South Moresby National Park Reserve, a 145,000 ha area set aside on 12 July 1988. A broader purpose is to draw attention to Sitka spruce's ecological niche in north temperate rain forest ecosystems, beyond its role as a producer of high-quality wood. For this reason, the first half of the book is devoted to the biology and ecology of naturally occurring Sitka spruce. The second half focuses on its management as a renewable and sustainable resource.

Sitka spruce is a prominent tree in certain ecosystems along the western coast of North America (Plate 1). The common name, Sitka, is derived from the name of Sitka Island (now Baranoff Island), Alaska, where an early botanical specimen of this species was collected. Two of this spruce's other common names – coast spruce and tideland spruce – are descriptive of its relatively narrow natural range along the coast (Little 1953; Ruth 1958). On the immediate coast, in areas of brackish water or within the narrow band subjected to sea spray, Sitka spruce appears to benefit more than other conifers from the relatively high amounts of available calcium and magnesium (Krajina et al. 1982). The literature contains frequent reference to Sitka spruce's association with the fog belt zone close to the outer coast (Ruth 1954; Bodsworth 1970; Franklin and Dyrness 1973; Harris 1990) and, where it occurs in pure stands, with the salt spray zone on the immediate coastline (Cordes 1972; Nelson and Cordes 1972; Harris 1990). Sitka spruce's abundance on the coastal fringe, such as on western Graham Island, Queen Charlotte Islands (Plate 2), is typical of its occurrence in the northern parts of the Very Wet Hypermaritime Subzone (CWHvh) of the Coastal Western Hemlock Biogeoclimatic Zone (Pojar et al. 1988; Banner et al. 1993) (see Appendix 4 for names and abbreviations of British Columbia's biogeoclimatic units).

The natural range of Sitka spruce from south to north is about 2,900 km, but its maximum range from west to east is only about 400 km if one measures from the western edge of the Queen Charlotte Islands to the easternmost limit of Sitka spruce along the Bulkley River north of Smithers, British Columbia (Figure 1). If one excludes the Queen Charlotte Islands from this approximation of west-east range, the greatest west-to-east extent of Sitka spruce on the mainland is in the panhandle of southeastern Alaska, where it occupies a zone about 200 km wide when measured from the western shores of the islands to the inland limits of this spruce.

The southernmost Sitka spruce trees occur near Casper, Mendocino County, California (latitude 39°20′N). The species occurs north to Prince William Sound, to about latitude 61°00′N, and west as far as about longitude 155°00′W on the Alaskan coast opposite Kodiak Island (Figure 1). Based on observations 60 years ago, the westernmost naturally occurring Sitka spruce clump recorded by Griggs (1934) in Alaska was at Cape Kubugakli. Griggs thought that these spruce likely originated from seed blown from Kodiak Island, about 80 km across Shelikof Strait, because at that time the main stands of Sitka spruce on the mainland extended west only to Iniskin

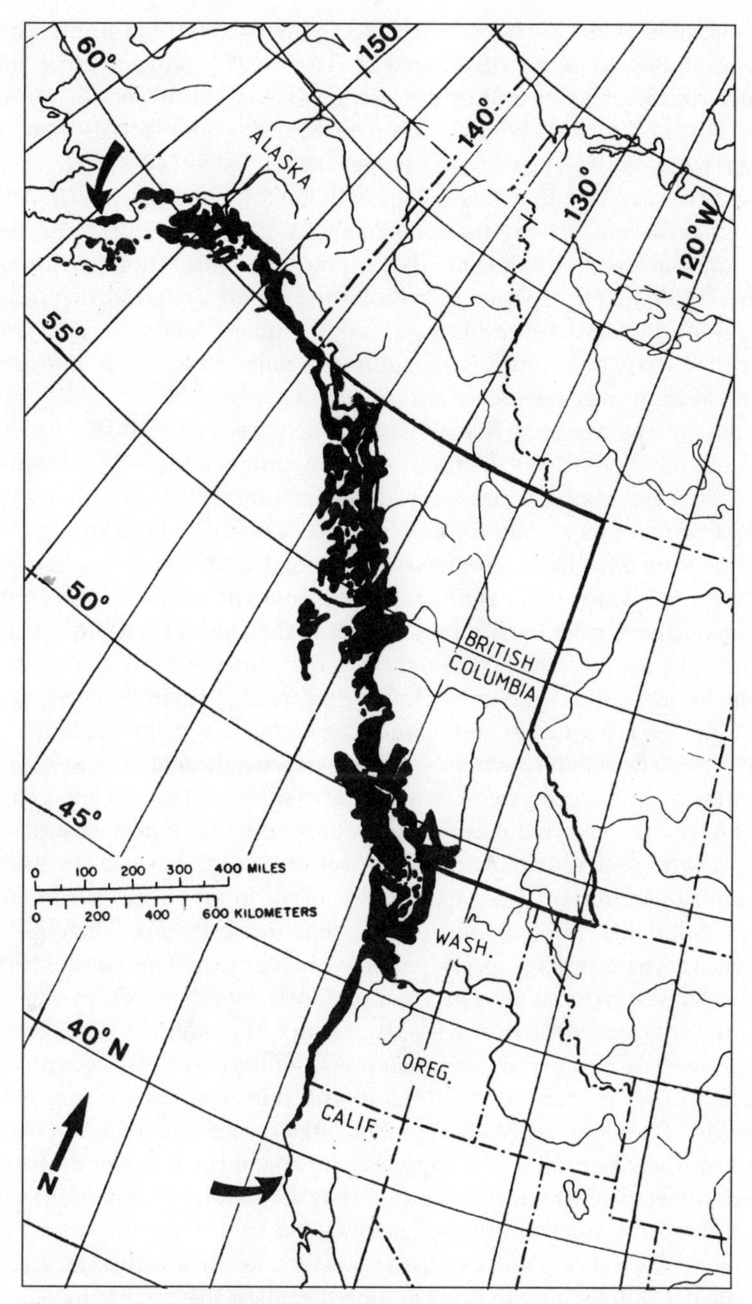

Figure 1. Natural range of Sitka spruce from near Casper, California (lat. 39°20′N), to Prince William Sound, Alaska, where its northern limit is near lat. 61°N and its western limit is at Cape Kubugakli (about long. 155°W) (adapted from Harris 1984). Arrows indicate southern and northwestern range limits.

Bay (latitude about 59°40′N, longitude 153°30′W), 160 km northeast of Kubugakli Bay. More recently, Farr and Harris (1983) indicated that Sitka spruce continues to expand to the west in Alaska, but no specific records were obtained for this book to compare with the observations made by Griggs 60 years ago. The record by Griggs (1934) that Sitka spruce were still alive 125 years after they were planted in 1805 by Russians on Unalaska Island, near Unalaska Bay (latitude about 53°50′N, longitude 166°30′W), 800 km southwest of Kubugakli Bay, supports the suggestion by Farr and Harris (1983) that the species is likely able to extend its range further southwest along the Alaska Peninsula towards the Aleutian Islands. Alaback (1996) suggested that Sitka spruce is expanding its range westward from Kodiak Island at an average rate of 1.6 km per century.

Sitka spruce is the state tree of Alaska, where, because relatively productive sites have a restricted distribution in a narrow coastal zone, it is one of the few species available for high-yield silviculture at such a northerly latitude. Except for its affinity for near-shore ecosystems, it faces strong competition from other native conifers that reforest disturbed areas. As described in Chapter 2, Sitka spruce is much more prominent in upland ecosystems in the northern parts of its range, including the Queen Charlotte Islands, than in the southern half of its natural range. These upland sites are very prone to debris slides and gully erosion (Figure 2), a circumstance that results in a high frequency and abundance of erosion-related disturbances that provide favourable surfaces for natural regeneration of spruce (Plate 3).

Recently various ecosystems in which Sitka spruce is a significant component have come to symbolize the last opportunities to protect examples of old-growth coniferous rain forests in western North America. In British Columbia magnificent Sitka spruce trees occur in all of the coastal areas where Aboriginal interests and other public interests have converged in recent years to encourage special protective measures for remnants of North American west coast rain forests. Examples are the southern part of Moresby Island in the Queen Charlotte Islands, Meares Island off the west coast of Vancouver Island, and the Carmanah Creek drainage basin on western Vancouver Island. In western Washington and northwestern Oregon, forest ecologists have emphasized that coastal Sitka spruce rain forests in which Sitka spruce is a prominent component are one of the very significant coniferous forests of the world, and that they are unique among the world's mesic temperate forests. Although mature Sitka spruce trees are not as large as coast redwood or giant sequoia in western North America, they command attention because they are among the tallest trees on earth.

Throughout most of its range, Sitka spruce is associated with western hemlock in stands that have some of the highest growth rates in North America. This spruce is a valuable commercial species for lumber, pulp, and special uses. The high strength-to-weight ratio and the resonant qualities of clear

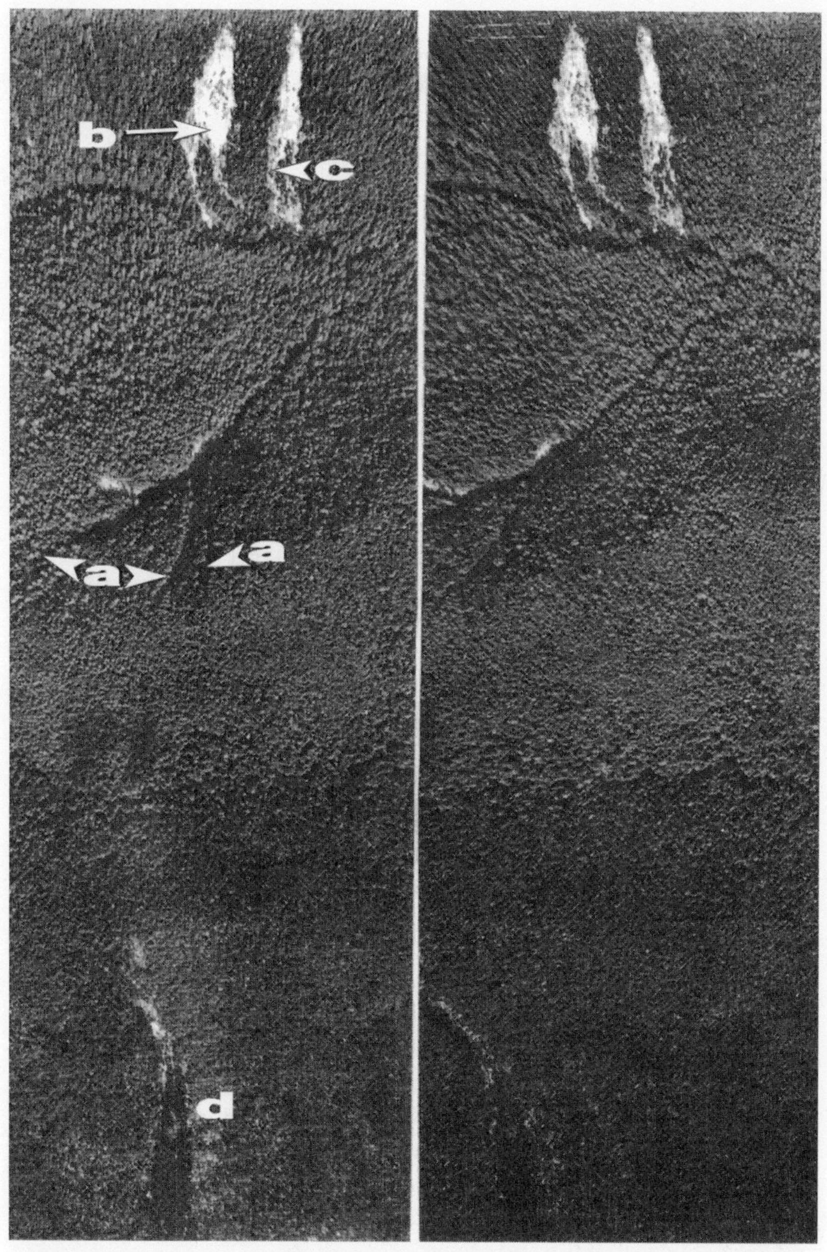

Figure 2. Stereogram of debris slides and gully erosion on Burnaby Island in the Queen Charlotte Islands, British Columbia, where Sitka spruce is a prominent component of forested slopes (Gimbarzevsky 1988; from British Columbia 1:12,000 colour negatives BCC 325, no. 41-42). Natural erosion occurs as steep V-notch gullies (a) that often extend into streams and debris slides (b, c, and d) that are in various stages of natural revegetation.

lumber are attributes that have traditionally made Sitka spruce wood valuable for specialty uses. Today, such uses include sounding boards for high-quality pianos, guitar faces, ladders, and components of experimental light aircraft.

One current example of the excellent value of high-quality Sitka spruce logs is the $850-1,100 per m^3 offered for 'Hi-grade' Sitka spruce on 28 June 1996 in the Vancouver area, at a time when the highest values for top-grade logs of other species were as follows: cypress (yellow-cedar), $570-910 per m^3; Douglas-fir, $525-850 per m^3; western hemlock/grand fir, $260-325 per m^3; and western redcedar, $267-300 per m^3 (Forest Industry Trader 1996). Another example is the high value-added use of Sitka spruce wood by a guitar manufacturer in east Vancouver, a company that in recent years employed 30 people and produced 15 guitars a day. The guitars are made from top-quality Sitka spruce stems purchased from log booms in the Fraser River. In some cases the company has paid as much as $700 for a single Sitka spruce log. The resulting premium guitars retail from $1,600 to $6,400 each, an excellent example of value-added use of a renewable and sustainable forest resource (Forest Renewal B.C. Newsletter 1996). Other products are oars, planking, masts, and spars for boats, and turbine blades for wind-energy systems (Harris 1984).

Because of these potential uses, and also because of its ecological adaptability to a wide variety of sites outside its natural range, Sitka spruce has become a commercially important introduced species in northern Europe and in several other regions of the world. Hermann (1987) provided a review of the significance of Sitka spruce in European forestry. Sitka spruce is particularly notable for its ability to grow very well on exposed moorland conditions with harsh climates. Countries in which Sitka spruce has been planted for commercial plantations, provenance trials, or amenity values include the following: **Australia** (Pederick 1979), **Belgium** (Nanson 1976), **Bulgaria** (Alexandrov 1993), **Chile** (Carrasco 1954), **Denmark** (Kjersgard 1976), **France** (Birot 1976), **Germany** (Kleinschmit and Sauer 1976), **Great Britain** (Macdonald 1931; Davies 1967, 1972; Pearce 1976; Hill and Jones 1978), **Iceland** (Benedikz 1976; Lines 1968), **Ireland** (O'Driscoll 1976a, 1976b), **Latvia** (Pirags 1976), **Netherlands** (Wilmes 1953; Kriek 1976), **New Zealand** (Weston 1957), **Northern Ireland** (Savill 1976), **Norway** (Bergan 1970; Magnesen 1976), **Poland** (Ilmurzynski et al. 1968), **Russia** (Kolomiets and Bogdanova 1979), **Sweden** (Karlberg 1961), **Turkey** (Eyuboglu 1986), and **Yugoslavia** (Pintaric and Mekic 1993).

Within Canada, but outside Sitka spruce's natural range, this species is known to have been successfully planted in Newfoundland (Khalil 1977; Richardson et al. 1984; Hall 1990) and in southwestern Nova Scotia (observations of senior author).

The preparation of this book began with recognition that a forest manager responsible for silvicultural interventions where Sitka spruce occurs naturally in coastal British Columbia could benefit from experience and information obtained from management of this species beyond its natural range. The available information was assessed in the following order of decreasing priority:

- British Columbia experience in documenting the biogeoclimatic and ecosystem characteristics of forest types in which Sitka spruce is a prominent component
- British Columbia experience in prescribed silvicultural interventions to manage Sitka spruce in relation to its adaptation to ecosystems where it occurs naturally, its resilience in recovering from disturbance, its stability and resistance to change, and the biological risks in encouraging this spruce for commercially viable and sustainable management
- North American west coast experience, from northern California to southeastern Alaska, which provides an additional knowledge base about the silvics and ecological characteristics of Sitka spruce in its natural range
- United Kingdom and other European experience in silviculture and management of plantation Sitka spruce.

Because of its widespread use internationally, Sitka spruce commands attention out of proportion with its somewhat restricted natural range along the west coast of Canada and the United States. Even if Sitka spruce were not a major component of the forest standing crop in western North America, however, this attention would be justified. As with other west coast trees, there are ancient Aboriginal links with this species (Turner 1979; Turner et al. 1990). Sitka spruce was involved in what may have been the first attempt to plant conifers in west coastal North America, when the Russians planted it in 1805 on Unalaska Island, in Alaska's Aleutian Islands chain. In British Columbia, Sitka spruce was among the conifers used in the very first production of nursery seedlings, in the late 1920s. It was also among the first trees logged in coastal British Columbia (Figure 3). During the Second World War, it played a vital economic and strategic role because of the high demand for its wood for aircraft construction (Figure 4). For several decades, Sitka spruce seed was an important British Columbia export, but now the United Kingdom and other countries of the European Economic Community (EEC) have enough registered seed stands to supply most of their needs; special permission must now be obtained to import Sitka spruce seed directly from British Columbia (T.C. Booth, pers. comm., May 1991).

To return to the original practical goal of the SMFRA project that led to this book – silvicultural ways to increase Sitka spruce wood production to

Figure 3. Harvesting of Sitka spruce began with the earliest logging in coastal British Columbia, as portrayed by this early scene, exact date and location unknown, from the Queen Charlotte Islands. (British Columbia Archives and Records Services, Photograph NA-05998, Catalogue no. FS-01972-0, photographer unknown)

Figure 4. Harvesting of large Sitka spruce reached a peak during the Second World War, when its wood was in high demand for aircraft construction. This spruce, 2.7 m in diameter at its base, was felled on the Queen Charlotte Islands in 1947. (British Columbia Archives and Records Services, Photograph I-029017, Catalogue no. 03366, B25-A0165, photographer unknown)

compensate for the removal of spruce-dominated ecosystems from the working forest by creation of a national park reserve – how can forest managers apply today's knowledge of Sitka spruce ecology? This question is the focus of the second half of this book. In the context of forested lands where harvesting is no longer an option because of the creation of the Gwaii Haanas/South Moresby National Park Reserve on the Queen Charlotte Islands, the silvicultural steps suggested here are one way to regain some of the allowable annual cut (AAC) that has been lost. This potential recovery of AAC can occur only if enhanced production through silviculture, sometimes referred to as the *allowable cut effect* (ACE), is taken into account. For example, the United States Forest Service uses ACE in determining AAC in the Mount Hood National Forest. However, in British Columbia there is reluctance to recognize ACE because of inadequate inventory data, limited available knowledge of baseline productivity, limited knowledge of yields of managed stands, and various unknown risks. One particular risk is the possibility that extra allowable cut today generated by more intensive silviculture practices may not be used by present tenure holders who carry out the work. This book does not address these issues comprehensively because they apply to many different ecosystems and to several major commercial tree species in coastal British Columbia. However, silvicultural steps that can be taken to increase the allowable cut effect in forests where Sitka spruce is a significant component are referred to in Chapter 3.

Sitka Spruce Supply in Its Natural Range

Compared with several other conifers with which it is associated in its natural range, Sitka spruce is not a major component of the timber supply of the Pacific Northwest and British Columbia. As a proportion of the total commercial timber supply, it is relatively more important in Alaska than further south. For example, the volume of Sitka spruce sawtimber on commercial forest land in the United States was estimated by Harris (1984) to be 71,341 million board feet, based on the International ¼-inch Rule. Conversion of this volume to metric units can be only approximate, but if one assumes 5.663 board feet (log scale) per cubic foot, the above volume estimate for Sitka spruce translates to 12,598 million cubic feet or 357 million m^3. Harris estimated that 89% (318 million m^3) of this volume was in Alaska and 11% (39 million m^3) was in Washington, Oregon, and California. The Council of Forest Industries of British Columbia (1990) estimated that the 1989 volume of mature standing Sitka spruce in commercial forests in British Columbia was 146 million m^3. Table 1 shows the approximate distribution of the North American Sitka spruce timber supply, based on early 1980s estimates for United States Sitka spruce (Harris 1984) and 1989 estimates for British Columbia (Council of Forest Industries of British Columbia 1990).

Table 1

Approximate distribution of North American Sitka spruce timber supply

Jurisdiction	Estimated volume of Sitka spruce timber supply (million m^3)	Percent of western North America Sitka spruce timber supply
Alaska	318	63.2
British Columbia	146	29.0
Washington, Oregon, and California	39	7.8
Total	503	100.0

Source: Estimates from Harris (1984) for the United States and the Council of Forest Industries of British Columbia (1990) for British Columbia

The estimated 146 million m^3 of Sitka spruce mature standing timber in British Columbia represents 4.76% of the 3,061 million m^3 total timber supply from all species on the coast (1989 estimate), and only 1.70% of the estimated 1989 provincial total of 8,588 million m^3 of mature timber supply from all forest species in all of British Columbia (Council of Forest Industries of British Columbia 1990). By comparison, Sitka spruce makes up more than 20% of the forest standing crop in Alaska's coastal western hemlock–Sitka spruce zone (Viereck and Little 1972).

When expressed in terms of 1989 log production, data compiled by the Council of Forest Industries indicate that production, in thousands of cubic metres, from species in the coastal portions of the Vancouver and Prince Rupert forest regions was as shown in Table 2.

These data indicate that Sitka spruce made up about 3.9% of the 28,313,000 m^3 produced by the top five coastal sources of logs in 1989.

Table 2

Log production, in thousands of cubic metres, from species in the coastal portions of the Vancouver and Prince Rupert forest regions in 1989

	Vancouver	Prince Rupert	Total coast
western hemlock	10,645	650	11,292
western redcedar	6,603	201	6,804
true firs	4,775	372	5,147
Douglas-fir	3,964	–	3,964
Sitka spruce	**925**	**181**	**1,106**
yellow cedar	932	14	946
hardwoods	432	–	432
cottonwoods	117	18	135
white pine	53	–	53
unspecified	34	4	38
lodgepole pine	21	2	23

Approach Used to Assemble and Select Information for This Book

In assembling information on the ecology and management of Sitka spruce in British Columbia, the highest priority was placed on research results from this province. The next highest priority was placed on information from Alaskan and Pacific Northwest ecosystems that support Sitka spruce, with preference being given to neither of the two data sources. It was assumed that information from Washington and Oregon would be very applicable to the southern part of Sitka spruce's range in British Columbia, approximately to latitude 51°30′N, south of which the Very Wet Hypermaritime Subzone of the Coastal Western Hemlock Biogeoclimatic Zone has a limited distribution near the outer coastline of Vancouver Island peninsulas. Northward from the north end of Vancouver Island (latitude 51°30′N), where the Very Wet Hypermaritime Subzone extends inland for a greater distance than it does further south, information from Alaskan data sources was considered to be more applicable to British Columbia than that from Washington and Oregon. Although this latitudinal distinction is not precise, it is an ecologically defensible guideline for the Sitka spruce manager concerned about the applicability of research results from south or north of any particular area of concern.

The lowest priority was given to research results from locations where Sitka spruce is an introduced species. These sources of information could not be totally ignored, because more is known about Sitka spruce's growth and responses to management from plantations in Europe than from natural stands of the species in British Columbia. Measured by numbers of published articles and reports, there is more information about Sitka spruce from locations where it is an introduced plantation species than from its natural range. The degree to which experience from Sitka spruce plantations in Europe can be applied to Sitka spruce management in western North America depends on the degree to which there are ecological parallels and equivalent management objectives. Thus we were very selective in including in this review ecological and management information based on European experience with Sitka spruce plantations. Most of the European information incorporated here deals with silvicultural response characteristics of Sitka spruce that are thought to have a sufficiently strong genetic basis to apply to the species wherever it occurs. We identify the geographic source of European Sitka spruce information summarized here.

This book acknowledges the public interest in Sitka spruce ecosystems that goes far beyond silviculture, crop planning, and the setting of allowable annual cuts. Sitka spruce's role in management for biodiversity, nontimber land uses, integrated resource management, representation in protected old-forest ecosystems, and sustainable forestry are all part of today's interest in ecosystems that contain Sitka spruce.

2
Biology and Ecology of Naturally Occurring Sitka Spruce

This chapter begins with a brief review of the history of Sitka spruce's taxonomy, natural variation, and morphology. Much of the chapter is devoted to biogeoclimatic relationships of ecosystems in which Sitka spruce can occur, because many forest management decisions today are based on treatment units defined at an ecosystem level within a biogeoclimatic classification of the management area. An example is the ecosystematic approach to a subunit plan for the Koprino River watershed on northwestern Vancouver Island (Klinka et al. 1980).

The concluding sections review published information on Sitka spruce regeneration; growth and productivity; nutrient relationships; soil-moisture relationships; chemical and physical damaging agents; fire relationships; insect and disease relationships; and some features of fish, mammalian, and bird habitats in spruce-dominated ecosystems. Additional information on the ecology of Sitka spruce can be found in several previously published reports (Day 1957; Ruth 1958; Harris 1964; Fowells 1965; Phelps 1973; Harris and Johnson 1983; Harris 1984, 1990).

Nomenclature and Taxonomy

Based on more detailed botanical accounts available elsewhere, the following subsections provide a brief summary of the taxonomic history of Sitka spruce. The genus name *Picea* is derived from the Latin *pix* or *picis,* meaning pitch. The original botanical use of this term was in relation to pitch-pine, but later *Picea* was applied to spruces as the genus name (Farrar 1995). The species name *sitchensis* means 'of Sitka,' referring to Sitka Island, Alaska, where the German naturalist Mertens collected specimens of this spruce in 1827. As outlined below, these Sitka Island specimens were used by the French botanist Bongard when he assigned the species name *sitchensis* to this tree.

History of Sitka Spruce Nomenclature

Sitka spruce first became known to science through Archibald Menzies, a

Scottish naturalist, who sailed as a surgeon aboard the HMS *Prince of Wales* in 1786 on a three-year voyage around the world, and who recorded this species on the shores of Puget Sound in 1787. Specimens of the foliage collected are in the British Museum in London, England (Fletcher 1976).

Sitka spruce was introduced into Great Britain by the Horticultural Society of London through David Douglas, who named it *Pinus menziesii* (Lambert 1832; Lindley 1833; Lindley and Gordon 1850). The seed of Sitka spruce was brought to England by Douglas's expedition, arriving in 1831 without Douglas, who had fallen victim to a fatal accident on the Sandwich Islands (Loudon 1838). Douglas's description of Sitka spruce must have arrived with the seed, but was probably not published until after Bongard (1832) had described it as *Pinus sitchensis,* from a collection made by Mertens on Sitka Island (now Baranoff Island).

One year later, Lindley (1833) also described the species and named it *Abies menziesii,* but the earlier (1827) collection of the same species by Mertens and the name *sitchensis* given by Bongard took precedence over this new name. According to Rafinesque (1832), Lewis and Clark noticed, but did not name, several fine 'fir' trees in the Columbia River area. These trees were named and characterized by Rafinesque in 1832, but he concluded that he had already described them in 1817 in his *Florula Oregonensis* as *Abies falcata.* Carriere (1855) was the first to ascribe the genus name *Picea* to what we now know as Sitka spruce, and Little (1944) felt justified in accepting *Picea sitchensis* as the correct scientific name for Sitka spruce.

Common names used for Sitka spruce include airplane spruce, coast spruce, Menzies spruce, silver spruce, tideland spruce, western spruce, and yellow spruce. The name 'golden spruce' has been used to refer to an unusual form of this species on the Queen Charlotte Islands, in which the outer sun foliage is a bright golden yellow, possibly indicative of a genetic mutant. In the early taxonomic literature, Sitka spruce varieties that have been recorded include *Picea menziesii crispa* and *Abies menziesii crispa* (Carriere 1855). These varieties differ from the species by having margins of cone scales more undulated and somewhat jagged and more extended; it is possible that all varieties are genetically identical to *Picea sitchensis*. The designation of varieties has been questioned taxonomically, as variation within the species' natural range is mainly clinal (C.C. Ying, pers. comm., March 1991).

Within Sitka spruce's natural range, one hybrid that has been recognized is *Picea glauca* × *Picea sitchensis* (Little 1953). This hybrid was collected by H.J. Lutz on the Kenai Peninsula, Alaska, and the holotype is deposited in the U.S. National Museum, Washington, D.C. Besides the Kenai Peninsula hybrids between Sitka and white spruce (*P. glauca*) reported by Little (1953) and referred to by Harris (1984), natural hybrids between these two spruces were also reported by Day (1957), Daubenmire (1968), Roche (1969), and Woods (1988) in the Skeena, Nass, and Bulkley river valleys in British Columbia.

Roche (1969) studied spruce populations in the Skeena River valley and concluded that spruces there are a result of introgressive hybridization between Sitka spruce and white spruce. In northwestern British Columbia, the common practice at present is to refer to 'Roche spruce' when foresters encounter trees in the field that appear to be hybrids between Sitka and white spruce or between Sitka and white and Engelmann spruce. Elsewhere in the literature, hybrids between Sitka and white spruce are sometimes referred to as 'Lutz spruce,' particularly by American foresters. Experimentally created hybrids between Sitka and white spruce have also been reported, mainly from work in Denmark (Hansen 1892, cited in Karlberg 1961; Fabricius 1926, cited in Larsen 1948; Schreiner 1937; Larsen 1945; Thaarup 1945; Langner 1952).

Other hybrids apparently exist outside the natural range of Sitka spruce. Wright (1955) listed the following: *Picea sitchensis* × *Picea engelmannii* (Cheng 1947; Johnson 1939), *Picea sitchensis* × *Picea omorika* (Johnson 1939; Eklundh 1943; Oksbjerg 1953), *Picea sitchensis* × *Picea excelsa* (Langner 1952), and *Picea sitchensis* × *Picea jezoensis* (Karlberg 1961). Wright (1955) suggested that in warmer pre-Pleistocene times Sitka spruce probably grew on the Aleutian Islands in the vicinity of *Picea jezoensis,* a species to which it is taxonomically similar. In addition to the hybrids listed above, Fowler (1983) successfully crossed Sitka spruce and black spruce and obtained viable seed, but natural hybridization between Sitka spruce and black spruce was not verified in the literature review for this book.

There has been research in British Columbia to assess growth responses of artificial crosses between Engelmann and Sitka spruce (Kiss 1989), in an attempt to develop better regeneration stock for reforestation in interior regions of the province. Based on the limited sampling of only one Sitka spruce male as a source of tested hybrids, it was concluded that Engelmann × Sitka spruce hybrids were not promising as a future source of reforestation stock for north-central British Columbia. However, several studies have noted that spruce hybrids exhibit better growth and survival than the parent species (Mitchell et al. 1974; Fowler 1983; Geburek and Krusche 1985; Sheppard and Cannell 1985).

Taxonomy of Sitka Spruce in Relation to Other British Columbia Spruces

Sitka spruce potentially overlaps in its geographic range with the three other spruces native to British Columbia. Based on spruce distribution maps prepared by Krajina et al. (1982), with updated information from Coates et al. (1994), Figure 5 shows locations where the mapped natural distribution of Sitka spruce overlaps with the mapped distribution of one or more of the interior spruces. These areas of potential overlap, which range from the Chilliwack River drainage basin near Hope, British Columbia (near latitude

49°N) to the Tagish Lake drainage basin near the Yukon–British Columbia border, almost at latitude 60°N, are not by themselves evidence of hybridization between Sitka spruce and interior spruces.

The overlapping ranges of British Columbia spruces (Figure 5) suggest that there are many geographic areas of potential hybridization between

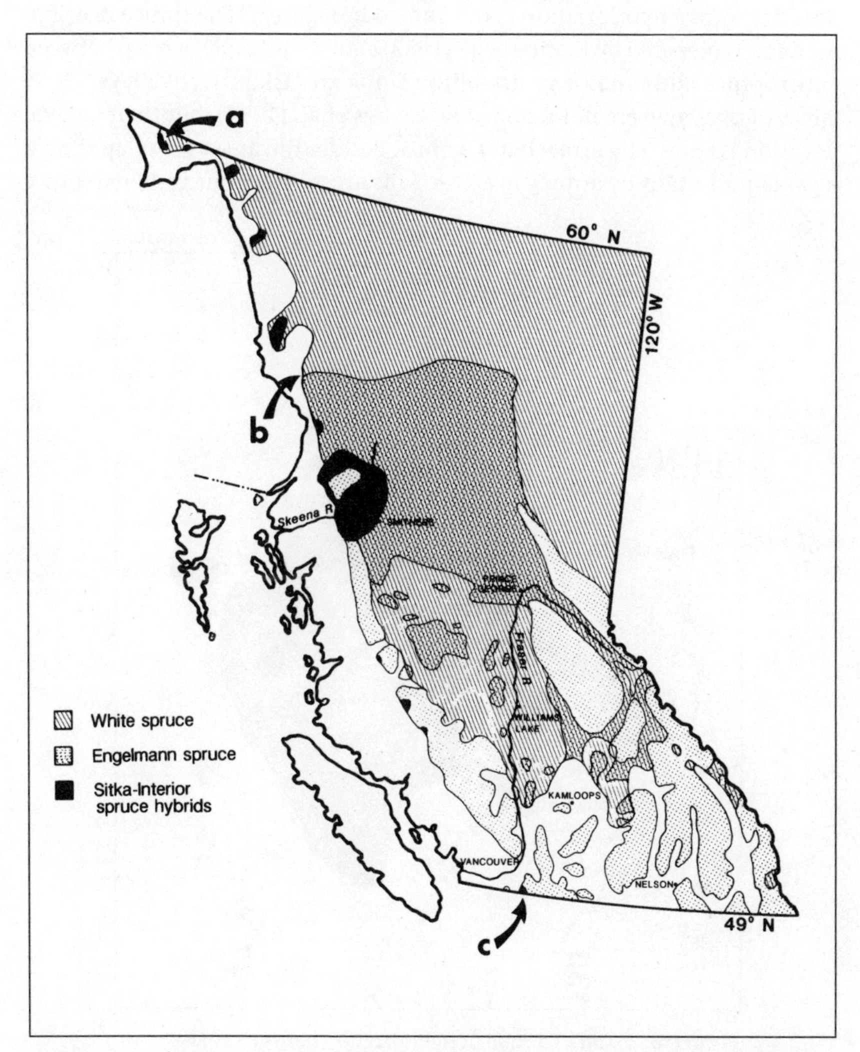

Figure 5. Locations in British Columbia (black areas) where the natural distribution of Sitka spruce overlaps with one or more of the interior species of spruce (Coates et al. 1994, as modified from Krajina et al. 1982). The northwesternmost and southernmost areas of overlap are shown by arrows **a** and **c**, respectively. Arrow **b** indicates the general location of two areas of overlap, one in the Unuk River valley and one in the Iskut River valley, which were too small to map at this scale.

Sitka spruce and any one of white spruce, Engelmann spruce, or black spruce. Conversely, hybrids may occur in areas where cartographic scales of species distributions do not show overlap between coastal and interior spruces. For example, there are reports, not yet confirmed by detailed taxonomic study, of Sitka spruce–Engelmann spruce hybrids in the vicinity of Whistler, British Columbia, at latitude approximately 50°06′N. This area of potential coast-interior spruce hybridization is not shown in Figure 5. The spruce distribution maps prepared by Krajina et al. (1982) showed possible overlap between Sitka spruce and interior spruce in the Unuk and Iskut river valleys. These areas of overlap were not mapped by Coates et al. (1994), but their general location is shown by arrow **b** in Figure 5. Besides the area of overlap shown near latitude 49°N by arrow **c** in Figure 5, Beamish and Stone gathered spruce

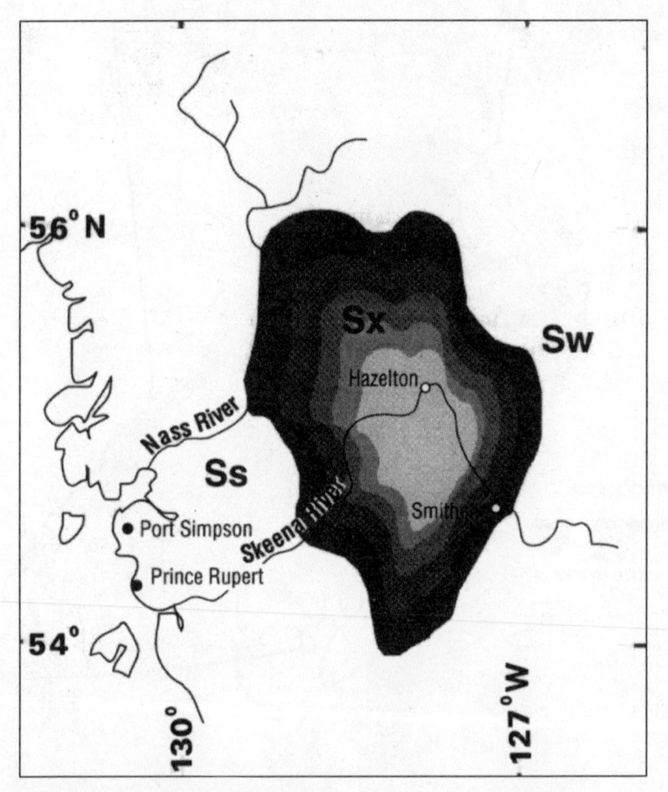

Figure 6. Estimated zone of hybridization between Sitka spruce (Ss), white spruce (Sw), and Roche spruce (Sx) in the Nass, Skeena, and Bulkley river areas of British Columbia (adapted from Woods 1988 and Coates et al. 1994). Boundaries are approximate until more information is available. The banding from light grey to progressively darker shades of grey indicates increasing probability of spruce hybridization.

(unpublished University of British Columbia herbarium specimen 7659) at an elevation of 2,000 m possessing characteristics that suggest a hybrid Sitka × Engelmann spruce. However, this specimen may simply be a variety of Engelmann spruce, which can be proven only by investigating several more specimens collected in the same locality.

When expressed in terms of British Columbia river valleys from north to south, Sitka spruce and interior spruces appear, because of geographic overlap, to have opportunities to hybridize in the following drainage basins: Tatshenshini, Tagish Lake headwaters area, Taku, Stikine, Iskut, Unuk, Nass, Skeena, Bulkley, Kimsquit, Bella Coola, Klinaklini, Cheakamus, and Chilliwack. Field personnel in these areas may face problems of spruce identification because of interbreeding, particularly in the Skeena, Nass, and Bulkley river valleys (Figure 6), where there is evidence of localized gradations in the probability of hybridization between Sitka and interior spruces (Coupé et al. 1982; Woods 1988; Coates et al. 1994; Grossnickle et al. 1996b; Morgenstern 1996).

Where interbreeding does not occur, British Columbia's four species of spruce can generally be distinguished on the basis of leaf, twig, and cone characteristics. The dichotomous key in the *Forestry Handbook for British Columbia* (Watts 1983), modified to agree with the criteria shown in Table 3 (Coates et al. 1994), indicates the following distinguishing characteristics:

A. Tips of cone scales rounded, stiff.
- B. Leaves 1-1.5 cm long, blunt generally [twigs usually velvety pubescent; cone egg-shaped, 3 cm long; margin of scales smooth or ragged]. **BLACK SPRUCE.**
- BB. Leaves longer, sharp [twigs smooth; cone to 6 cm long, margins smooth]. **WHITE SPRUCE.**

AA. Tips of cone scales wedge-shaped, flexible.
- C. Leaves very sharp, flat, bluish below [not crowded towards upper side of twig; cone to 9 cm, scales ragged]. **SITKA SPRUCE.**
- CC. Leaves 4-sided, less sharp [crowded towards upper side of twig; twigs sometimes pubescent; cone to 5 cm]. **ENGELMANN SPRUCE.**

Morphological features for distinguishing between Sitka, white, and Engelmann spruce are summarized in Table 3. For those interested in more detail, the most definitive research on the taxonomy and genecology of the genus *Picea* in British Columbia is the work by Roche (1969), with more recent suggestions by Yeh and Arnott (1986) for criteria to differentiate Sitka spruce and white spruce.

Table 3

Morphological features for distinguishing between Sitka, white, and Engelmann spruces

Character	Sitka spruce	White spruce	Engelmann spruce
Young twigs	smooth, shiny	smooth and shiny, usually not hairy	hairy (occasionally smooth)
Needles (leaves)	somewhat flattened	4-angled	4-angled
Cones	6.0-9.0 cm long	2.5-6.0 cm long	4.0-5.0 cm long
Scales			
morphology	rounded, finely irregularly toothed, somewhat stiff; longer than broad; narrower than Engelmann spruce	elliptical, rounded to blunt, margin smooth; stiffer than Engelmann spruce; broader than long	blunt to sharp-pointed, finely irregular, wavy margin (ragged); longer than broad
mean free scale length	longer than white spruce, and narrower than Engelmann spruce (4.0-5.0 mm)	short (1.0-2.0 mm)	longer than white spruce, slightly longer than Sitka spruce (up to 6.3 mm)
free scale percentage	greater than white spruce (24-34%)	small (8-16%)	greatest (30-40%)

Hybrids show intermediate characteristics.

Sources: Coates et al. (1994), based on information from Douglas (1975, unpublished), van Barneveld et al. (1980), Coupé et al. (1982)

There has been substantial research on criteria for laboratory identification of Sitka spruce and its hybrids (Hanover and Wilkinson 1970; Copes and Beckwith 1977; Yeh and Arnott 1986; Sigurgeirsson et al. 1990; White et al. 1993; Sutton et al. 1991, 1994; Grossnickle et al. 1996a, 1996b). Copes and Beckwith (1977) indicated that starch gel electrophoretic techniques, using apparatus similar to that described by Conkle (1972), can be used to identify stands in which various levels of hybridization have occurred between white spruce and Sitka spruce. The presence of white spruce genes in Sitka spruce populations can usually be detected by using a tetrazolium oxidase isoenzyme. Tests have confirmed that the introgression demonstrated by isoenzymes agrees with the hybridity detected by chromatographic methods (Hanover and Wilkinson 1970) as well as the results obtained for the same population by morphological study (Daubenmire 1968). Although electrophoretic techniques can be useful for identifying species and populations in which introgressive hybridization has occurred, the initial detection of stands suspected of containing hybrids is still more easily done with traditional taxonomic methods (Copes and Beckwith 1977). Joint work by the Western Forest Genetics Association and the International Union of Forestry Research Organizations (IUFRO) typifies the significance given to Sitka spruce as an important genetic resource (King and El-Kassaby 1990; Meier 1990; Pollard and Portlock 1990; Van de Sype and Roman-Amat 1990; Ying 1990).

Apart from criteria for spruce hybrid differentiation, Sitka spruce has been an important component of recent work by several forest genetics researchers in British Columbia (Ying 1990; Chaisurisri and El-Kassaby 1993a, 1993b, 1994a, 1994b; Chaisurisri et al. 1992, 1993, 1994a, 1994b).

Natural Variation of Sitka Spruce

Phenotypes of Sitka spruce (used here to mean visible tree characteristics that are produced by interaction of genotype and environment) have been frequently observed but have been little emphasized in the literature. A recorded example occurs along the coastal strip at Long Beach, between Ucluelet and Tofino, British Columbia, where wind-shaped Sitka spruce trees grow on islands, dunes, and bluffs facing the ocean (Stanek and Krajina 1964). In this environment, the highest order of Sitka spruce branches is arranged in a pattern that provides the least resistance to wind and maximum protection from sandblasting. Another example was recorded by Cary (1922), who distinguished two general types of Sitka spruce forms. One is a bottomland type, where the trees, although large and tall, are characterized by large buttressed bases, limbiness, and comparatively short length of stem clear of branches. An example of this type, from the Queen Charlotte Islands in British Columbia, is portrayed in Figure 7. The largest specimens of Sitka spruce are to be found on such sites. The second form of Sitka spruce is that which Cary referred to as the 'slope type,' which includes trees that are large and tall, without buttressed bases, and with long lengths of stem clear of branches. An example of this form from the Queen Charlotte Islands is shown in Figure 8.

A third type, the 'upper slope' type, was suggested by Cary for Alaskan examples of shrublike Sitka spruce that occur near the altitudinal limit of tree growth. Interestingly, at its southern latitudinal limits in Mendocino County, California, Sitka spruce also occurs as shrubby specimens. The extent to which these shrublike phenotypes at the northern and southern latitudinal extremes of the species' range represent genotypes cannot be determined without progeny tests. Where there is heavy browsing by deer,

Figure 7. Typical fluted form of old-growth bottomland Sitka spruce, at lower Gregory Creek, Graham Island, Queen Charlotte Islands, British Columbia. (Photograph by R.B. Smith)

Figure 8. Typical unfluted form of slope-type old-growth Sitka spruce at about 300 m elevation, near Kaisun, on the northwest coast of Moresby Island, Queen Charlotte Islands, British Columbia. (Photograph by T. Chatwin)

there may also be compact and dwarfed Sitka spruce that is very different in shape from surrounding unbrowsed spruce, as shown in Plate 4 for a Sitka spruce on a landslide scar on the Queen Charlotte Islands.

Sitka spruce has probably occupied its present range since the Pleistocene (Hultén 1937). Its distribution in a narrow coastal belt from California to Alaska leads to the assumption of different genotypes or provenances in response to climatic differences (Betts 1945; Laing 1951; Day 1957; Vaartaja 1959; Schober 1962; Phillips 1963; Lines and Mitchell 1966; Mergen and Thielges 1967; Daubenmire 1968; Roche 1969; Yeh and Rasmussen 1985). Sitka spruce exhibits geographic variation in height growth, wood quality, response to day length, seed characteristics, karyotype, morphological events, stem and needle coloration, susceptibility to frost (Robak 1957; Burley 1965a, 1965b, 1965c, 1965d, 1966a, 1966b, 1966c; Clark 1965), tracheid length (Dinwoodie 1963), and chemical properties (Forrest 1975, 1980; von Rudloff 1978; Yeh and El-Kassaby 1980).

Provenance research has been under way in North America and in many European countries during the last 50 years, but most provenance work involving Sitka spruce is relatively recent (Lines 1967). By 1972 a total of 84 sources from the range of Sitka spruce had been sampled (Fletcher 1976). From these collections, 10 provenances that were representative of Sitka spruce's distribution were selected for international provenance tests, participated in by the following countries: Belgium (Nanson 1976), Denmark (Kjersgard 1976), France (Birot 1976), Germany (Kleinschmit and Sauer 1976), Great Britain (Pearce 1976), Iceland (Benedikz 1976), Ireland (O'Driscoll 1976a, 1976b), Latvia (Pirags 1976), Netherlands (Kriek 1976), Northern Ireland (Savill 1976), and Norway (Magnesen 1976).

A recent analysis of Sitka spruce nursery stock revealed an inverse relationship between all growth parameters (averaged over all nursery locations) and latitude of seed source (Burley 1976). The underlying pattern was a linear north-south trend of increased growth and increased growing period with decreasing latitude of seed source. There was a quadratic relationship between growth (averaged over all provenances) and latitude of nursery location. The highest growth values were associated with Sitka spruce seed collected between latitudes 52°N and 54°N. Provenance × location interaction effects were small, but some provenances and locations contributed more than others. The best general growth and the greatest contribution to interaction was provided by the seedlot from Big Qualicum River, Vancouver Island, in combination with certain nursery locations, particularly in Norway.

Khalil (1977) reported the results of trials in Newfoundland with 12 provenances from British Columbia, Alaska, Oregon, and Washington, five years after planting 2+2 seedlings at nine locations. Resistance to winter damage and height growth were the best criteria for evaluating provenances, and eight of the provenances were recommended for further trials. Sitka spruce is not currently planted in Newfoundland, however.

After 10 years of trials in southern British Columbia (Project EP 702), involving 38 provenances from Alaska to southern Vancouver Island, Sitka spruce has shown strong north-south and coast-inland trends in growth and hardiness; provenances from southern latitudes and of outer coastal origin are faster-growing but less hardy than the northern and inland provenances. On the basis of these results, the B.C. Ministry of Forests has recommended the introduction of seeds from the Oregon and Washington coast for low-elevation test plantations on Vancouver Island and the Queen Charlotte Islands. For inner coastal valleys and high-elevation sites, the continued use of local seeds is suggested. Further details, based on the *Seed and Vegetative Material Guidebook* released under the Forest Practices Code of British Columbia Act (Ministry of Forests and B.C. Ministry of Environment, Lands and Parks 1995i), are provided under 'Stock Selection' in Chapter 3.

In other Sitka spruce provenance trials in British Columbia, involving populations with a latitudinal range from Oregon to Alaska, northern and inland provenances tended to flush earlier than those from the south.

During the period of observation, from September 1973 to May 1977, no cold injury was recorded at test locations on Vancouver Island or the Queen Charlotte Islands, where populations from the Oregon and Washington coasts were substantially more vigorous than local or northern ones (Illingworth 1978). It should be emphasized that the introductions of Oregon and Washington provenances to the Queen Charlotte Islands were for operational trials only, not for reforestation. Roche and Fowler (1975) point to areas outside the natural range of Sitka spruce where tests have shown heritable geographic variation of early growth, dormancy, and frost susceptibility; they suggested that form, vigour, and wood properties can probably also be genetically manipulated. In trials in the United Kingdom, involving seed from a latitudinal range similar to that described above (Oregon to Alaska), Lines and Mitchell (1966) discerned similar flushing dates for all seed origins, but major variations in cessation of growth. This makes Sitka spruce from southern origins prone to autumn frost damage in the United Kingdom, and restricts the areas in which these faster-growing provenances can be used (T.C. Booth, pers. comm., May 1991).

Variations in height growth and frost resistance among the provenances reported in IUFRO (1976, 1984) were highly significant, but the ranking varied with the test locations. These variations are generally attributable to the date of growth cessation rather than rate of growth. The date of growth cessation is controlled largely by decreasing photoperiod (Illingworth 1976). Besides the IUFRO provenance trials described by Illingworth for British Columbia, there have been some trials to identify Sitka spruce provenances that are resistant to the white pine weevil (Hartman 1974; White 1986). Spruce observed by W. Stanek (pers. comm., 1985) in the Kitimat-Terrace area, which he thought to be hybrids between Sitka and interior spruce, are suggested to be good genetic material for further testing in relation to white pine weevil susceptibility.

Sitka Spruce Tree Morphology

For the forest manager interested in detailed information on morphological features of Sitka spruce, basic references are Harlow and Harrar (1958), Phelps (1973), Hosie (1990), and Farrar (1995). Some features of Sitka spruce morphology are shown in Figure 9.

The following subsections focus only on certain aspects of the morphology of Sitka spruce that may be of interest to a silviculturist involved with the management of Sitka spruce in its natural range, with emphasis on roots, stems, bark, twigs and branches, tree crowns, cones, seeds, and maximum tree sizes.

Roots

Sitka spruce has a relatively high root spread/crown width ratio, as shown

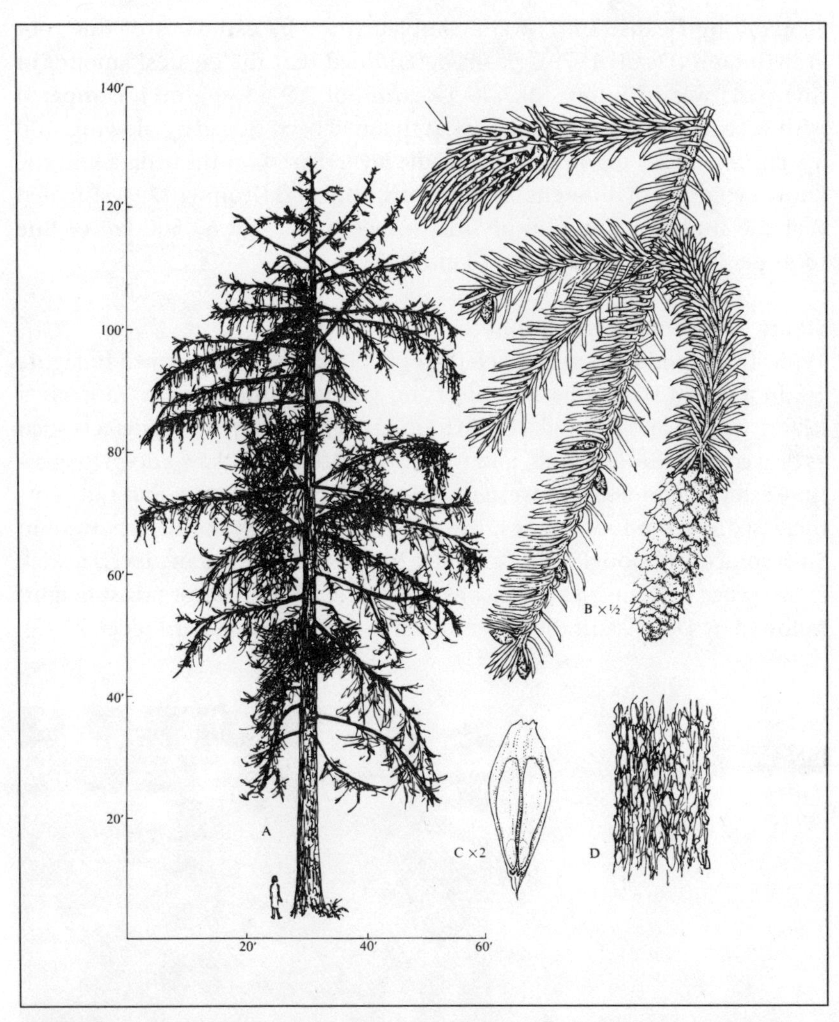

Figure 9. Some features of Sitka spruce morphology. A – typical tree form and branching habit (the vertical and horizontal axes, shown in feet, represent a tree height of about 39 m and a crown width of about 17 m). B – branch with mature cone and with gall on tip of branchlet (indicated by arrow) caused by Cooley spruce gall aphid (*Adelges cooleyi*). C – seed-bearing cone scale. D – surface pattern and texture of bark. (Sketches by L. Bohm, as published in Taylor and Taylor 1973)

in the following ranking by Smith (1964) of several British Columbia tree species from highest to lowest root spread/crown width ratio: lodgepole pine, Sitka spruce, ponderosa pine, Douglas-fir, western hemlock, western redcedar, and red alder.

As with most other tree species, there is limited information on fine roots of Sitka spruce. Studies of coarse root production in young plantations in

Scotland by Deans (1981) were complemented by estimates of fine root growth rates (Deans 1979). Deans determined that the greatest amount of fine roots was in humus (9.7-19.1 cm/mL or 2.9-6.9 mg/mL), compared with 9.3 cm/mL or 2.6 mg/mL in peat that had been ridged by plowing, and 5.5 cm/mL or 2.0 mg/mL in the needle layer. Based on these data and the known volumes of different soil horizons, Ford and Deans (1977) estimated that the studied plantation of Sitka spruce had about 67,500 km of fine roots per hectare 11 years after planting.

Stems, Bark, and Stem Epiphytes

Typical bark and branch characteristics of Sitka spruce are shown in Figure 10. In a recent study of stem defects on young plantations on Vancouver Island, Omule and Krumlik (1987) found that the frequency of defects such as broken tops, crookedness, and forking increased in Sitka spruce, Douglas-fir, western hemlock, and western redcedar as site moisture and nutrients increased. Of the four species, the greatest frequencies of defects were in Sitka spruce and Douglas-fir. On average, over all three spacing levels tested, Sitka spruce had the most branches (10) at the whorl nearest breast height, followed by Douglas-fir (8), western hemlock (6), and western redcedar (5).

Figure 10. Typical bark and branch characteristics of Sitka spruce, near Big Qualicum River, Vancouver Island, British Columbia. (Photograph by R.B. Smith)

For comparable spacings, branch diameters were slightly larger in Sitka spruce and Douglas-fir than in western hemlock and western redcedar.

From his exhaustive literature review, Minore (1979) ranked Sitka spruce second only to red alder in its receptivity to epiphytes. According to Minore, western hemlock, Douglas-fir, western white pine, western redcedar, black cottonwood, and bigleaf maple all have stem surfaces that are less receptive to epiphytes than those of Sitka spruce. This ranking differs from that of some observers who have suggested that bigleaf maple is the Pacific Northwest tree species most receptive to epiphytes because stemflow and throughfall that pass through the base-rich maple foliage encourage epiphytic growth on stems of maple (Kirk 1966). Aside from the mosses, lichens, and fungi that grow on Sitka spruce stems and on spruce logs on the forest floor, conspicuous vascular species of epiphytes on spruce include *Selaginella* spp. in the Olympic Peninsula in Washington (Kirk 1966) and *Polypodium scouleri* in Pacific Rim National Park on Vancouver Island (Cordes and MacKenzie 1972). In the latter location, *Frullania nisquallensis* is a prevalent epiphyte on Sitka spruce stems, along with *Polypodium scouleri.*

The sapwood of Sitka spruce is creamy white to light yellow and blends gradually into the heartwood, which is pinkish yellow to brown. Sapwood in mature trees typically ranges from 7.5 to 15.0 cm wide. The wood has a fine, uniform texture, generally straight grain, with no distinct taste or odour and relatively few resin ducts. The annual rings are distinct, with a band of lighter-coloured earlywood that shades gradually into a narrower band of darker latewood (Harris 1984).

Sitka spruce wood is classed as moderate in weight, stiffness, hardness, resistance to shock, shrinkage, bending, and compressive strength. It ranks high in strength on a weight basis. The average specific gravity is 0.37, based on green volume and ovendry weight. The average density of air-dry wood (12% moisture content) is 0.45 g/cm^3. Young, dense managed stands exhibit slow shedding of dead branches and thus poor self-pruning. Consequently, wood from such stands generally lacks the knot-free, slow-growth characteristics of old-growth trees (Harris 1984).

Needles, Twigs, and Branches

Sitka spruce needles stand out from all sides of the twigs. They are linear, up to 27 mm long, and flat to oval in cross section; they are slightly keeled. Dry needles show dorsally and ventrally protruding ribs, and range in colour from pale green to clay-brown; needles take on a pale bluish colour in response to nitrogen fertilization or high nitrogen availability (Plate 5). Needles are generally very stiff and acuminate at the end; there are 6-11 (average 8) rows of stomata in each of two ventral stomatal bands, and 1-5 (average 3) rows of stomata in each of the two dorsal stomatal bands. The ratio of length to breadth of leaves varies from one herbarium specimen to

Figure 11. Typical needle orientation and bud shape of Sitka spruce. Scale at top is 15 cm.

another, but averages about 13. Typical needle orientation and bud shape of Sitka spruce are shown in Figure 11. Twigs of Sitka spruce are stout, stiff, and hairless, and vary from light to dark brown. Twigs are typically rough after needle drop because of the peglike bases of the needles (Viereck and Little 1972).

For most genera of vascular plants, lateral shoots originate from meristems in the axils of leaves, but in a few genera, including *Picea,* shoots develop from meristems apparently differentiated from internodal cortical tissues (Ford 1985). For Sitka spruce specifically, most of the current knowledge about the early stages of branch development comes from studies in plantations rather than natural stands. These studies indicate that a bud of Sitka spruce extends in one year, and in the next year acts as a source of whorl and interwhorl branches (Cochrane and Ford 1978). Each bud may produce a similar sequence of branches, but, as they become submerged in the canopy, branches gradually fail to produce first their own interwhorls and then whorl branches; finally they fail to elongate at all.

Longman (1985) noted that evidence for aging may be readily observed in conifers from changes in the number, vigour, and type of vegetative shoots on successively older branches. For instance, at the tops of mature Sitka spruce grafts, and on first-order shoots on the main branches, most or all of the terminal buds contain preformed shoots and lateral bud primordia in winter. The total number of buds with preformed shoots per branch increases from the 1-year-old branches at the top of the spruce trees to the 4-year-old branches below, but below this level the number decreases because an increasing proportion of the buds contain only a living apex and bud scales. Failure to form preformed shoots is particularly true of third- and fourth-order shoots, which are mostly weak and fail to produce lateral bud primordia (Longman 1985).

Cochrane and Ford (1978) found that, during canopy development, there was considerable variability in the production, dispersion, and extension of branches along the main stem of the tree. Branch dispersion followed consistent rules from year to year, however. For example, whorl branches were always arranged in a spatially regular (not random) fashion around the stem, irrespective of their numbers; interwhorl branches were absent immediately below their distal whorl and above their proximal whorl. The angles between the vertical main stem and the branches increased from the top of the tree towards the base of the crown, especially for interwhorl branches (Ford 1985).

The development of new branches on previously branch-free lower stems, referred to as epicormic branching (Plate 6), is a characteristic of Sitka spruce (Isaac 1940; Ruth 1958). Thinning or other disturbance that permits increased light exposure of the stem can stimulate growth of epicormic sprouts, which can develop from either dormant or adventitious buds (Stone and Stone 1943). Such branching can increase harvest costs and lower the value of timber for veneer, but epicormic branches do not lower the utility of Sitka spruce for structural lumber or pulpwood. Herman (1964) noted that moderate or heavy thinning will stimulate epicormic branching of Sitka spruce. However, he also concluded that most Sitka spruce in a 100-year-old stand had some epicormic branches before release, and that new sprout production was slight in a subsequent 10-year period following low-intensity thinning. From observations in the Cascade Head Experimental Forest near Otis, Oregon, Herman (1964) concluded that exposing 110-year-old Sitka spruce stems to increased solar radiation by cutting a logging road right-of-way resulted in increased growth of existing epicormic branches and production of new ones, but the changes were not spectacular.

Crowns

As shown by three adjacent Sitka spruce trees in Figure 12, growing near Bowser, eastern Vancouver Island, there can be considerable tree-to-tree variation in branch angle, branch length, and crown form. Open-grown Sitka spruce frequently have exceptionally wide crowns relative to tree height (Figure 13). For stereo airphoto interpretation, the large, jagged crowns of Sitka spruce make it relatively easy to distinguish this tree from its companion species (Hegg 1967).

Day (1957) and Phelps (1973) noted that variations in Sitka spruce crown characteristics result from differences in growing conditions and from the

Figure 12. Branch angle, branch length, and crown form vary greatly in Sitka spruce, as shown by these three adjacent trees near Bowser, Vancouver Island, British Columbia, suggesting substantial variation in the genetic composition of Sitka spruce.

Figure 13. Open-grown individuals of Sitka spruce are characterized by exceptionally wide crowns relative to tree height.

incidence of root rot. On favourable sites, the crowns of well-developed trees are broadly conical and the foliage is dense; on shallow soil and where the roots are diseased, the crowns are thin and sparse. Sometimes the crowns of mature trees are narrowly spired, which Day (1957) suggested was indicative of a stony soil, a soil deficiency in nutrients, or a shallow soil with a high water table. From observations in Great Britain, Hamilton (1969) found that narrow-crowned trees were more efficient producers of stem biomass than wide-crowned trees.

Taylor and Taylor (1973) referred to 'golden spruce' on the Queen Charlotte Islands, where two such trees were known to occur near the Yakoun River southwest of Port Clements. The largest and best known of these spruces was about 50 m tall, with a diameter at breast height (dbh) just over 2 m, and was estimated to be 400 years old (Plate 8). Unfortunately, this special tree, revered by the Haida people, was felled by a vandal in January 1997. In these unusual trees, the outer foliage that is exposed to the sun is bright golden yellow; the inner foliage is green but not as dark as a normal Sitka spruce. By examining chlorophyll content, Scott (1969) determined that the sun foliage needles contained 40% more carotene but 5 times less chlorophyll compared with normal sun foliage of Sitka spruce. Attempts have been made to propagate this unusual Sitka spruce variety by grafting and by cuttings; Oscar Sziklai, now retired from the Faculty of Forestry, University of British Columbia, was successful with both methods, but regeneration of the variety from seeds has not been tested (Taylor and Taylor 1973).

Cones

Eremko et al. (1989) summarized Sitka spruce cone production characteristics as follows:

reproductive cycle	2 years
cone length	5-10 cm
cone-bearing age (collectable quantities)	25-40 years
cones per hectolitre	4,000-6,000
periodicity	3-4 years
viable seeds per hectolitre of cones	194,270
position of cones in crown	top one-third
ease of cone detachment	moderate
plantable trees per hectolitre of cones	
bareroot	83,000
container	67,000

Sitka spruce cones hang vertically from short stalks. They are cylindrical and light orange-brown, and fall off at maturity. The cones are oblong, sometimes curved, and can be up to 10 cm long, although they are often shorter

Figure 14. Typical cone and seed shape of Sitka spruce. Scale at bottom is 15 cm.

(Figure 14); the ratio of length to breadth in dry, open cones averages 1.9. Cone scale bracts are acuminate, finely serrate, or toothed; the ratio of scale length to bract length averages 2.9. Cone scales are rounded to wedge-shaped at the apex, with margins erose to finely toothed.

In monoecious conifers, typical characteristics are as follows: a leading shoot and apical parts of vigorous upper branches that are strongly vegetative, with little or no cone formation; female cones that occur principally in the upper and apical parts of the middle crown; and male cones that occur especially in the middle and apical parts of the lower crown (Longman 1985). Longman (1985) demonstrated that when an adult graft of Sitka spruce was bark-girdled and hormonally stimulated to produce reproductive buds, most of the female cones occurred on 4-year-old branches but there were also some on younger branches right up to the basal section of the terminal shoot. Most of the male cones were formed on 4- and 5-year-old branches. In the upper part of the tree, both female and male cones were found on the weaker branches, between the main pseudowhorl groups, whereas lower down the tree they occurred mostly on the main branches. Within a branch, female cones tended to be nearer the tips than male cones. In both sexes, there was a distinct tendency for cones on more apical, vigorous shoots to be in lateral positions, whereas on less vigorous, more basal sections, terminal cones were predominant (Longman 1985).

Large within-species variation in flowering is typically found between individuals of Sitka spruce. For example, observations over eight years on 25 mature grafts and cuttings showed that two grafted clones never flowered, 20 clones flowered occasionally to varying extents, whereas three clones flowered every year. Also, certain grafted clones produced predominantly one sex, as has been reported for other conifers. Pronounced genetic control of cone production in Sitka spruce has also been demonstrated in progeny trials (Longman 1978, 1985).

Seeds

Sitka spruce seeds are brown and, including the wing, measure up to 12 mm long (Viereck and Little 1972). Sitka spruce is a prolific seed producer. It has been noted to drop more than 14.5 kg of seed per hectare in an old-growth stand; a good seed crop may occur as often as every 3 or 4 years. In southeastern Alaska, researchers have noted that some seed is produced every year and there is a good cone crop every 5-8 years. Although Sitka spruce seeds are relatively small and have a correspondingly slow rate of seedfall, most seeds fall within the stand from which they originated. There is little correlation between Sitka spruce seed weight and latitude, although there are indications that seed from northern stands is slightly heavier than seed from southern sources (Cleary et al. 1978).

Based on observations in Alaska, Minore (1979) indicated that Sitka spruce has larger average cone crops than western hemlock during most years. In contrast, Ruth and Berntsen (1955) found western hemlock to be a more prolific seed producer than Sitka spruce in coastal Oregon. In terms of seed crop frequency, Sitka spruce tends to be less frequent than western hemlock (Meyer 1937). Sitka spruce seeds mature earlier than western hemlock seeds in southern Alaska (Harris 1969).

The following seed collection standards for Sitka spruce were suggested by Eremko et al. (1989):

filled seeds per half-cone	7
cone colour	yellowish-brown
storage tissue	opaque and firm
seedcoat	should occupy 90% of the cavity
seed wing	golden brown with darker strip along one edge
cleanliness	less than 5% debris and unacceptable cones

Maximum Dimensions of Mature Trees

Typical and maximum dimensions of Sitka spruce can be best portrayed by comparison with other conifers of the Pacific Northwest. Such data were assembled by Waring and Franklin (1979) and are reproduced in Table 4. Of the 16 conifers listed in Table 4, only Douglas-fir and coast redwood have typical heights exceeding those of typical Sitka spruce; in terms of

Table 4

Typical and maximum ages and dimensions of Sitka spruce compared with other conifers of the Pacific Northwest

	Typical			Maximum[a]	
Species	Age (yr)	Diameter (cm)	Height (m)	Age (yr)	Diameter (cm)
silver fir	> 400	90-110	44-55	590	206
noble fir	> 400	100-150	45-70	> 500	270
Port Orford cedar	> 500	120-180	60	–	359
Alaska yellow-cedar	> 1,000	100-150	30-40	3,500	297
western larch	> 700	140	50	915	233
incense-cedar	> 500	90-120	45	> 542	368
Engelmann spruce	> 400	> 100	45-50	> 500	231
Sitka spruce	> 500	180-230	70-75[b]	> 750	525
sugar pine	> 400	100-125	45-55	–	306
western white pine	> 400	110	60	615	197
ponderosa pine	> 600	75-125	30-50	726	267
Douglas-fir	> 750	150-220	70-80	1,200	434
coast redwood	> 1,250	150-380	75-100	2,200[c]	501[d]
western redcedar	> 1,000	150-300	> 60	> 1,200	631
western hemlock	> 400	90-120	50-65	> 500	260
mountain hemlock	> 400	75-100	> 35	> 800	221

Source: Waring and Franklin (1979)

[a] Besides these maximum ages, Worrall (1990) indicated that bristlecone pine may reach almost 5,000 years, and provided circumstantial evidence that subalpine larch may exceed 3,000 years. If the latter is confirmed, the subalpine larch trees examined by Worrall in Manning Provincial Park, British Columbia, would be the oldest known trees in Canada.

[b] While not typical, the 'Carmanah Giant' Sitka spruce is 95 m tall (Stoltmann 1993).

[c] Worrall (1990) suggests that coast redwood can exceed 3,000 years.

[d] The maximum diameter recorded for coast redwood by Harlow and Harrar (1958) was 609 cm.

maximum stem diameter, only coast redwood and western redcedar have greater girth than the 525 cm recorded for Sitka spruce.

The 1996-97 *National Register of Big Trees* (American Forests 1996) lists two Sitka spruce trees as 'co-champions' for the United States. Their recorded dimensions are shown in Table 5.

The circumferences listed in Table 5 translate into breast height diameters of about 5.7 m for the Olympic Peninsula spruce and 5.4 m for the spruce near Seaside, Oregon. Large trees in the west coast rain forest are difficult to measure accurately, and different dimensions are often recorded when different researchers measure the same tree. For example, Harris (1990) recorded the Seaside spruce as 5.1 m in diameter at breast height and 65.8 m tall, with a crown spread of 28.3 m. By comparison, mature Sitka spruce in Alaska's Tongass National Forest average 48.7 m in height and 1.0-1.5 m

Table 5

Dimensions of the two largest Sitka spruce trees in the United States

Location	Circumference at breast height (m)	Height (m)	Crown spread (m)
Olympic National Forest, Washington	17.96	62.8	28.3
Near Seaside, Oregon	17.09	58.2	29.3

Source: 1996-97 *National Register of Big Trees* (American Forests 1996)

in diameter at breast height. Viereck and Little (1972) indicated that the largest Sitka spruce known in the United States at that time, near Forks, Washington, had a trunk diameter of 5.4 m and a height of 75.6 m. In this book it is assumed that this is the same spruce that was recorded as 62.8 m tall in a 1987 remeasurement (American Forests 1996).

Recent measurements indicate that Sitka spruce larger than the Oregon or Washington records are present on Vancouver Island. Based on information published in 1993, the Carmanah Creek valley contains British Columbia's finest Sitka spruce trees. The 'Carmanah Giant' Sitka spruce, of uncertain age, is 95.7 m tall and has a circumference of 9.58 m at breast height (MacMillan Bloedel Ltd. 1989). Stoltmann (1993) reported a crown spread of 14.0 m for this tree. It is presently thought to be the tallest tree in Canada and also the tallest Sitka spruce in the world. Beese (1989) estimated the ages of other Sitka spruce in the Carmanah Creek valley that were not as tall as the Carmanah Giant (Table 6).

The estimated maximum ages for the largest spruces are from ring counts of the outer 30-40 cm, plus an estimate of the number of rings to the tree's centre based on the growth rate at the inner end of the core. Because actual growth rates would be as good as or better than this during the early life of the tree, this method tends to overestimate age.

Most of the spruce in the protected area in the Carmanah Creek valley are 1.5-2.5 m in diameter, placing the estimated origin of the main canopy at 200-300 years ago. The scattered veteran spruce are probably about 500 years

Table 6

Ages of several large Sitka spruce in the Carmanah Creek valley other than the 'Carmanah Giant'

Diameter at 1.3 m (cm)	Circumference (m)	Age (years)
89	2.8	150
165	5.2	200 maximum
223	7.0	260 maximum
312	9.8	570 maximum

Source: Beese (1989)

old, although individual specimens, such as the Carmanah Giant, could be older. The ages of Sitka spruce 1.6-2.3 m in diameter in the adjacent Walbran valley were 250-300 years, as determined by ring counts of stumps. The

Table 7

Tree height data from selected locations on Vancouver Island

Location	Date	Species	Height (m)
Nimpkish Island	July 1983	Douglas-fir	72.6
Proposed Ecological		Douglas-fir	80.8
Reserve (16 ha)		Douglas-fir	72.0
		Douglas-fir	79.3
		Douglas-fir	71.3
		Douglas-fir	74.7
		Douglas-fir	78.7
		Douglas-fir	68.9
			(average) 74.4
Tahsish River	July 1983	Sitka spruce	66.8
Proposed Ecological		Sitka spruce	69.2
Reserve (40 ha)		Sitka spruce	68.6
		Sitka spruce	76.2
		Sitka spruce	67.1
		Sitka spruce	74.4
		western hemlock	75.6
			(average) 71.0
Puntledge River	August 1987	Douglas-fir	73.8
Strathcona Provincial		Douglas-fir	76.5
Park		Douglas-fir	64.3
		Douglas-fir	64.3
		Douglas-fir	76.2
		Douglas-fir	68.6
		Douglas-fir	75.6
		Douglas-fir	83.2
			(average) 72.9
Carmanah Creek	April 1988	Sitka spruce	71.3
		Sitka spruce	80.5
		Sitka spruce	78.0
		Sitka spruce	74.1
		Sitka spruce	70.1
		Sitka spruce	75.9
		Sitka spruce	73.4
		Sitka spruce	75.9
			(average) 74.9

Source: Heritage Forests Society and Sierra Club of Western Canada (1988)

maximum ages of Sitka spruce 75-80 m tall in the South Fork of the Hoh River valley in Olympic National Park were approximately 260 years (McKee et al. 1982). These data generally support the age estimates for all but the largest veteran spruce in the Carmanah Creek valley. Some estimates have claimed that the largest Carmanah Sitka spruce are 700-1,000 years old. Beese (1989) believed that, because of the extremely productive site, the trees may be younger than they appear. If so, this fact highlights the exceptional character of the site, with growth rates perhaps unrivalled in Canada.

The dominant tree heights of old-growth Sitka spruce are similar to those of old-growth Douglas-fir. Some recent tree height data from selected locations on Vancouver Island are shown in Table 7. Table 8 shows a comparison between Carmanah Creek valley spruce and other large Sitka spruce currently recorded in British Columbia.

Current information, summarized above from Stoltmann (1993) and American Forests (1996), indicates that the spruces with the largest diameters recorded to date are in Washington and Oregon, and the tallest are in British Columbia, on Vancouver Island. These implied north-south differences in the tree form (height/diameter relationship) of very large Sitka spruce have not been confirmed or explained. If bigness were expressed in terms of stem volume or aboveground biomass, and if presence or absence of wind-broken treetops as an influence on tree height were taken into account, there may be little difference in the bigness of Sitka spruce in the portions of its north-south range from southern British Columbia to northern Oregon.

Geographic Distribution and Ecological Characteristics of Ecosystems Supporting Sitka Spruce

The ecological characteristics of Sitka spruce that define and influence its natural distribution in western North America can be summarized by the following points emphasized by Klinka et al. (1989):

- a shade-intolerant, submontane to montane, Pacific North American evergreen conifer
- occurs in hypermaritime to maritime cool mesothermal climates on nitrogen-rich soils
- grows poorly on moisture-deficient and nutrient-deficient soils
- occurrence increases with increasing latitude and precipitation and decreases with increasing elevation and continentality
- forms pure, open-canopy stands along the outer coast on sites affected by ocean spray and brackish water, and in advanced stages of primary succession on floodplains
- usually associated with black cottonwood, western hemlock, or western redcedar

- most productive on fresh and moist, very nutrient-rich soils within very wet cool mesothermal climates
- characteristic of hypermaritime mesothermal forests.

The first subsection below, 'Latitudinal and Altitudinal Distribution of Sitka Spruce,' begins with a description of the distribution of Sitka spruce within its natural range in western North America. The subsection 'An Overview of Sitka Spruce and Sitka Spruce–Western Hemlock Forest Types in Coastal British Columbia' provides an overview of British Columbia ecosystems that contain Sitka spruce. A typical mature Sitka spruce ecosystem is described in 'An Example of a Typical Mature Sitka Spruce Ecosystem,' based on recent documentation of ecosystems in the Carmanah Creek valley, Vancouver Island. 'Sitka Spruce in Sparsely Treed Wetland Ecosystems' describes a Sitka spruce swamp ecosystem that is common in the Very Wet Hypermaritime CWH Subzone from the northern tip of Vancouver Island northward to southeastern Alaska.

This is followed by the subsection 'Summary of Biogeoclimatic Zones, Subzones, Variants, and Site Series Where Sitka Spruce Occurs in British Columbia.' The subsection 'An Example of a Typical Ecosystem Where Sitka

Table 8

Comparison of Carmanah Creek valley Sitka spruce with other large Sitka spruce currently recorded in British Columbia

Location	Circumference at breast height (m)	Approx. diameter at breast height (m)	Height (m)	Crown spread (m)
Carmanah Pacific Provincial Park, Vancouver Island	9.58	3.05	95.7	14.0
Nasparti River, Vancouver Island	13.74	4.37	60.1	26.8
Meares Creek, Meares Island	13.70	4.36	48.8	21.6
West Walbran Creek, Vancouver Island	11.28	3.59	78.0	18.3
Dolomite Narrows, Gwaii Haanas/South Moresby National Park Reserve, Queen Charlotte Islands	14.78	4.70	?	?

Source: Guide to the Record Trees of British Columbia (Stoltmann 1993)

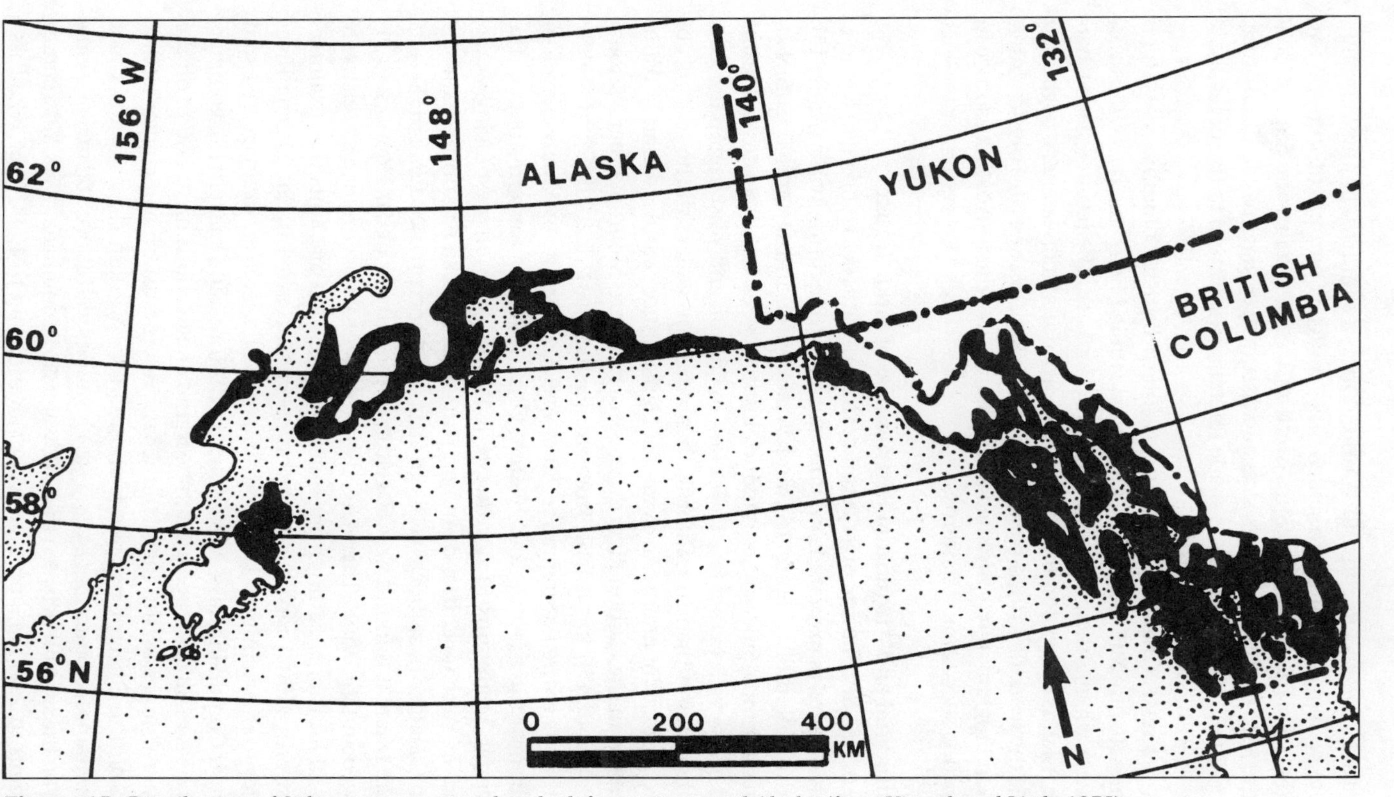

Figure 15. Distribution of Sitka spruce–western hemlock forests in coastal Alaska (from Viereck and Little 1975).

Spruce Occurs in Oregon and Washington' is a summary of plant communities or ecosystem units in which Sitka spruce is prominent in these two states. The last subsection, 'Sitka Spruce's Tree Associates and Successional Sequences,' describes several successional sequences that involve Sitka spruce, concluding with notes on its shade tolerance in relation to its successional role in west coast ecosystems.

It is stressed that ecological succession in British Columbia ecosystems dominated by Sitka spruce has not been studied in as much detail as ecosystems where other west coast conifers dominate. The recent Queen Charlotte Islands successional studies by Forestry Canada researchers, beginning with the work described by Smith et al. (1984) for the Fish-Forestry Interaction Program, have provided new insights into the role of Sitka spruce in recurring successional patterns.

Latitudinal and Altitudinal Distribution of Sitka Spruce

The overall distribution of Sitka spruce in North America is shown in Figure 1. More detailed maps of its distribution are provided here for Alaska (Figure 15), British Columbia (Figure 16), and Washington and Oregon (Figure 17).

In California, Sitka spruce is found close to the sea at the mouths of streams and in low valleys that face the ocean. In Oregon and Washington it occurs further inland than in California, extending up valleys to the foothills of the Cascade Mountains. An example of Sitka spruce's eastward limit in Washington is in the Carbon River drainage of northern Mount Rainier National Park, about 170 km inland from the Pacific coast (Franklin et al. 1988). Within its entire natural range, Sitka spruce attains its maximum development near the coast in northwestern Oregon, on the western slopes of the Olympic Mountains of northwestern Washington, on west-facing watersheds of Vancouver Island, and on parts of the Queen Charlotte Islands.

In Canada, Sitka spruce has its greatest inland occurrence along the Skeena River, where it is found near Hazelton in the Interior Cedar-Hemlock (ICH) biogeoclimatic zone. There are also reports of its occurrence at relatively inland locations, dating from some of the earliest accounts. For example, Sudworth (1908) reported that this species occurred in the Skeena River valley to the vicinity of Hazelton, and in the Fraser River drainage eastward to near Coquihalla Pass. According to Whitford and Craig (1918), commercial sizes of Sitka spruce used to be found at elevations up to 300 m on the western slopes of the mainland Coast Mountains, on Vancouver Island, and on the Queen Charlotte Islands. On the western slopes of the Coast Mountains, it sometimes extends to elevations well above 300 m. For example, on the summit between the Coldwater and Coquihalla rivers, Sitka spruce extends to 1,000 m, and it grows at elevations up to 1,500 m at Taku Pass.

On Vancouver Island, prevailing winds from the west can carry Sitka spruce seed over some ridges of the mountains, and if there is a favourable site, the species may become established at a relatively high altitude. Stanek (pers. comm., 1985) noted that Sitka spruce trees about 60 m tall and with diameters of 1.6 m have been reported at elevations of about 1,000 m on Vancouver Island. On the British Columbia mainland, Stanek also observed Sitka spruce in 1962 on benches of the North Alouette River and in the southwest corner of the Malcolm Knapp Research Forest near Haney that were up to 1.5 m in diameter just above their fluted bases. In the same area, Sitka spruce over 1 m in diameter just above the fluting, about 300 years

Figure 16. Natural range of Sitka spruce in British Columbia (from Krajina et al. 1982).

old, were present on the peak west of Gwendoline Lake, about 750 m above sea level. Perhaps an indication that Sitka spruce will have a long-term role in future forest management is the fact that the species is scattered throughout second-growth forests in much of the Lower Mainland of British Columbia.

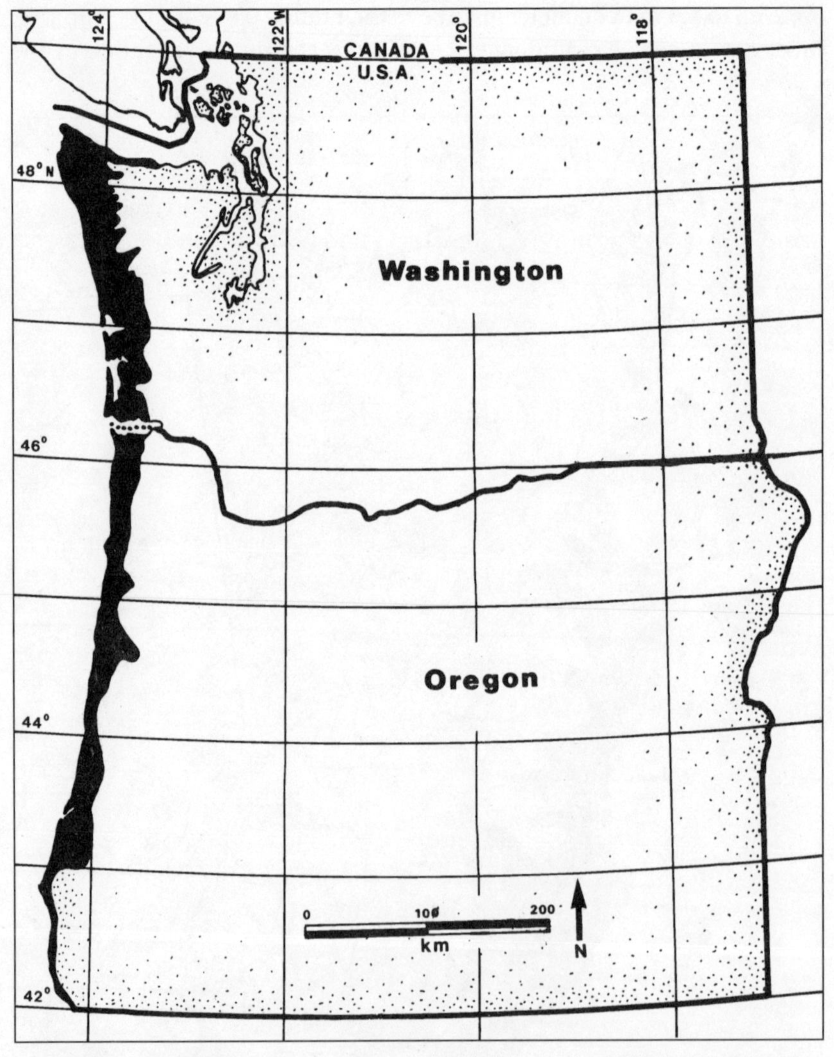

Figure 17. The Sitka spruce zone (black area) in western Washington and Oregon (from Franklin and Dyrness 1973). As shown in Figure 1, Sitka spruce's natural range in Washington extends further inland to about long. 121°30'W, east of Puget Sound, but its main zone of occurrence is in the north-south coastal band shown above.

Most of the remaining old-growth Sitka spruce stands in British Columbia are in the northern coastal region extending from about the northern end of Vancouver Island to the vicinity of Portland Canal and including the Queen Charlotte Islands. Sitka spruce in this region commanded the attention of early researchers; Bier et al. (1946), Foster and Foster (1951), and Day (1957) all reported on Sitka spruce distribution on Graham Island and Moresby Island. In the more accessible parts of its range in British Columbia, the best opportunity to view stately, old-growth Sitka spruce now is generally around designated protected areas such as parks or picnic sites. Some of the key areas of protected Sitka spruce in British Columbia are: Gwaii Haanas/South Moresby National Park Reserve, Pacific Rim National Park, Cape Scott Provincial Park, the vicinity of Lake Victoria near Port Alice on Vancouver Island, Naikoon Provincial Park on Graham Island, and the campground area near the southeastern end of Nitinat Lake.

An Overview of Sitka Spruce and Sitka Spruce–Western Hemlock Forest Types in Coastal British Columbia

A recent inventory of old-growth forests in coastal British Columbia by Roemer et al. (1988) provided a concise overview of ecosystems characterized by the presence of Sitka spruce. Their description, with minor modifications, is reproduced here.

The Sitka spruce forest type occurs along the length of the British Columbia coast, but in its typical form is restricted to a narrow shoreline fringe characterized by a hypermaritime climate accentuated by sea spray and frequent fogs. The belt widens from south to north, so that in the vicinity of the Queen Charlotte Islands and the adjacent mainland coast, Sitka spruce extends further inland and further up the mountain slopes than in the vicinity of Vancouver Island. Typical associated tree species are western hemlock on well-drained soils, western redcedar in swampy and alluvial environments, and black cottonwood and red alder on alluvial floodplains and landslide surfaces.

Understorey vegetation varies greatly according to habitat exposure, crown closure, and stand age. Major shrub species include *Rubus spectabilis, Gaultheria shallon, Vaccinium parvifolium, Oplopanax horridus,* and *Ribes bracteosum.* Of these shrubs, *Rubus, Oplopanax,* and *Ribes* are most prominent on alluvial sites, whereas *Gaultheria* and *Vaccinium* are more abundant on upland sites. On the Queen Charlotte Islands, a shrub layer that would normally be dominant in that area's biogeoclimatic variants is often very suppressed because of heavy browsing by Sitka deer (*Odocoileus hemionus sitkensis*).

The herb layer also varies greatly. Typical species are *Maianthemum dilatatum, Polystichum munitum,* and *Tiarella trifoliata,* but *Blechnum spicant* and *Moneses uniflora* are typical for poorer and upland sites, while *Athyrium filix-femina* dominates in rich alluvial sites. Dense *Calamagrostis nutkaensis* occurs under Sitka spruce on exposed marine headlands. The main mosses are *Kindbergia oregana* and *Rhytidiadelphus loreus,* accompanied by *Plagiomnium* spp. on richer sites.

Typical coastal-fringe habitats for Sitka spruce include sand, shingle, and rocky beaches; headlands; delta terraces; and steep colluvial windward slopes. On sites less severe than these near-coastal habitats, Sitka spruce achieves its best growth on floodplain terraces, which may occur a considerable distance inland along major rivers. This natural occurrence of Sitka spruce on a wide range of sites indicates an inherently wide ecological amplitude that allows this species to grow successfully on a wide range of climatic and soil conditions when planted as an exotic species in Europe. The key feature limiting its occurrence appears to be its requirement for seedbeds that have little competition from other species. Examples of desirable seedbeds are rotten logs, landslides, and exposed mineral soil near streams, rivers, and tidal flats. High light intensity is another requirement for the establishment of Sitka spruce.

Sitka spruce–western hemlock forest types occur along the entire British Columbia coast, but on the south coast are restricted to a relatively narrow subzone on the west side of Vancouver Island, especially on river terraces. North of Vancouver Island, ecological conditions favourable to Sitka spruce are more common than on Vancouver Island. On the Queen Charlotte Islands and the northern mainland coast, inland from the Alaska Panhandle, Sitka spruce forests may dominate upland till landscapes as well as moist but well-drained colluvial and fluvial landforms. Sitka spruce and western hemlock codominate these stands – the spruce because of very large and old but sometimes scattered trees, the hemlock because of high numbers and total volume (Plate 9). In these ecosystems, associated tree species include western redcedar, amabilis fir, yellow-cedar, mountain hemlock, and red alder. Typically, about 25% of the total basal area of these old-growth stands consists of standing dead snags in various stages of deterioration (Smith et al. 1986). Shrub layers contain ericaceous species on upland sites, whereas *Oplopanax horridus* and *Rubus spectabilis* dominate on rich sites of floodplains. Herb layers are dominated by ferns and may be lush in alluvial forests. The bryophyte layer shows high coverage and is rich in species.

Roemer et al. (1988) pointed out that the best growth of Sitka spruce is on fertile, well-drained soils of floodplains, where massive, tall, clean-boled spruce are dominant in open stands with well-developed understoreys. By contrast, on upland till sites western hemlock dominates the main canopy, whereas spruce tends to occur as scattered emergents. Sitka spruce may not be a true climax species in such forests, but long-lived individuals persist, often as dominants. Young spruce regenerate, commonly on decaying wood, and regain the canopy through gaps resulting from death, blowdown, and landslides (see Figure 23). The Sitka spruce–western hemlock forest type is also common on surfaces of former landslides following an earlier successional stage of red alder (Plates 10 and 11). Typical soils are Orthic and Gleyed Humo-Ferric and Ferro-Humic Podzols (often with folic phases), with the addition of Dystric and Sombric Brunisols on fluvial, colluvial, and marine parent materials. There are also Sitka spruce swamp forests with western redcedar, red alder, and *Lysichiton americanum* on Gleysols and organic soils (Roemer et al. 1988).

An Example of a Typical Mature Sitka Spruce Ecosystem

If site association (Pojar et al. 1987) is taken as the basic unit for describing forest ecosystems and is understood to mean an area that will support the same 'climax' plant species in a relatively homogeneous environment, then the Carmanah Creek valley site associations described by Beese (1989) serve as a typical description of a Sitka spruce ecosystem. The protected area for Sitka spruce in the Carmanah Creek valley is within the Submontane Variant (CWHvm1) of the Very Wet Maritime CWH Subzone. Most of the remainder of the Carmanah Creek watershed is also within this variant, except slopes above about 600 m elevation, which are within the Montane Variant (CWHvm2) (Banner et al. 1990).

The CWHvm1 (submontane) variant is characterized by the presence of amabilis fir, western hemlock, western redcedar, and varying amounts of Sitka spruce. The variant has a long growing season and high rainfall. Many areas receive over 3,500 mm of annual precipitation, most of which is in the form of rain. The CWHvm2 (montane) variant is characterized by the presence of yellow-cedar, with correspondingly less western redcedar and more amabilis fir than the submontane variant. It has a shorter, cooler growing season than the submontane variant, with winter precipitation in the form of snow (Klinka et al. 1991).

The following description of the Carmanah Creek valley Sitka spruce ecosystem is adapted from Beese (1989). Stands of large Sitka spruce occur on alluvial deposits in the Carmanah Creek floodplain and on glaciofluvial and glaciolacustrine terraces. Soils are generally well drained and range in texture from clayey or sandy silts to gravelly sands, interbedded in layers of varying thickness on the alluvial plain (Rollerson 1989).

Most of the Carmanah Creek floodplain is occupied by the BaCw–Foamflower Site Association (Silviculture Interpretations Working Group 1994). The shrub layer is dominated by *Rubus spectabilis, Vaccinium parvifolium,* and *Vaccinium alaskaense. Ribes bracteosum* is common on river banks. The lush herbaceous layer includes the following: *Polystichum munitum, Athyrium filix-femina, Dryopteris expansa, Blechnum spicant, Disporum hookeri, Maianthemum dilatatum, Galium triflorum, Urtica dioica, Tolmiea menziesii, Trillium ovatum, Petasites palmatus, Viola glabella,* and several grasses.

Aruncus dioicus is a characteristic herb on river banks. The most abundant mosses are *Hylocomium splendens* and *Rhytidiadelphus loreus*. *Kindbergia oregana, Plagiothecium undulatum, Leucolepis menziesii,* and *Pogonatum* spp. are also common. The liverwort *Pellia neesiana* is found in moist depressions.

The moist, nutrient-rich soils of this ecosystem make it very productive for tree growth. The mull humus form typifies sites with rapid incorporation of organic matter into the upper mineral soil by microfaunal activity. Sitka spruce achieves its best growth on these alluvial sites, which have an abundant supply of nutrients, flowing groundwater, and a long growing season.

The HwBa–Deer Fern Site Association occurs on well-drained terraces on the western edge of lower Carmanah Creek. This association is dominated

Table 9

Some comparative features of Sitka spruce and western hemlock that control their distribution and growth

Sitka spruce	Western hemlock
• Very long sustained height growth	• Shorter sustained height growth
• Needs moist, competition-free seedbed with light; can be nurse logs or soil	• Can become established on low light seedbeds with competition
• Can grow well on mineral soils, if the exchange capacity is high	• Grows on moist organic seedbeds as well as moist mineral soils
• Can tolerate salt spray	• Low tolerance to salt spray
• Responds well to rich soils and N and P fertilizer additions	• Nutrition in organic soils often adequate; erratic record of response to fertilizers
• Narrow niche or ecological amplitude in natural forests due to specialized natural regeneration requirement on rich soils	• Wide niche and wider amplitude in natural forests due to unspecialized natural regeneration requirement and advance-growth reproduction strategy
• When planted, actually has wide ecological amplitude in terms of moisture, nutrition, and climate	• When planted as container stock grown in organic soils, has a narrower nutritional amplitude
• Has restrictive tolerance to frost; grows only on the coast of British Columbia	• Can tolerate frost and cold better; grows in the moist interior of British Columbia
• Subject to weevil attack but is only an occasional host for dwarf mistletoe; resistant to flooding	• Is a primary host for dwarf mistletoe but is not subject to weevil attack; weak resistance to flooding

by *Vaccinium parvifolium, V. alaskaense,* and *Blechnum spicant,* with a moss layer of *Hylocomium splendens* and *Rhytidiadelphus loreus.* Some *Menziesia ferruginea* and *Gaultheria shallon* are also present. Small areas of *Lysichiton americanum* are found in poorly drained depressions. *Sphagnum* and *Pellia* are common in these wet areas (Beese 1989).

It should be noted that the restricted occurrence of Sitka spruce in natural forests, where it competes with western hemlock, may provide the best indication of where Sitka spruce should be planted. When planted in situations without severe competition, Sitka spruce has a wide ecological amplitude (see Table 9).

Sitka Spruce in Sparsely Treed Wetland Ecosystems

Besides its role in certain closed-canopy forest ecosystems, Sitka spruce is also represented in coastal wetland ecosystems that contain only scattered individual trees. Banner et al. (1986) described such ecosystems, referring to them as part of the Pacific Oceanic Wetland Region. This particular regional landscape classification terminology is not used here because it is assumed that Banner and co-workers were simply referring to lowlands and lower mountain slopes of the outer British Columbia coast from northern Vancouver Island to southeastern Alaska, which in this text is referred to as the Very Wet Hypermaritime CWH Subzone (CWHvh). Wetlands cover nearly as much area as closed forest in this region, not only because the topography is generally subdued but also because the cool, wet climate promotes organic soil formation.

The four main classes of wetlands are bogs, fens, marshes, and swamps; bogs are more common than the others. Bog forest and bog woodland support shore pine (*Pinus contorta* var. *contorta*) as well as western redcedar and yellow-cedar, but shore pine is not common in conifer swamps, which are frequent but localized in this region. There are three general types of conifer swamps: *Thuja plicata* seepage swamps, *Picea sitchensis* floodplain swamps, and *Chamaecyparis nootkatensis* montane swamps. Similar types of treed swamps occur in southeastern Alaska (Banner et al. 1986). The description of the Sitka spruce–dominated wetland by Banner et al. (1986) can be summarized as follows. Their summary of climatic data from 12 locations in the northern part of the Very Wet Hypermaritime CWH Subzone (CWHvh) is reproduced in Table 10.

Sitka spruce swamps occur on fluvial fans and terraces, and are characterized by Gleysols and Mesisols, with Histomor and Histomoder humus forms (Banner et al. 1986). Selected soil characteristics are reproduced in Table 11. These swamps develop on level or very gently sloping valley-bottom terrain. Sitka spruce swamps have a less stagnant hydrology than western redcedar swamps, resulting in better soil aeration, higher nutrient status, and better tree productivity.

Table 10

Climate statistics for the northern part of the Very Wet Hypermaritime Subzone (CWHvh) of the Coastal Western Hemlock (CWH) Biogeoclimatic Zone, British Columbia

Characteristic	Value (range)[a]
Number of stations	12
Mean elevation (m)	25 (3-89)
Mean annual temperature (°C)	7.8 (6.7-8.5)
Mean temperature May to Sept. (°C)	11
Mean temperature June to Aug. (°C)	13
Mean temperature warmest month (°C)	14
Mean extreme maximum temperature (°C)	30 (23-39)
Mean temperature Nov. to Feb. (°C)	3.5
Mean temperature coldest month (°C)	2
Mean extreme minimum temperature (°C)	−16 (−11 to −24)
Mean number of days of frost	43
Mean frost-free period (days)	205 (156-272)
Mean number of months with mean monthly temperature greater than 10°C	4 (4-5)
Mean number of months with mean monthly temperature below 0°C	0 (0-1)
Mean annual growing degree-days (based on 5°C)	1,337 (1,148-1,592)
Mean annual total precipitation (mm)	2,378 (1,152-4,387)
Mean annual total snowfall (cm)	96 (51-155)
Driest month	July
Mean total precipitation driest month (mm)	91 (43-151)
Wettest month	Oct.
Mean total precipitation wettest month (mm)	334 (198-625)
Mean number of days with measurable precipitation	217 (210-254)
Mean total potential evapotranspiration March to Oct. (mm)	87 (49-156)

Source: Banner et al. 1986. Data from Environment Canada (1980) for the following 11 stations: Bonilla Island, Cape St. James, Ethelda Bay, Langara, McInnes Island, Masset, Ocean Falls, Prince Rupert, Prince Rupert Airport, Tasu Sound, Tlell.

[a] Values are means except for ranges of data in parentheses.

Table 11

Selected soil profile characteristics for shallow open water, western redcedar swamp, and Sitka spruce swamp in the northern part of the Very Wet Hypermaritime Subzone (CWHvh) of the Coastal Western Hemlock (CWH) Biogeoclimatic Zone, British Columbia

Wetland type	Horizon	Depth (cm)	pH	% C	% N	Pyro-phosphate sol. index	% ash	Bulk density	me/100 g Ca	Mg	K	Al	% sand	% silt	% clay	Calorific value cal/g
Shallow open	Om	0-28	3.8	34.1	1.85	21.4	34.0	0.07	4.86	3.36	1.55	4.66	–	–	–	4,162
water	Oh	28-52	3.7	17.3	1.03	18.3	67.3	0.36	3.00	0.69	0.51	4.00	–	–	–	2,450
	IICg	130+	4.0	0.4	–	–	–	–	0.71	1.48	0.39	0.72	91.7	5.8	2.5	
Western redcedar	Fy	0-5	4.2	19.7	0.67	19.0	60.9	0.15	15.54	3.36	1.98	1.67	–	–	–	2,166
swamp	Om1	6-60	3.6	41.7	1.59	41.8	24.2	0.13	7.08	2.07	0.58	8.33	–	–	–	5,063
	Om2	60-105	3.3	48.3	1.41	33.9	10.7	0.14	11.16	2.76	0.48	7.66	–	–	–	5,427
	Om3	105-130	3.6	38.9	1.26	47.9	26.9	0.20	7.25	3.36	0.39	9.32	–	–	–	4,443
	IIAhg	130-150	3.9	4.7	0.14	–	–	–	1.50	0.30	0.37	3.33	86.1	14.7	2.2	
Sitka spruce	Fyi	4-0	4.0	19.1	0.80	18.9	65.4	–	9.00	1.58	1.36	2.99	–	–	–	2,064
swamp	Ahg	0-55	4.0	8.4	0.48	–	–	–	3.78	0.69	0.60	4.66	59.1	5.4	32.6	

Source: Banner et al. (1986)

Table 12

Prominence values for species in various strata of Sitka spruce swamp, based on 9 sampled stands in the northern part of the Very Wet Hypermaritime Subzone (CWHvh) of the Coastal Western Hemlock (CWH) Biogeoclimatic Zone, British Columbia

Strata	Species	Prominence values[a]
Overstorey trees	*Thuja plicata*	150
	Tsuga heterophylla	211
	Picea sitchensis	306
	Alnus rubra	30
Understorey trees	*Thuja plicata*	28
	Tsuga heterophylla	178
	Picea sitchensis	11
	Alnus rubra	17
Tree saplings and seedlings/stunted trees	*Tsuga heterophylla*	220
	Thuja plicata	25
	Picea sitchensis	30
Shrubs	*Menziesia ferruginea*	102
	Gaultheria shallon	10
	Vaccinium alaskaense	37
	Vaccinium ovalifolium	12
	Vaccinium parvifolium	97
Herbs/dwarf shrubs	*Lysichiton americanum*	108
	Blechnum spicant	54
	Cornus unalaschkensis	46
	Coptis aspleniifolia	10
	Rubus pedatus	28
	Maianthemum dilatatum	25
	Veratrum viride	2
Mosses, liverworts, and lichens	*Rhytidiadelphus loreus*	191
	Hylocomium splendens	180
	Sphagnum girgensohnii	65
	Pellia neesiana	22
	Rhizomnium glabrescens	61
	Stokesiella oregana	45
	Leucolepis menziesii	29
	Conocephalum conicum	26
	Scapania bolanderi	34
	Pogonatum alpinum	32
Average cover (%) of strata	Tree	75
	Shrub	27
	Herb/dwarf shrub	40
	Moss/liverwort/lichen	85

Source: Banner et al. (1986)

[a] mean cover × frequency

Stands are typically patchy, with irregular and uneven canopies and frequent windthrow. Sitka spruce and western redcedar dominate, while western hemlock and sometimes amabilis fir form a subcanopy layer. Red alder typically occurs in canopy openings. Sitka spruce exhibits greater dominance and much better growth in these swamps than in western redcedar swamps.

The moderately well developed shrub layer typically has abundant tree regeneration (dominated by western hemlock) and scattered clumps of *Vaccinium alaskaense, Menziesia ferruginea, Rubus spectabilis,* and *Oplopanax horridus. Lysichiton americanum* characterizes the moderately developed herb layer. The well-developed bryophyte layer is dominated by *Pellia neesiana, Conocephalum conicum, Sphagnum girgensohnii,* and *S. squarrosum. Rhizomnium glabrescens* dominates in wetter, periodically flooded depressions, and *Rhytidiadelphus loreus, Hylocomium splendens,* and *Kindbergia oregana* on drier, elevated microsites.

Prominence values (mean cover × frequency) for species in various strata of nine sampled Sitka spruce swamps were compiled by Banner et al. (1986) and are reproduced in Table 12.

Summary of Biogeoclimatic Zones, Subzones, Variants, and Site Series Where Sitka Spruce Occurs in British Columbia

To portray Sitka spruce's ecological and forestry role in British Columbia in relation to biogeoclimatic zones, it is necessary to summarize the geographic distribution of the zones where this spruce can occur. The distribution of 14 biogeoclimatic zones in British Columbia is shown in Plate 7. Sitka spruce's natural occurrences and growth potential in British Columbia are most closely linked with the CWH zone, the distribution of which is shown in Figure 18. The biogeoclimatic zone with the second greatest probability of Sitka spruce or Sitka spruce hybrid occurrence is the near-coastal part of the Interior Cedar-Hemlock (ICH) Zone, which is shown in Figure 19.

To portray Sitka spruce's potential distribution altitudinally and latitudinally in British Columbia, Figures 20 and 21 show elevational profiles of the CWH zone, the main zone of Sitka spruce occurrence, from the latitude of the Queen Charlotte Islands (about 53°20′N) south to the southern part of Vancouver Island (about 49°15′N).

On a local scale that would typically include 4-6 site series along a topographic gradient of differing soil moisture availability, Sitka spruce's most characteristic occurrence is at the more moist, lower-elevation end of the topographic gradient, as shown in a generalized sketch for the CWH zone (Figure 22).

The earliest investigators of Sitka spruce (Cary 1922) noted that this species occurs in a region characterized by an abundance of rainfall, frequent fog, and temperatures moderated by proximity to the sea. Climatic normals

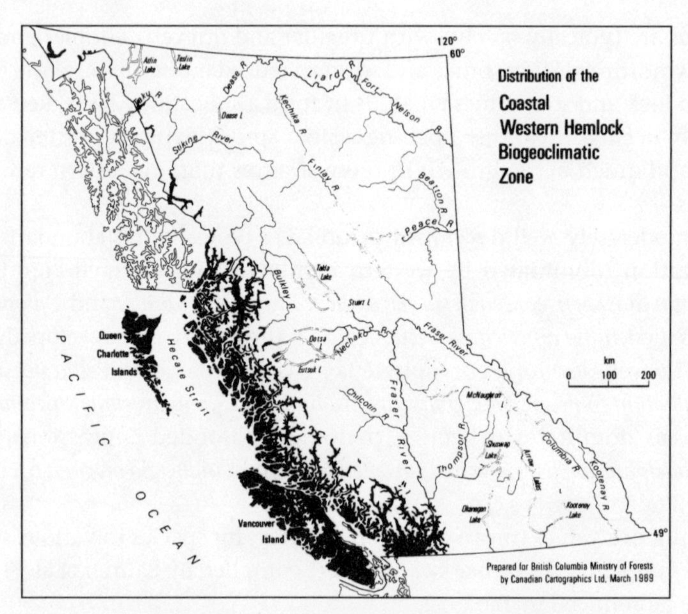

Figure 18. Distribution of the Coastal Western Hemlock Biogeoclimatic Zone in British Columbia (from Pojar et al. 1991), the main zone of Sitka spruce occurrence in the province.

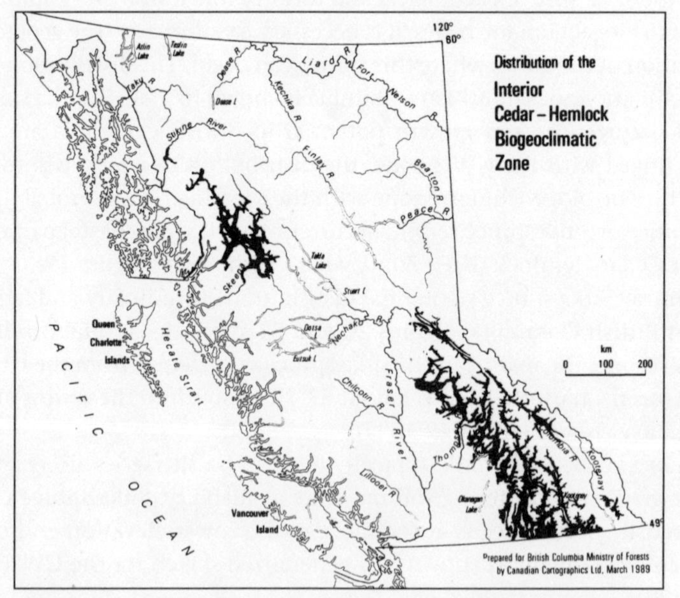

Figure 19. Distribution of the Interior Cedar-Hemlock Biogeoclimatic Zone in British Columbia (from Ketcheson et al. 1991), in which Sitka spruce occurs only in the near-coastal part of the zone, mainly along the Skeena and Nass River valleys.

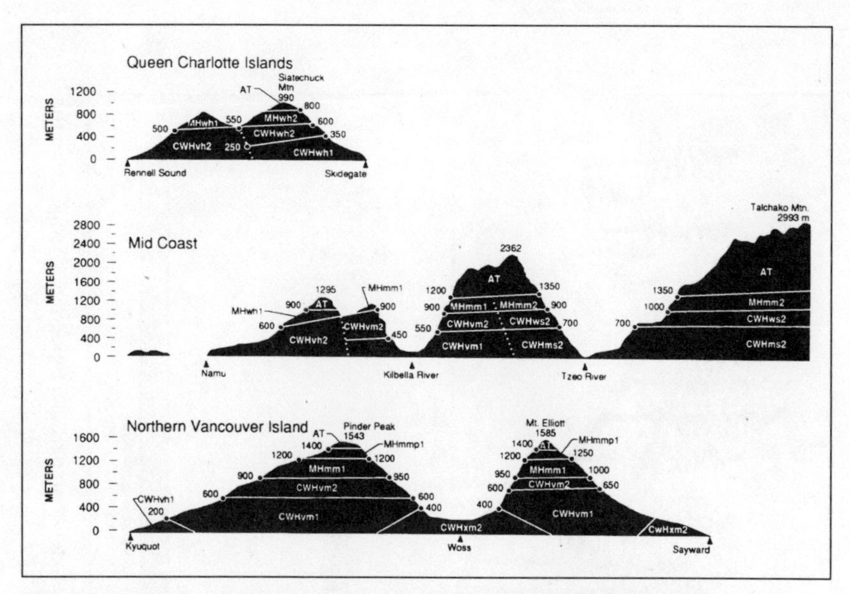

Figure 20. Elevational profiles of the Coastal Western Hemlock Biogeoclimatic Zone, the main zone of Sitka spruce occurrence, at latitudes 53°20′N (Queen Charlotte Islands), 51°50′N (mid coast), and 50°15′N (northern Vancouver Island) (from Green and Klinka 1994). There is limited occurrence of Sitka spruce in the Mountain Hemlock Biogeoclimatic Zone at the latitudes from mid-coast north to the Queen Charlotte Islands.

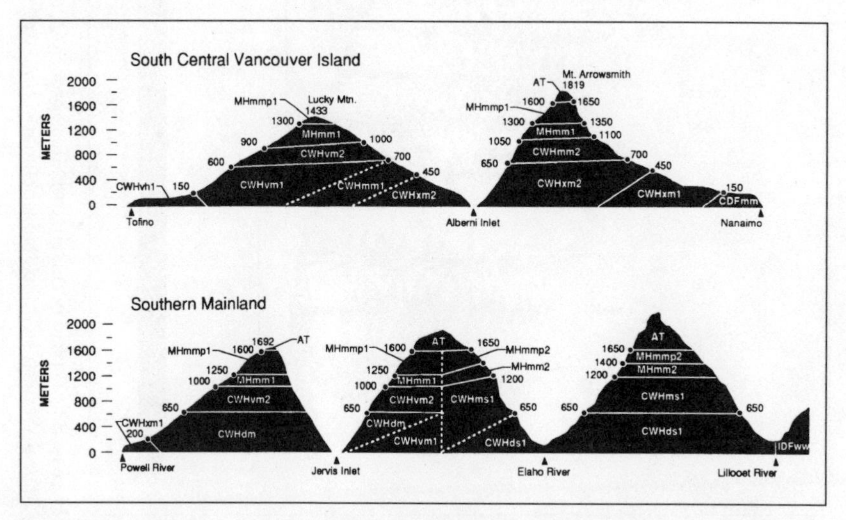

Figure 21. Elevational profiles of the Coastal Western Hemlock Biogeoclimatic Zone, the main zone of Sitka spruce occurrence, at latitudes 49°15′N (south-central Vancouver Island) and 49°50′N (southern mainland) (from Green and Klinka 1994).

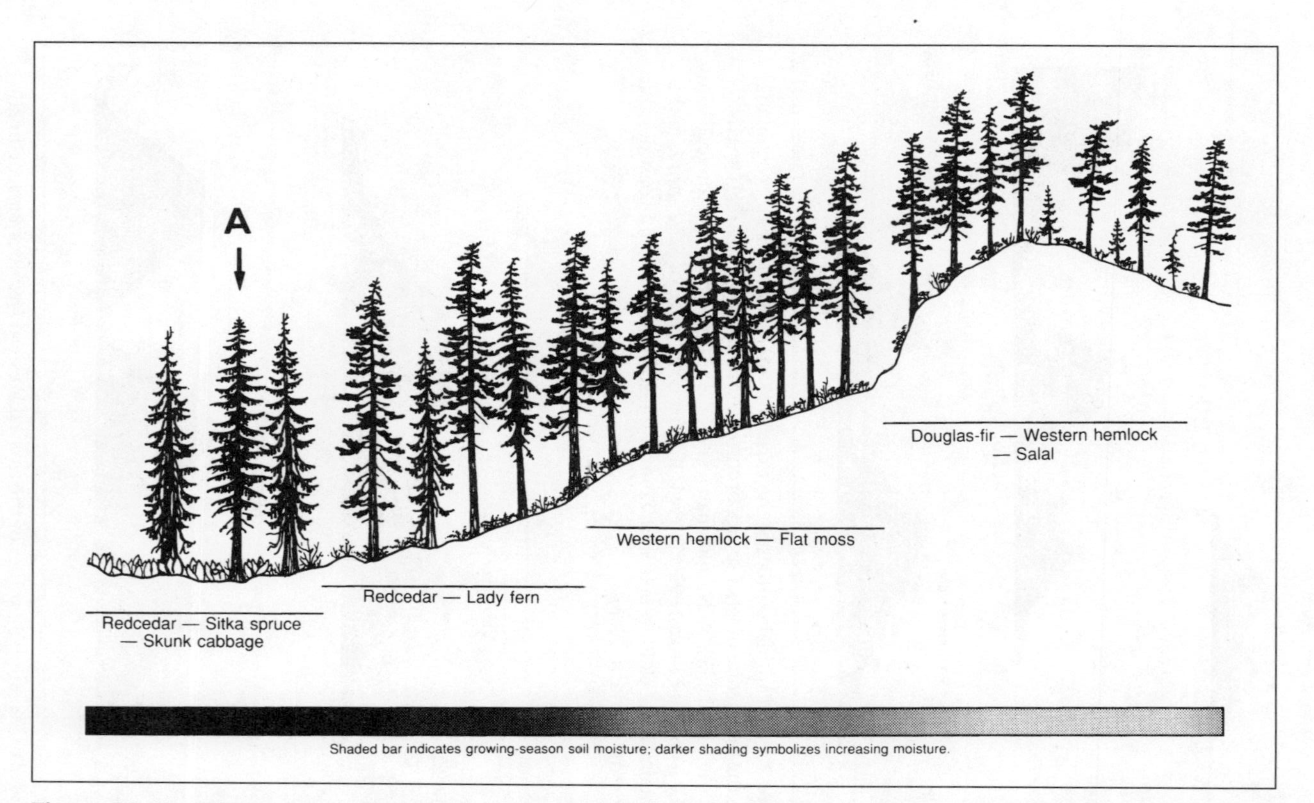

Figure 22. Typical occurrence of Sitka spruce (A) in a relatively moist, lower-elevation part of a topographic sequence in the Coastal Western Hemlock Biogeoclimatic Zone (from Pojar et al. 1991).

are summarized in Table 13 for nine weather stations in the Submontane Variant (CWHvm1) on western Vancouver Island. This is the biogeoclimatic variant in which Sitka spruce reaches its greatest prominence relative to other west coast conifers on the west coast of Vancouver Island.

Sitka spruce occurs naturally, or has silvicultural potential as a regeneration species, on a large number of site series in coastal British Columbia. Criteria for the ecological acceptability of tree species on a site-series basis have been defined by the Silviculture Interpretations Working Group (1994), and are summarized here to portray the breadth of Sitka spruce's potential silvicultural role. Site series in which Sitka spruce is considered ecologically acceptable (as either a primary, secondary, or tertiary species) are listed, in order of decreasing productivity (site index) classes, for the

Table 13

Means and standard deviations (in parentheses) of climatic characteristics in a portion of the Submontane Variant (CWHvm1) on western Vancouver Island, British Columbia

Climatic characteristic	Mean (standard deviation)
Number of climatic data sets	9
Mean annual precipitation (mm)	3,016 (70.8)
Mean temperature of the coldest month (°C)	4.6 (0.37)
Index of continentality	1 (0.2)
Mean radiation during growing season (Ly)	40,300 (310)
Mean precipitation April to September (mm)	774 (48.8)
Mean temperature of the warmest month (°C)	14.2 (0.72)
Accumulated degree-days over 5.6°C	1,390 (145.5)
Frost-free period (days over 0°C)	263 (11.4)
Mean precipitation of the driest month (mm)	83 (6.1)
Mean precipitation of the wettest month (mm)	445 (20.5)
Mean annual temperature (°C)	9.2 (0.43)
Number of months with mean temperature over 10°C	5.1 (0.93)
Number of months with mean temperature below 0°C	0.0 (0.0)
Water surplus (mm)	2,485 (73.4)
Water deficit (mm)	0 (0.0)
Number of months with water deficit	0.0 (0.0)
Maximum snow depth (cm)	0.0[a] (0.0)
Number of months with snow	0.0[a] (0.0)
Potential evapotranspiration (mm)	531 (10.7)
Actual evapotranspiration (mm)	531 (10.7)
Ratio actual to potential evapotranspiration (%)	100 (0.0)
Actual evapotranspiration April to September (mm)	485 (8.6)

Source: Klinka et al. (1979)

[a] Although snow occasionally occurs on western Vancouver Island, for the period of climatic records analyzed by Klinka and co-workers (years of data records unspecified), no snow was recorded in these 9 climatic data sets.

Table 14

Site series in the Vancouver Forest Region in which Sitka spruce has potential as a primary, secondary, or tertiary species, listed in order of decreasing productivity (site index) classes

Biogeoclimatic zone, subzone, or variant[a]	Site series name (no.)[a]	Site index, median ht (m) at b.h. age 50[b]
CWHdm; CWHxm; CWHds1; CWHds2; CWHvm1	Ss–Salmonberry (08,09)	34[c]
CWHvm1	BaCw–Salmonberry (07)	34
CWHds2	Cw–Devil's club (07)	34
CWHwh1; CWHvh1; CWHvh2	Ss–Lily-of-the-valley (07,08)	34
CWHwh1	PlYc–Sphagnum (11)	– (34)[d]
CWHwh1	CwSs–Skunk cabbage (12)	– (37)[d]
CWHwh1; CWHvh1; CWHvh2	HwSs–Lanky moss (01,04)	29[c] (31)[d]
CWHwh1; CWHvh1; CWHvh2	CwSs–Foamflower (05,06)	29[c] (36)[d]
CWHms1; CWHms2; CWHmm1	Ss–Salmonberry (07,08)	29[c]
CWHwh1	CwSs–Conocephalum (06)	29[c] (32)[d]
CWHvh1; CWHvh2	CwSs–Sword fern (05)	29
CWHvm1; CWHvm2	HwBa–Blueberry (01)	29
CWHvm1; CWHvm2	CwHw–Sword fern (04)	29
CWHvm1; CWHvm2	BaCw–Foamflower (05)	29
CWHvm1; CWHvm2	HwBa–Deer fern (06)	29
CWHvm1	BaSs–Devil's club (08)	29
CWHvm2	BaCw–Salmonberry (07)	29
CWHwh1	CwSs–Sword fern (03)	29 (31)[d]
CWHvh1; CWHvh2	CwSs–Devil's club (07)	29 (20)[d]
CWHwh1; CWHvh1; CWHvh2	Ss–Sword fern (16,17)	29
CWHds2	Cw–Solomon's seal (05)	29
CWHms2	BaCw–Oak fern (04)	29
CWHms2	BaCw–Devil's club (06)	29
IDFww[e]	CwFd–Vine maple (05)[e]	29
IDFww[e]	Cw–Devil's club–Lady fern (06)[e]	29
CWHwh2	CwSs–Conocephalum (04)	22[c]
CWHwh2	CwSs–Foamflower (03)	22[c]
CWHwh2	HwSs–Lanky moss (01)	22[c]
CWHvm2	CwYc–Skunk cabbage (11)	22
CWHwh1	CwSs–Salal (02)	22 (30)[d]
CWHwh1	CwHw–Salal (04)	22 (28)[d]
CWHvh1; CWHvh2	CwHw–Salal (01)	22
CWHwh1; CWHvh1; CWHvh2	Ss–Kindbergia (14,15)	22
CWHwh1; CWHvh1; CWHvh2;	Ss–Reedgrass (15,16)	22
CWHdm; CWHds1; CWHds2; CWHmm1; CWHvm1; CWHxm	CwSs–Skunk cabbage (12)	22
IDFww[e]	CwSs–Skunk cabbage (07)	22

►

◀ *Table 14*

Biogeoclimatic zone, subzone, or variant[a]	Site series name (no.)[a]	Site index, median ht (m) at b.h. age 50[b]
CWHwh1; CWHvh1; CWHvh2	Ss–Salal (13, 14)	< 22
CWHms2; CWHvh1; CWHvh2	CwSs–Skunk cabbage (11,13)	< 22
CWHwh1; CWHvh1; CWHvh2	Ss- Slough sedge (17, 18)	< 22
CWHvh1; CWHwh1	CwYc–Goldthread (11)	– (21, 32)[d]
MHwh	HmSs–Blueberry (01)	< 22
MHwh	SsHm–Reedgrass (03)	< 22
MHwh	YcHm–Twistedstalk (05)	< 22
MHwh	YcHm–Hellebore (07)	< 22
MHwh	YcHm–Skunk cabbage (09)	< 22
CWHds2; CWHvm1	Act–Red-osier dogwood (09, 10)	?[f]
CWHvh1; CWHwh1	Ss–Pacific crab apple (19)	?
CWHvh1; CWHvh2; CWHwh1	Ss–Trisetum (08, 09)	?

Source: Site series names are from Silviculture Interpretations Working Group (1994) and Green and Klinka (1994); site index data are from Green and Klinka (1994) and Consortium of Thrower-Blackwell-Oikos (1995).

[a] Site series, and the subzones or variants in which they occur, are listed for site series in which the Silviculture Interpretations Working Group (1994) recorded Sitka spruce either in the name of the site series or as a primary, secondary, or tertiary species in the guidelines for tree species selection. The site series number recorded in parentheses in the middle column of this table is the number assigned to a particular site series in given subzones/variants by the Silviculture Interpretations Working Group (1994). In some cases there are two numbers in parentheses; this indicates that a given site series has a different official number in different biogeoclimatic variants. See Appendix 4 for full names of abbreviated biogeoclimatic units and for tree species whose abbreviations begin the name of each site series.

[b] The Sitka spruce site index values (column 3) assigned to a particular site series (column 2) in the designated subzones/variants (column 1) are based on median site index values shown for Sitka spruce for site classes I, II, and III in Figure 12 of Green and Klinka (1994). For this purpose, Sitka spruce site index class I was recorded as 34 m at breast-height age 50; class II as 29 m; and class III as 22 m. Green and Klinka (1994) gave no numerical value for the class IV site index for Sitka spruce; where their handbook identified site series where Sitka spruce would likely occur as site class IV, the site index in column 3 is recorded as < 22, to indicate that it would be less than the class III site index.

[c] These are cases where a particular site series supports Sitka spruce of a given site index in one biogeoclimatic variant but the same site series in a different biogeoclimatic variant has a lesser or greater median site index.

[d] The Site Index–Biogeoclimatic Ecosystem Classification (SIBEC) project, British Columbia Ministry of Forests, resulted in assembly of Sitka spruce site index data by the Consortium of Thrower-Blackwell-Oikos (1995). The SIBEC Sitka spruce site index estimates that were classed as 'highly reliable' and 'reliable,' and that are assignable to specific site series (excluding the Thrower-Blackwell-Oikos Groupings as site associations), are recorded in parentheses in column 3 of this table. In the CwYc–Goldthread site series, site index is 21 m in the CWHvh1 variant and 32 m in the CWHwh1 variant.

[e] Sitka spruce is not a typical component of the Interior Douglas-Fir (IDF) Biogeoclimatic Zone. Mention of Sitka spruce by the Silviculture Interpretations Working Group (1994) for these three site series of the IDFww subzone is an indication that Sitka spruce could be an acceptable tertiary choice for tree species selection in silvicultural planning in that IDF subzone.

[f] A question mark indicates a site series for which Green and Klinka (1994) did not assign a Sitka spruce site index class in a particular biogeoclimatic variant.

Table 15

Site series in the Prince Rupert Forest Region in which Sitka spruce has potential as a primary, secondary, or tertiary species, listed in order of decreasing productivity (site index) classes

Biogeoclimatic zone, subzone, or variant[a]	Site series name (no.)[a]	Site index, median ht (m) at b.h. age 50[b]
CWHvh2	CwSs–Devil's club (07)	34
CWHvh2	Ss–Lily-of-the-valley (08)	33
CWHvm1	Ss–Salmonberry (09)	33[c]
CWHvm1	Act–Red-osier dogwood (10)	32[c]
CWHwm	Ss–Salmonberry (05)	32[c]
CWHwm	Act–Red-osier dogwood (06)	30[c]
CWHvm1	BaCw–Foamflower (05)	30[c]
CWHvh2	HwSs–Lanky moss (04)	30
CWHvh2	CwSs–Foamflower (06)	30
CWHvh2	CwSs–Sword fern (05)	29
CWHvh2	Ss–Trisetum (09)	29
CWHvm1	BaSs–Devil's club (08)	29[c]
CWHvm1	HwBa–Blueberry (01)	28[c]
CWHvm1	HwBa–Deer fern (06)	28[c]
CWHvm2	BaSs–Devil's club (08)	28[c]
CWHvm2	BaCw–Foamflower (05)	26[c]
CWHvm1	CwHw–Sword fern (04)	25
CWHwm	SsHw–Devil's club (04)	25
CWHwm	SsHw–Oak fern (03)	24
CWHwm	HwSs–Blueberry (01)	23
CWHvm2	HwBa–Blueberry (01)	21[c]
CWHvm2	HwBa–Deer fern (06)	21[c]
CWHvm1	CwSs–Skunk cabbage (14)	19
CWHwm	HwSs–Step moss (02)	17
CWHwm	Ss–Skunk cabbage (09)	15
CWHwm	Hw–Sphagnum (08)	11
CWHvm2	CwHw–Sword fern (04)	?[d]
CWHvm2	BaCw–Salmonberry (07)	?
CWHvm2	CwYc–Skunk cabbage (11)	?
CWHvh2	CwSs–Skunk cabbage (13)	?
CWHvh2	Ss–Salal (14)	?
CWHvh2	Ss–Kindbergia (15)	?
CWHvh2	Ss–Reedgrass (16)	?
CWHvh2	Ss–Sword fern (17)	?
CWHvh2	Ss–Slough sedge (18)	?
CWHvh2	Ss–Pacific crab apple (19)	?
MHwh	HmSs–Blueberry (01)	?
MHwh	SsHm–Reedgrass (03)	?

▶

◀ *Table 15*

Biogeoclimatic zone, subzone, or variant[a]	Site series name (no.)[a]	Site index, median ht (m) at b.h. age 50[b]
MHwh	YcHm–Twistedstalk (05)	?
MHwh	YcHm–Hellebore (07)	?
MHwh	YcHm–Skunk cabbage (09)	?

Source: Site series names are from Silviculture Interpretations Working Group (1994) and Banner et al. (1993); site index data are from Banner et al. (1993).

[a] Site series, and the subzones or variants in which they occur, are listed for site series in which the Silviculture Interpretations Working Group (1994) recorded Sitka spruce either in the name of the site series or as a primary, secondary, or tertiary species in the guidelines for tree species selection. The site series number recorded in parentheses in the middle column of this table is the number assigned to particular site series in given subzones/variants by the Silviculture Interpretations Working Group (1994). See Appendix 4 for full names of abbreviated biogeoclimatic units and for tree species whose abbreviations begin the name of each site series.

[b] The Sitka spruce site index values (column 3) assigned to a particular site series (column 2) in the designated subzones/variants (column 1) are derived from linear measurement of the placement of Sitka spruce on the site index bar graphs shown in the interpretations table on pages 7.32, 7.36, 7.38, 7.40, 7.42, and 7.46 of Banner et al. (1993). Banner and co-workers stressed that these site index estimates are based on very limited, unpublished Ministry of Forests data.

[c] These are cases where a particular site series supports Sitka spruce of a given site index in one biogeoclimatic variant but the same site series in a different biogeoclimatic variant has a lesser or greater estimated site index.

[d] A question mark indicates a site series for which Banner et al. (1993) did not assign a Sitka spruce site index class in a particular biogeoclimatic variant.

Vancouver Forest Region (Table 14) and the Prince Rupert Forest Region (Table 15). Table 16 provides a similar listing of site series, also in the Prince Rupert Forest Region, in which Sitka spruce hybrids have silvicultural potential according to the same criteria.

Sitka spruce has silvicultural potential in 38 different site series in the Vancouver Forest Region (Table 14) and in 33 different site series in the Prince Rupert Forest Region (Table 15). In the Prince Rupert Forest Region, there is also potential for Sitka spruce hybrids to have primary, secondary, or tertiary ecological acceptability on 25 different site series in the CWH and ICH zones (Table 16). In these tables, some site series are listed more than once because they can occur as two different productivity (site index) classes in different biogeoclimatic variants. Also, considering the entire British Columbia distribution of Sitka spruce and its natural hybrids, the total number of site series where this spruce is ecologically acceptable is somewhat less than the 38 + 33 + 25 shown in Tables 14, 15, and 16, respectively. This is because some wide-ranging site series (such as Ss–Salmonberry and Act–Red-osier dogwood) are listed in each of the three summary tables.

Table 16

Site series in the Prince Rupert Forest Region in which Sitka spruce hybrids have potential as a primary, secondary, or tertiary species, listed in order of decreasing productivity (site index) classes.

Biogeoclimatic zone, subzone, or variant[a]	Site series name (no.)[a]	Site index, median ht (m) at b.h. age 50[b]
MHwh	YcHm–Twistedstalk (05)	?[c]
CWHws1; CWHws2	Ss–Salmonberry (07)	30
CWHws1; CWHws2	Act–Red-osier dogwood (08)	30
CWHws1; CWHws2	BaCw–Devil's club (06)	25
ICHmc1	ActSx–Dogwood (05)	25
ICHmc1	HwBl–Devil's club (04)	23
ICHmc1[a]	HwBa–Devil's club–Lady fern (03)	23
ICHmc2	CwHw–Devil's club–Oak fern (04)	23
ICHmc2	Sx–Devil's club–Lady fern (05)	23
ICHmc2	ActSx–Dogwood (06)	23
ICHmc2	SxEp–Devil's club (54)	23
ICHmc2	HwCw–Oak fern (03)	22
ICHmc1[a]	HwBa–Oak fern (02)	22
ICHmc2	SxEp–Thimbleberry–Hazelnut (52)	21
ICHmc1	Hw–Azalea–Skunk cabbage (06)	20
ICHmc1[a]	HwBa–Bramble (01)	19
ICHmc2	CwSx–Horsetail–Skunk cabbage (07)	18
CWHws1; CWHws2	HwBa–Bramble (01)	?
CWHws1; CWHws2	BaCw–Oak fern (04)	?
CWHws1; CWHws2	HwBa–Queen's cup (05)	?
CWHws1; CWHws2	CwSs–Skunk cabbage (11)	?
ICHvc; ICHwc	HwBl–Devil's club (01, 04)	?
ICHvc; ICHwc	Hw–Step moss (02, 03)	?
ICHvc; ICHwc	Sx–Devil's club (03, 05)	?
ICHvc; ICHwc	Sx–Horsetail (06, 08)	?
ICHvc; ICHwc	ActSx–Dogwood (05, 06)	?
ICHvc	Sx–Devil's club–Dogwood (04)	?
ICHwc	HwBl–Oak fern (01)	?
ICHwc	HwSx–Blueberry–Sphagnum (07)	?

Source: Site series names are from Silviculture Interpretations Working Group (1994) and Banner et al. (1993); site index data are from Banner et al. (1993).

[a] Site series, and the subzones or variants in which they occur, are listed for site series in which the Silviculture Interpretations Working Group (1994) recorded Sitka spruce hybrids either in the name of the site series or as a primary, secondary, or tertiary tree that should be considered in the guidelines for tree species selection. The site series number recorded in parentheses in the middle column of this table is the number assigned to particular site series in given subzones/variants by the Silviculture Interpretations Working Group (1994). In some cases there are two numbers in parentheses; this indicates that a given site series has a different official number in different biogeoclimatic variants. See Appendix 4 for full names of abbreviated biogeoclimatic units and for tree species whose abbreviations begin the name of each site series.

[b] The hybrid Sitka spruce site index values (column 3) assigned to a particular site series (column 2) in the designated subzones/variants (column 1) are derived from linear measure-

ment of the placement of Sitka spruce hybrids on the site index bar graphs shown in the interpretations table on pages 7.46, 7.54, 7.56, 7.58, and 7.60 of Banner et al. (1993). Banner and co-workers stressed that these site index estimates are based on very limited, unpublished Ministry of Forests data.

[c] A question mark indicates a site series for which Banner et al. (1993) did not assign a hybrid Sitka spruce site index class in a particular biogeoclimatic variant.

One of the distinctive features of Sitka spruce's natural distribution is its prominence on sites exposed to ocean spray near coastal shorelines. This narrow band of Sitka spruce's natural range can be used to depict how some of the site series listed in Tables 14 and 15 are indicative of different soil nutrient and soil moisture conditions. For the ocean spray zone, Banner et al. (1990) gave the following edatopic ranking to several key site series in which Sitka spruce is the dominant tree.

On very poor to medium nutrient regimes

- Slightly dry sites: Ss–Salal (rocky headlands and beach plains)
- Fresh sites: Ss–Kindbergia (old beach plains)

On medium-rich to very rich soil nutrient regimes

- Slightly dry sites: Ss–Reedgrass (rocky headlands, colluvium, and old dunes)
- Fresh sites: no Ss association represented
- Moist sites: Ss–Sword fern (marine terraces and adjacent scarps)
- Very moist sites: no Ss association represented
- Wet sites: Ss–Slough sedge (sites with a greatly fluctuating water table) and Ss–Pacific crabapple (sites affected by brackish water)

To summarize Sitka spruce's occurrence in different biogeoclimatic units on a latitudinal basis, Table 17 lists biogeoclimatic zones, subzones, and variants in which Sitka spruce can be expected to occur in a transect from the outer coast eastward to the inland geographic extent of Sitka spruce, at each degree of latitude from 49°00′N to 59°00′N, based on data from Clement (1984), Nuszdorfer et al. (1985), Houseknecht et al. (1987), Pojar et al. (1988), Banner et al. (1993), and Green and Klinka (1994). The nomenclature for subzones and variants has been amended to match the symbols and names used by the Silviculture Interpretations Working Group (1994), as listed in Appendix 4.

Biogeoclimatic Units Where Sitka Spruce Hybrids Occur

The Silviculture Interpretations Working Group (1994) considered Sitka spruce hybrids to have primary, secondary, or tertiary ecological acceptability in certain variants of only the CWH and ICH zones of the Prince Rupert Forest Region. These hybrids, Roche or Lutz spruce (Sx) (Watts 1983; Coates

Table 17

Biogeoclimatic zones, subzones, and variants in which Sitka spruce can be present in the Vancouver and Prince Rupert forest regions, at each degree of latitude from 49°N to 59°N

Latitude	Zone[a]	Subzone[a]	Variant[a]
49°N	CWH	Very Wet Hypermaritime	CWHvh1
Ucluelet, Vancouver Island,	CWH	Very Wet Maritime	CWHvm1
east to Chilliwack Lake on the	CWH	Very Wet Maritime	CWHvm2
mainland	CWH	Moist Maritime	CWHmm2
	CWH	Very Dry Maritime	CWHxm
	CDF	Moist Maritime	CDFmm
	CWH	Dry Maritime	CWHdm
	CWH	Moist Submaritime	CWHms1
50°N	CWH	Very Wet Hypermaritime	CWHvh1
Kyuquot, Vancouver Island,	CWH	Very Wet Maritime	CWHvm1
east to Princess Louisa Inlet	CWH	Very Wet Maritime	CWHvm2
on the mainland	CWH	Moist Maritime	CWHmm2
	CWH	Moist Maritime	CWHmm1
	CWH	Very Dry Maritime	CWHxm
	CDF	Moist Maritime	CDFmm
51°N	CWH	Very Wet Hypermaritime	CWHvh1
Bramham Island on outer	CWH	Very Wet Maritime	CWHvm1
coast east to the head of	CWH	Very Wet Maritime	CWHvm2
Bute Inlet on the mainland	CWH	Moist Submaritime	CWHms2
	CWH	Wet Submaritime	CWHws2
	CWH	Moist Submaritime	CWHms1
	CWH	Dry Submaritime	CWHds1
52°N	CWH	Very Wet Hypermaritime	CWHvh2
Kunghit Island, southern	CWH	Very Wet Maritime	CWHvm1
Queen Charlotte Islands,	CWH	Very Wet Maritime	CWHvm2
east to South Bentinck Arm	CWH	Wet Submaritime	CWHws2
on the mainland	CWH	Moist Maritime	CWHmm1
53°N	CWH	Very Wet Hypermaritime	CWHvh2
Englefield Bay, Moresby Island,	CWH	Wet Hypermaritime	CWHwh2
central Queen Charlotte Islands,	CWH	Wet Hypermaritime	CWHwh1
east to the head of Mathieson	CWH	Very Wet Maritime	CWHvm2
Channel on the mainland	CWH	Wet Submaritime	CWHws2
54°N	CWH	Very Wet Hypermaritime	CWHvh2
Beresford Bay, NW Graham	CWH	Wet Hypermaritime	CWHwh2
Island, Queen Charlotte Islands,	CWH	Wet Hypermaritime	CWHwh1
east to Kitimat on the mainland	CWH	Very Wet Maritime	CWHvm2

►

◀ *Table 17*

Latitude	Zone[a]	Subzone[a]	Variant[a]
55°N Pearse Island on Portland Inlet east to the vicinity of Cedarvale, Skeena River valley	CWH CWH CWH CWH ICH	Very Wet Hypermaritime Wet Hypermaritime Wet Submaritime Wet Submaritime Moist Cold	CWHvh2 CWHwh2 CWHws1 CWHws2 ICHmc2
56°N Bear River valley near Stewart	CWH	Wet Maritime	CWHwm
57°N Stikine River valley	CWH	Wet Maritime	CWHwm
58°N Whiting River valley	CWH	Wet Maritime	CWHwm
59°N	Within British Columbia at this particular latitude, there are no distinct areas where a particular subzone is associated with Sitka spruce, but this species does occur at several locations north of 58°N in British Columbia in the valleys of the Taku, Chilkat, and Tahini rivers, where Pojar et al. (1988) mapped small occurrences of the Wet Maritime (CWHwm) Subzone, in which there is potential overlap with Sitka spruce's distribution as mapped by Krajina et al. (1982).		

Source: Key information sources were Clement (1984), Nuszdorfer et al. (1985), Pojar et al. (1988), Banner et al. (1993), and Green and Klinka (1994). Names of subzones and variants have been amended to follow the nomenclature used by Klinka et al. (1991) and the Silviculture Interpretations Working Group (1994), as reproduced in Appendix 4.

[a] Biogeoclimatic subzones and variants are listed in the order in which they would be encountered on a transect from the outer coast eastward to the inland geographic limit of Sitka spruce at any given latitude. Each subzone or variant is listed only once even though any variant can occur at several locations along a line of latitude, such as CWHvh2, which at lat. 53°N occurs on the western slopes of both Moresby Island and Princess Island.

et al. 1994), occur mainly in the Nass and Skeena river valleys, on the site series listed in Table 16 (Houseknecht et al. 1987; Banner et al. 1993). However, Pojar (1983) and Coates et al. (1994) indicate that Sitka spruce hybrids are also present in the Mountain Hemlock (MH) Zone in that part of British Columbia, but their presence in this subalpine zone is uncommon to rare. Based on work by Houseknecht et al. (1987) and a 1:500,000 map prepared by Pojar et al. (1988), Sitka spruce hybrids can be expected to occur in the

ICHmc2 variant of the Moist Cold ICH Subzone (ICHmc) but not in the Very Wet Cold ICH Subzone (ICHvc). In the judgment of the Silviculture Interpretations Working Group (1994), however, Sitka spruce has silvicultural potential in any of the following: ICHmc1, ICHmc1a, ICHmc2, ICHvc, and ICHwc (Table 16).

In the Prince Rupert Forest Region, Sitka spruce hybrids have also been recorded in the Very Wet Cold Subzone (ESSFwv) of the Engelmann Spruce–Subalpine Fir (ESSF) Zone (Yole et al. 1989). In this subzone, spruce of potential Sitka spruce hybrid origin could occur on any of the following site series that are not listed in Table 16: BlHm–Azalea; BlHm–Feathermoss; Bl–Oakfern–Heron's bill; Bl–Devil's club–Lady fern; Bl–Valerian–Sickle moss; and Bl–Horsetail–Glow moss.

Biogeoclimatic Units Where Sitka Spruce Is Uncommon or Rare

Compared with Sitka spruce's abundance in the northern part of the CWH zone and its common occurrence in the southern part of the CWH zone, this species is uncommon in all subzones of the Coastal Douglas-fir (CDF) and Mountain Hemlock (MH) zones, and in coastal transition subzones (ICHmc and ICHvc) of the ICH zone (Coates et al. 1994). The review by Coates and co-workers acknowledged the possible occurrence of Sitka spruce in coastal portions of the Alpine Tundra Biogeoclimatic Zone. The circumstances of such occurrences were not documented by Coates and co-workers, but one would expect this spruce to reach altitudinal treeline only at the latitude of the Queen Charlotte Islands and northward from there.

Field observations by Yole et al. (1989) indicated that at latitude 53°00'N there is evidence that Sitka spruce occurs substantially further inland than shown by Krajina et al. (1982) in the range map for this species. Based on the latter reference, Table 17 lists Sitka spruce extending inland, at latitude 53°00'N, to approximately the head of Mathieson Channel (longitude about 128°00'W). However, Clement (1984) referred to abundant Sitka spruce on floodplains of the Kimsquit River valley, above the head of Dean Channel (longitude about 127°00'W), which is about 60 km further inland than the generalized eastern range limit mapped by Krajina et al. (1982). At this latitude (53°00'N), therefore, Sitka spruce would also occur in the Submontane Moist Maritime Coastal Western Hemlock Variant (CWHmm1) and in the Montane Wet Submaritime Coastal Western Hemlock Variant (CWHws2). On the Kimsquit River floodplain, the main sites in which Sitka spruce is prominent are those dominated by *Oplopanax horridus, Gymnocarpium dryopteris,* and *Plagiomnium insigne,* which are sites similar to those listed by Green and Klinka (1994) as the Ss–Salmonberry site series.

Listing Sitka spruce in the Interior Douglas-fir (IDF) Zone in Table 14 may seem anomalous. Sitka spruce does not naturally occur in the few areas of the Vancouver Forest Region that are classified as part of the Wet Warm IDF

Subzone (IDFww). The Silviculture Interpretations Working Group (1994) listed Sitka spruce as a potential tertiary species in the IDFww subzone, on the assumption that it may have some silvicultural role for regeneration purposes in the parts of the IDF zone that come closest to the coast. Within the Vancouver Forest Region, the IDFww subzone occupies relatively small areas in the valleys from Spuzzum to Boston Bar and downstream to Keefers, from Pemberton downstream to Birken, upstream from Pemberton along the Lillooet River valley, in a small portion of the Similkameen River valley on the east side of Manning Provincial Park, and in the portion of the Skagit River valley that lies west of Manning Provincial Park.

In general, Sitka spruce can be considered uncommon or rare in the biogeoclimatic subzones, variants, or site series where it has been recorded in hybrid form. The biogeoclimatic units where Sitka spruce hybrids are most likely to occur are summarized in the previous subsection.

An Example of a Typical Ecosystem Where Sitka Spruce Occurs in Oregon and Washington

In Washington, as in British Columbia, there are major differences between Sitka spruce ecosystems in the coastal *Picea sitchensis* zone (Franklin and Dyrness 1973) and valley-bottom Sitka spruce–western hemlock ecosystems further inland. The latter are typified by stands on terraces of the South Fork Hoh River, Olympic National Park (Franklin 1982; Franklin et al. 1982), and elsewhere in the Olympic National Forest (Henderson et al. 1989).

The Twin Creek Research Natural Area exemplifies Sitka spruce forests as they occur under the 'rain forest' conditions in river valleys on the west side of the Olympic Peninsula (Franklin et al. 1972). This 40 ha natural area occupies gentle topography on river terraces in the Hoh River valley. Elevations range from about 130 to 195 m. A wet, mild, maritime climate prevails. Precipitation is heavy, but less than 10% falls during summer months. The following climatic data are from the Forks weather station located approximately 32 km northwest of the natural area (Franklin et al. 1972):

mean annual temperature	9.5°C
mean January temperature	3.7°C
mean July temperature	15.4°C
mean January minimum temperature	0.2°C
mean July maximum temperature	21.5°C
average annual precipitation	2,974 mm
June through August precipitation	214 mm
average annual snowfall	348 cm

The forest within the Twin Creek Research Natural Area is a mixture of Society of American Foresters forest cover types 225, Western Hemlock–

Sitka Spruce, and 223, Sitka Spruce, with the latter being dominant (Society of American Foresters 1980). The two units within the natural area are mosaics of Sitka spruce and western hemlock forest of varying ages and sizes interspersed with open areas dominated by *Acer circinatum* and occasionally *Acer macrophyllum*. Both the spruce and the hemlock are present in all size classes ranging up to a maximum of 330 cm dbh for spruce and 150 cm dbh for hemlock. Mature trees of either species attain heights of 60 m.

Forest stands in the Twin Creek Research Natural Area have relatively rich and well-developed understoreys. *Acer circinatum, Vaccinium ovalifolium, V. parvifolium, Rubus ursinus,* and *R. spectabilis* are the most common species in the shrub layer. The major herbaceous species are *Oxalis oregana, Polystichum munitum, Tiarella unifoliata, Carex deweyana, Trisetum cernuum, Rubus pedatus, Montia sibirica, Athyrium filix-femina,* and *Gymnocarpium dryopteris*. *Polystichum* and *Oxalis* are the most important herbs. The 'rain forest' region of the western Olympic Peninsula is famous for an abundance of cryptograms. Some of the more common ground species are *Kindbergia oregana, Hypnum circinale, Rhytidiadelphus loreus, Leucolepis menziesii, Hylocomium splendens,* and *Plagiomnium insigne*. Common epiphytes are *Pseudoisothecium stoloniferum, Porella navicularis, Rhytidiadelphus loreus, Radula bolanderi, Frullania nisqualensis, Scapania bolanderi,* and *Ptilidium californicum*.

Some other examples of plant communities or ecosystems in which Sitka spruce is a component in Oregon and Washington are summarized in Table 18.

Sitka Spruce's Tree Associates and Successional Sequences

Although Sitka spruce occurs in pure stands, particularly where ocean spray is common, overall its most common associate is western hemlock. This combination is designated as the Western Hemlock–Sitka Spruce forest type in the Pacific Northwest Region of the United States (Harris and Johnson 1983). Within this type, tree species composition changes with latitude. For example, Douglas-fir is an important associate of Sitka spruce in Oregon and Washington, whereas western hemlock is a more frequent associate than Douglas-fir along much of coastal British Columbia. At the latitude of southeastern Alaska, western redcedar is commonly associated with Sitka spruce. Shore pine is an occasional associate throughout the range of Sitka spruce. Red alder and black cottonwood are common associates throughout the range, except on the Queen Charlotte Islands, where cottonwood is absent. At latitudes north of Vancouver Island, mountain hemlock and yellow-cedar are associated with Sitka spruce at higher elevations.

Succession involving Sitka spruce in particular ecosystems is summarized in the subsections that follow. An overview of successional stages typical of even-aged stands of western hemlock can be summarized as follows, based on Deal and Farr (1994).

A major disturbance, such as clearcutting or a large windthrow area, that

Table 18

Plant communities or ecosystem units in which Sitka spruce occurs in Oregon and Washington

Location	Forest type	Community or ecosystem	Reference
Diamond Point Res. Nat.Area on Washington coast (lat. 46°29′N)	second growth	*Polystichum munitum* type with Sitka spruce dominant	Franklin et al. 1972
	Sitka spruce–western hemlock	*Gaultheria shallon* type with hemlock dominant	
Neskowin Crest Res. Nat. Area on Oregon coast (lat. 45°05′N)	Sitka spruce–western hemlock	*Rubus–Sambucus–Menziesia Polystichum munitum–Gaultheria shallon*	Franklin et al. 1972
Quinalt Res. Nat. Area in Olympic Mountains (lat. 47°50′N)	Sitka spruce–western hemlock	*Vaccinium parvifolium–Polystichum munitum–Oxalis oregana*	Franklin et al. 1972
Twin Creek Res. Nat. Area in Olympic Mountains (lat. 47°50′N)	Sitka spruce and Sitka spruce–western hemlock	*Vaccinium ovalifolium–V. parvifolium–Oxalis oregana–Polystichum munitum*	Franklin et al. 1972
Northern Oregon coast	Sitka spruce–western hemlock	*Tsuga–Picea–Gaultheria shallon–Blechnum spicant*	Franklin et al. 1972
Oregon inland from coast	Sitka spruce–western hemlock	*Tsuga–Picea–Oplopanax horridus–Athyrium filix-femina*	Franklin and Dyrness 1973

removes all or most of the overstorey, is followed by rapid growth of existing forbs, herbs, and shrubs; advance conifer regeneration; and the germination of new conifers. This *stand-initiation stage* (Oliver and Larson 1990) typically lasts 15-30 years in the hemlock-spruce forests that Alaback (1982a, 1982b, 1982c, 1984) and Alaback and Herman (1980) have documented in southeastern Alaska. During this stage, tree crowns expand to form a dense canopy that eventually shades out understorey vegetation.

The next stage is what Oliver and Larson (1990) call the *stem-exclusion stage;* it follows canopy closure and lasts 50-100 years. During this stage, there is little or no understorey vegetation other than mosses on the forest floor (Alaback 1982a; Harcombe 1986). This stage is followed by the slow reinvasion of shrubs, herbs, forbs, and conifers as light gradually increases at the understorey level. In the absence of another major disturbance, the overstorey develops into a multi-aged old-growth stand over several hundred years (Deal et al. 1991).

These typical hemlock-spruce successional stages can be significantly changed by silvicultural events. For example, in a southeastern Alaska study by Deal and Farr (1994), thinning delayed the stem-exclusion stage in young stands; in older stands, thinning added a new initiation phase and hastened the onset of the *understorey-reinitiation stage*. Stands less than 30 years old still had conifers originating from the stand-initiating disturbance, and these understorey conifers rapidly expanded to fill the available growing space provided by thinning. Such expansion excluded new conifer germinants. In contrast, older stands that were in the stem-exclusion stage when they were thinned had little, if any, residual conifer understorey. In this circumstance, at least in southeastern Alaska, new hemlock and spruce germinants became established abundantly within the first 2-3 years after thinning. The degree to which these relationships between thinning and subsequent succession holds true in more southerly biogeoclimatic zones within Sitka spruce's natural range is not well known.

The shortest successional sequences leading to relatively pure Sitka spruce stands are on sites where bare mineral soil has been exposed. Examples of such sites are areas of recent glacial recession in Alaska (Ruth 1958), landslide scars on the Queen Charlotte Islands, British Columbia (Smith et al. 1986), recently deposited alluvium (Ruth 1958), and soils exposed by blowdown during winds of hurricane force. At present some of the most productive stands in which Sitka spruce is a significant component on northern Vancouver Island are those that regenerated after a 1908 hurricane; stand volumes on such sites 80 years after blowdown were as high as 1,500 m^3/ha (S. Joyce, pers. comm., Nov. 1989).

Succession on Alluvial Ecosystems in British Columbia

Alluvial ecosystems are the most productive Sitka spruce sites and must be considered separately for management purposes. It is therefore important to understand successional relationships in these dynamic ecosystems. Day (1957) recognized three stages of development in alluvial sites on the Queen Charlotte Islands. These stages are characterized not only by changes in forest type but also by distinct differences in soil texture and structure, amounts of silt deposition, and soil genesis.

The first stage, identified by Day (1957) as *meadow forest,* is characterized by the initial stabilization of alluvium, although there continues to be regular silt deposition. Typically this stage is dominated by Sitka spruce and red alder, with few western hemlock and no cedar present. On the mainland, this stage is dominated by black cottonwood, with groups of western redcedar and Sitka spruce appearing as the cottonwood develops a full canopy. It is common to find the spruce established on logs or other raised microsites, as indicated by regeneration growing in relatively straight lines. Eventually such stands develop to contain the largest specimens of Sitka spruce and western redcedar.

By the second stage of development, silt deposition has raised the alluvium to a level that is submerged only by very high floods. At this stage, humus begins to develop. In general, western hemlock increases in significance once active silt deposition ceases. On the Queen Charlotte Islands, red alder is absent or uncommon by this stage. Western hemlock becomes increasingly common as an understorey to Sitka spruce. The latter continues to regenerate, but poorly. On the mainland, this stage is characterized by a mixture of black cottonwood, Sitka spruce, and western redcedar, with a dense understorey of shrubs (Day 1957).

The third stage is characterized by alluvial terraces above current flood levels. These sites have well-developed humus, and soils have leached horizons. The ecosystem is dominated by mixed stands of western hemlock and Sitka spruce, in which western hemlock is present in all age classes and is the species with the most numerous stems. However, Sitka spruce remains the outstanding dominant species because of its large size. By this stage, Sitka spruce regeneration is absent or sparse, occurring only on downed tree stems and on the upturned root systems of windthrown trees. In contrast, western hemlock regeneration is abundant and widespread on many microsites. With time, the large spruce dominants die and disappear from the canopy, and are succeeded by western hemlock as the climax species (Day 1957).

Succession Involving Sitka Spruce on Landslides in British Columbia

In some parts of coastal British Columbia, scars created by slope failures are known to be suitable sites for vegetational succession that leads to development of Sitka spruce stands. Successional processes involving Sitka spruce on landslide scars (Figure 23) have been best documented for the Queen Charlotte Islands (Smith et al. 1986), but similar processes, although not

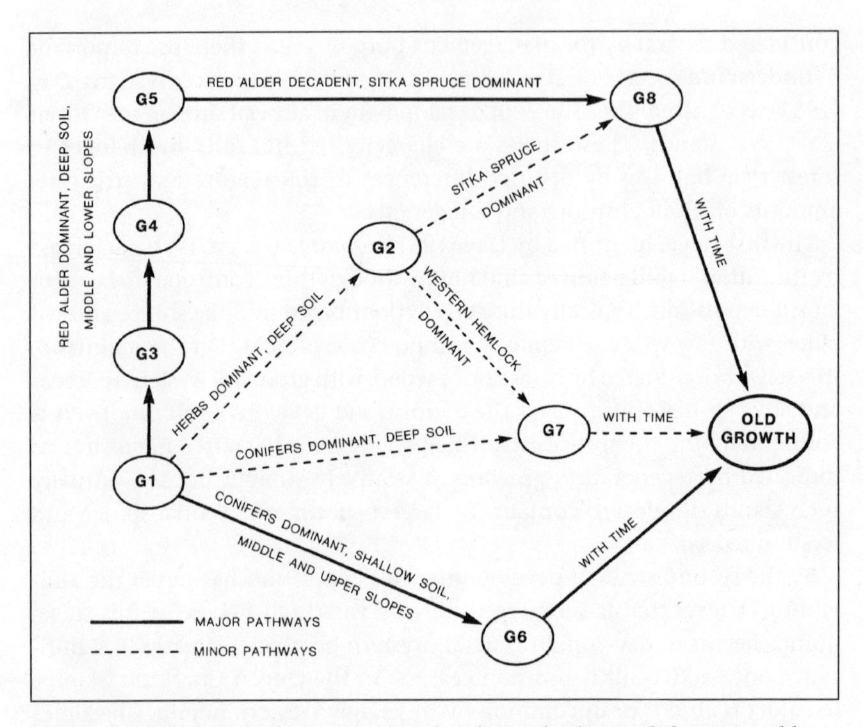

Figure 23. Major and minor successional pathways leading to Sitka spruce old growth on landslides in the Queen Charlotte Islands, British Columbia (from Smith et al. 1986).

Note that in this legend, low shrubs = shrubs or trees under 2 m tall; tall shrubs = shrubs or trees 2-10 m tall.

G1 **Initial stage**: Vegetative cover is generally less than 10% and age of landslide is generally less than 10 years. Western hemlock, western redcedar, Sitka spruce, and yellow-cedar occur sporadically as low shrubs or seedlings with a cover of less than 1%. Revegetation is enhanced by the presence of fallen trees, logs, and stumps, and is retarded by surface soil erosion.

G2 **Herbs dominant, deep soil**: This early stage is dominated (15-40% cover) by ferns, grasses, rushes, fireweed, thistle, and other herbs and bryophytes. Red alder, western hemlock, western redcedar, and Sitka spruce occur in the low shrub layer, but with only up to 5% total cover.

G3 **Red alder dominant, young age**: Found typically on middle and lower portions of slide scars, this group is dominated by red alder in the tall or low shrub layers, with other plants sparsely scattered underneath the dense red alder or in patches in small openings. Western hemlock, Sitka spruce, and western redcedar are present in the low shrub layer but with cover less than 5%. Total vegetative cover ranges from 60-100%, and the ages of slides are generally less than 10 years.

G4 **Red alder dominant, intermediate age**: A later stage of G3, this group is found on slides 14-45 years of age. Red alder averaging 10 m in height contributes a vegetative cover of about 60%. Sitka spruce dominates the understorey tall shrub layer. Western hemlock is usually shorter than spruce; western redcedar is relegated to the low shrub layer as a result of severe deer browsing. Thimbleberry and huckleberry are common, but herbs and a very diverse assemblage of bryophytes make up most of the non-tree vegetation. Total plant cover averages 86%.

G5 **Red alder dominant, old age**: A later stage of G4 occurring on slides 60-102 years of age in which decadent red alder is still dominant (50% cover) but in which Sitka spruce is common (10-40% cover) as a tall shrub or small tree just under the main canopy of red alder. Western hemlock and western redcedar tend to be reduced in height and cover relative to Sitka spruce. Herb cover is markedly less than for G4, but the cover of bryophytes normally exceeds 60%.

G6 **Conifer dominant, shallow soil**: The substrate of largely coarse rubble and bare bedrock on these middle and upper slide sites keeps total plant cover below 40%. Red alder is not a component. Western hemlock or Sitka spruce is dominant but, despite slide ages up to 48 years, usually less than 2 m in height. Herb cover is generally less than 5%, and bryophyte cover on most plots is less than 10%.

G7 **Western hemlock dominant, deep soil**: Relatively uncommon, these stands (35-38 years of age) are dominated by western hemlock (50% cover) and Sitka spruce (< 20%) in the tall and low shrub category. Western redcedar is severely browsed and has a cover of less than 1%. Red alder stems are absent or occur in insignificant numbers even on the lower half of the slides. Shore pine is an unusual component of some of these stands. Shrubs, herbs, and bryophytes contribute less than 20% to the total plant cover, which usually amounts to more than 90%.

G8 **Sitka spruce dominant, deep soil**: These stands on relatively old (48-155 years) slides developed either from stands originally devoid of red alder or from stands in which red alder has largely disappeared. Sitka spruce contributes most to the 90% tree cover, but western hemlock may also be abundant as a generally shorter tree. Understorey plants are less abundant, making up less than 3% of the vegetative cover.

quantified, can be expected to be common in all coastal mountainous areas that lie within the Wet and Very Wet Hypermaritime subzones of the CWH zone. Black cottonwood is sometimes a competitor of Sitka spruce on landslides on the British Columbia mainland, but cottonwood does not occur on the Queen Charlotte Islands, leaving spruce with a very prominent role in plant community succession on landslide areas (R.B. Smith, pers. comm., October 1993). As described elsewhere in this book, a complicating factor in ecological succession on the Queen Charlotte Islands is the heavy browsing by high densities of Sitka deer, which reduces shrub cover and may accelerate normal succession patterns from alder and other shrubs to Sitka spruce and other conifers.

Based on aerial photographic analysis of 8,328 debris slides, debris avalanches, debris flows, and debris torrents on the Queen Charlotte Islands, Gimbarzevsky (1988) estimated that the average number of slope failures per unit area was greatest (1.50 failures per square kilometre) in the Very Wet Hypermaritime Subzone (CWHvh). Second in number of slope failures were the MH and AT zones, which, combined, had an average of 1.00 failure per square kilometre. In the portions of the Queen Charlotte Islands that are now classified as the Montane Variant (CWHwh2) and the Submontane Variant (CWHwh1) of the Wet Hypermaritime CWH Subzone (Pojar et al. 1988), Gimbarzevsky estimated that there were 0.53 slope failures per square kilometre.

Recent landslide surfaces satisfy the mineral soil requirements for above-average natural restocking of Sitka spruce. However, the subsequent growth and dominance of Sitka spruce established in this way depends on such factors as the position of establishment on the landslide scar, the presence and abundance of red alder, and the origin of the parent material (Smith et al. 1986).

Smith et al. (1986) described two major and several minor pathways for vegetation development on landslide surfaces (Figure 23). All of these originate on exposed bedrock, rubble, finer mineral soil, woody mixtures, or mixtures of these substrates. The first major pathway is dominated by red alder and occurs on the relatively deep soiled, moist middle and lower portions of landslide scars, particularly when associated with fine sedimentary, calcareous, or soft volcanic bedrock. In the early stages of this successional sequence, red alder dominates, with Sitka spruce, western hemlock, and western redcedar relegated to an understorey role and low cover values (Plate 10). The growth rate of both western hemlock and Sitka spruce in these young alder-dominated stands is about one-half that for the same species in stands originating after logging. By 30 years of age, red alder continues to dominate, but with greatly reduced numbers of stems per hectare (G4 in Figure 23). Of the conifers, Sitka spruce tends to fare best, although it is still largely confined to the understorey. At 60-100 years, red alder cover begins to decrease due to the death of branches and whole trees (Plate 11; see also G5 in Figure 23). Sitka spruce in particular approaches red alder in height and increases markedly in total cover. Western hemlock is usually shorter and maintains a lower percentage of cover than Sitka spruce.

The second major successional sequence described by Smith et al. (1986) occurs especially on the middle and upper portions of the landslide scars (G6 in Figure 23). The substrate consists largely of shallow, coarse rubble and exposed bedrock. The tree cover is relatively sparse, as it is restricted to pockets of organic and mineral materials and fissures in the bedrock (Figure 24). Red alder is not normally a component of this sequence, but is occasionally replaced by Sitka alder (*Alnus viridis* subsp. *sinuata*). Western hemlock and Sitka spruce are the dominant tree species. Shore pine is less commonly present (Figure 25). Western redcedar and yellow-cedar are severely suppressed by deer, and, with poor site conditions, they constitute a low proportion of the tree cover. Suppression of cedar and other vegetative competition by browsing deer is thought to be of major successional significance on the Queen Charlotte Islands, potentially favouring Sitka spruce regeneration over cedar regeneration over the long term.

In the synoptic surveys by Smith et al. (1986), a considerably less commonly observed sequence involved establishment of a high stocking of Sitka spruce (Figure 26), western hemlock, or mixtures of these conifers on relatively deep soil (G7 and G8 in Figure 23). This establishment of spruce and/or hemlock appears to be initiated directly and rapidly on disturbed surfaces, or sometimes after only a short period of domination by herbs (G2 in Figure 23).

Figure 26 shows an example of a pure Sitka spruce stand that developed on the middle portion of an approximately 48-year-old landslide, on Moresby Island, Queen Charlotte Islands, without a preceding development of red alder canopy. When photographed 48 years after the landslide's occurrence, the diameters of the largest spruce shown in Figure 26 were 40-45 cm; in

Figure 24. Young Sitka spruce established on the severely scoured upper portion of an approximately 21-year-old landslide scar, near Long Inlet, Graham Island, Queen Charlotte Islands, British Columbia. (Photograph by R.B. Smith)

Figure 25. Young Sitka spruce (some severely browsed), western hemlock, and shore pine on an exposed landslide surface, with the original Sitka spruce–western hemlock old-growth stand in the background, on Burnaby Island, Queen Charlotte Islands, British Columbia. (Photograph by R.B. Smith)

Figure 26. Pure Sitka spruce stand on the middle portion of an approximately 48-year-old landslide scar, near Sachs Creek, Moresby Island, Queen Charlotte Islands, British Columbia. (Photograph by R.B. Smith)

the sampled portion of the stand, diameters averaged 25 cm and ranged from 8 to 47 cm (R.B. Smith, pers. comm., Nov. 1990).

Averaged over all sequences, the highest cover of western hemlock and shore pine occurred on slide scars associated with granitic bedrock, and the highest cover of Sitka spruce on slides was associated with fine sedimentary bedrock. Smith et al. (1986) expressed tree cover values as ratios of one species to another on different types of bedrock. There appeared to be two different groups of bedrock influencing tree cover ratios. One group, composed of granitic, coarse conglomerate and hard volcanic types had alder/conifer ratios ranging from 0.1 to 0.3, alder/hemlock ratios from 0.2 to 0.8, and Sitka spruce/hemlock ratios from 0.4 to 1.3. The other group (soft volcanic, calcareous, fine sedimentary, and mixtures) had consistently higher ratios: alder/conifer from 0.8 to 1.8, alder/hemlock from 2.4 to 12.2, and Sitka spruce/hemlock from 1.6 to 5.4.

After tree removal, the gradual decay of tree roots often predisposes soils on steep slopes to failure. The rate of root strength deterioration following tree removal varies by species (Sidle 1991). Roots of Sitka spruce and western hemlock decay the least rapidly compared with several other species for which data are available (radiata pine, beech, coastal Douglas-fir, Rocky Mountain Douglas-fir), as documented by Sidle (1991) and Ziemer and Swanston (1977). This is a fortunate circumstance for landslide-prone slopes where Sitka spruce occurs. Sidle's root strength model, when adapted to different silvicultural systems (clearcutting, partial cutting, shelterwood cutting, and thinning), indicates that vegetation and site conditions that promote rapid root strength deterioration and slow regrowth may depress net rooting strength over several management rotations. Progressively shorter clearcut or partial cut rotations can cause a steady temporal decline in root strength. Longer intervals between initial and final shelterwood cuttings promote greater rooting strength than short intervals.

Successional Sequences in Washington and Oregon

Franklin and Dyrness (1973) described successional patterns in the *Picea sitchensis* zone that occurs along the entire length of the Washington and Oregon coastline (Figure 17). They acknowledged that the Sitka spruce zone could be considered a variant of the *Tsuga heterophylla* (western hemlock) zone, which is how it was treated in British Columbia, where Krajina (1969) and subsequent developers of the biogeoclimatic classification system did not recognize a distinct Sitka spruce zone. Some successional patterns recognized by Franklin and Dyrness (1973) are summarized below.

Early successional trends following fire or logging in the *Picea sitchensis* zone of Washington and Oregon are similar to those encountered in the *Tsuga heterophylla* zone. In the Sitka spruce zone, however, there is a stronger tendency towards development of dense shrub communities dominated by *Rubus spectabilis, Sambucus racemosa* subsp. *pubens* var. *arborescens,* and *Vaccinium* spp. There are two major kinds of seral forest stands in the Sitka spruce zone: (1) coniferous, containing varying mixtures of Sitka spruce, western hemlock, and Douglas-fir; and (2) broadleaf, dominated by red alder. In many cases, alder overtops conifer regeneration, resulting in pure or nearly pure *Alnus* forest. Replacement of alder by other tree species is often very slow, even though alder is a relatively short lived species. This is partially because of the dense shrubby understoreys of *Rubus spectabilis.* Successional sequences have not been thoroughly studied, although it appears that red alder can be replaced by any one of the following: semi-permanent brushfields; Sitka spruce released from a suppressed state (Franklin and Pechanec 1968); or western redcedar or western hemlock. Franklin and Dyrness (1973), based on unpublished work by Fonda, also identified a sere for river terraces on the western Olympic Peninsula in which red alder is replaced initially by Sitka spruce, black cottonwood, and bigleaf maple.

Succession in most mature conifer forest types in the Washington and Oregon Sitka spruce zone is towards replacement of mixed Sitka spruce, western redcedar, western hemlock, and Douglas-fir forests by hemlock. Western hemlock is apparently more shade-tolerant than Sitka spruce and dominates the reproduction in old-growth forests. Krajina (1969) reached a similar conclusion in coastal British Columbia. Hines (1971) found that Sitka spruce was not reproducing at all in the *Tsuga-Picea/Gaultheria/Blechnum* community type, and only about one-third of the *Tsuga-Picea/Oplopanax/Athyrium* stands had spruce seedlings, although hemlock was reproducing well. Hines concluded that the spruce is perpetuated, if at all, by natural openings created by windthrow or overstorey mortality. However, since Sitka spruce, western redcedar, and Douglas-fir are all long-lived species, even very old stands usually retain at least some of the original representation of these species. On moist to wet sites, western redcedar and, in some cases, Sitka spruce are a part of the climax forest along with western hemlock (Franklin and Dyrness 1973). Although Sitka spruce is commonly considered to be seral to western hemlock, Taylor (1990) presented evidence that in the Pacific Northwest, spruce may persist through gap-phase regeneration after small-scale disturbances. Gaps of 800-1,000 m^2 appear to be large enough for Sitka spruce to persist in these forests.

Many investigators have noted the successional role of rotting logs, referred to as 'nurse logs,' which often support hundreds of hemlock, spruce, and redcedar seedlings. Some of these seedlings survive, and their roots eventually reach mineral soil. The consequences are often readily visible in forests as lines of mature trees growing along the remains of the original nurse logs. Heilman (1990) noted that for the special conditions of poorly drained or periodically flooded soils, large fallen trees can be particularly important as nurse logs for the establishment and growth of tree seedlings. Often the most abundant and vigorous tree seedlings occur on fallen, decomposed logs, particularly in the more acidic soils of the coastal Sitka spruce zone, where seedling root development is better on decomposed wood than in mineral soil. On rich sites with intense vegetative competition, moist rotten logs provide suitable microsites for Sitka spruce germination, survival, and growth.

In general, Sitka spruce is commonly considered as a subclimax species because it is overtaken by western hemlock (Ruth 1958). There are exceptions, however. In the Twin Creek Research Natural Area, Olympic Peninsula, Franklin et al. (1972) described circumstances in which Sitka spruce does coexist with western hemlock as a climax species. They attributed this to the special conditions found in 'rain forest' valleys of the western Olympic Peninsula, particularly the relatively open nature of many of the stands and the selective grazing of hemlock seedlings by elk.

On the Olympic Peninsula, Fonda (1974) described succession on four terrace levels of different ages in the Hoh River valley, ranging from recent gravel bars to glaciofluvial terraces formed during the last glaciation period. Alder and willow were followed by spruce–bigleaf maple–black cottonwood on recent deposits, by spruce-hemlock on terraces of intermediate age, and by hemlock stands on older terraces. Although Sitka spruce is recognized as a seral species along all of its coastal geographic range, its status on alluvial sites, especially lower terraces, may be a special-case 'climax' (Franklin and

Dyrness 1973). Climax species are normally considered to be those that are self-perpetuating without major disturbance and are sufficiently shade-tolerant to regenerate under a forest canopy. In the case of Sitka spruce, which typically grows in association with western hemlock, seedbed conditions may be more influential than shade tolerance differences in determining the degree to which Sitka spruce regenerates as an understorey species. For example, western hemlock seems to need organic matter as a germination medium, whereas Sitka spruce does not. In most cases, for Sitka spruce to regenerate sufficiently there must be periodic disturbance to create larger canopy openings that favour spruce over the more shade-tolerant hemlock and amabilis fir.

Succession on Coastal Dunes

Where Sitka spruce occurs on coastal sand dunes, the successional sequence is very different from those summarized above. Based on unpublished thesis work by Wiedemann (1966), Franklin and Dyrness (1973) described a sere that begins on wet and stable bare sand in Oregon coastal dunes. Any one of three pioneer stages – marsh (*Carex obnupta* and *Potentilla anserina*), rush meadow (*Juncus mertensianus* subsp. *gracilis* and *Trifolium wormskjoldii*) or meadow (*Lupinus littoralis* and *Festuca rubra*) – can develop to a wet shrub community of *Salix hookeriana* and *Myrica californica*. The latter eventually develops into a pine-spruce forest, in which the dominant shrubs are joined by *Pinus contorta* and *Picea sitchensis* to form a very dense vegetation cover. The final state is a Sitka spruce forest in which there is little or no shrub layer.

Succession in Coastal Marshlands

There are no open coastal salt marshes in the Pacific Northwest, but marshlands do occur in portions of estuaries that are characterized by low salinity, large areas of soft sediments at high tide levels, and low wave energy. Like marshes elsewhere, Pacific Northwest salt marshes have very high annual plant production rates, a significant fraction of which is exported to the rest of the estuarine portion of the ecosystem as plant detritus (Maser et al. 1988). The role of trees in estuarine salt marshes is not widely documented in the natural range of Sitka spruce, but Eilers (1975) examined this question in marshes of the Nehalem River estuary in Oregon and found that the relative mobility of large driftwood trees and logs in different parts of a marsh influence successional processes. A Sitka spruce–alder–willow wetland forest community dominates the highest portions of the salt marsh, slowly invading it as the infilling of the marsh proceeds. This forest advances in some places by active colonization of stable piles of driftwood trees and logs left in the marsh by past storms.

Large drifted trees embedded in the marsh are colonized by terrestrial plant species that are unable to grow directly on the marsh soil because of salinity. Most of the Sitka spruce in the high marsh grow on nurse logs, and few of the spruce roots extend into the marsh soil. In other parts of the marsh exposed to winter storm waves, the forest edge periodically retreats, mainly because accumulations of drifted trees batter against standing trees at abnormally high tides (Eilers 1975; Maser et al. 1988).

On open coastal shorelines, away from estuaries, ocean-borne driftwood serves to stabilize coastal dune fronts. Sitka spruce is among the colonizing species that grow on or adjacent to stranded driftwood logs. Particularly in backshore areas, these nurse logs not only provide beach stability but also shelter, shade, moisture, and nutrients, all of which encourage the development of woody vegetation (Maser et al. 1988).

Shade Tolerance and Light Relationships

Minore (1979) examined information from a large number of published sources and compiled a comparative table of shade tolerance – the capacity of a tree species to survive in light of low intensity. The main tree species occurring in the Pacific Northwest and British Columbia were ranked as follows, in order of decreasing shade tolerance:

amabilis fir, western hemlock, western yew
western redcedar, yellow-cedar, mountain hemlock
subalpine fir
grand fir
Sitka spruce
white spruce
western white pine
Douglas-fir
Engelmann spruce, red alder
lodgepole pine
western larch
ponderosa pine

In the list above, species on the same line are not necessarily equal in shade tolerance; data are insufficient to separate them (Minore 1979). As indicated earlier, Sitka spruce is more shade-tolerant than white and Engelmann spruce but less shade-tolerant than its main associate, western hemlock. The relative ratings of Sitka and white spruce were also confirmed in tests in controlled environments (Brix 1972). He found that Sitka spruce was less productive than white spruce under high light and high temperature conditions. Also under controlled-environment tests, Tinus (1970) found that Sitka spruce was light-saturated at lower light intensities than red alder. Krueger and Ruth (1969) found that red alder's photosynthetic rate was also higher than that of Sitka spruce at high light intensities.

Minore (1979) indicated that Sitka spruce is relatively sensitive to a shortened photoperiod, but it has a longer optimum photoperiod than western redcedar (Malcolm and Caldwell 1971). When Vaartaja (1959) used the ratio of height growth under very long days to height growth under very short days, he found the following species sequence of decreasing ratios: Engelmann spruce > Sitka spruce > lodgepole pine > ponderosa pine > Douglas-fir > western redcedar > mountain hemlock.

Regeneration

Reproduction by seed is the main method of natural regeneration in Sitka spruce, but vegetative reproduction by layering has been recorded in both natural stands and plantations (Cooper 1931; Harris 1990). As with Engelmann or white spruce, layering is typically triggered when live lower branches hang close enough to the ground to come in contact with mineral soil or humus. For Engelmann and white spruce, layering is considered to be insignificant for establishing and maintaining closed forest stands (Coates et al. 1994), and the same appears to be true for Sitka spruce. Harris (1990) indicated that layering in Sitka spruce is most likely to occur on very moist sites at the edges of bogs. In locations where Sitka spruce may occur near treeline in the northern parts of its natural range, layering can be stimulated by snow accumulations that press low branches into contact with the ground, as happens with *Abies lasiocarpa* and *Picea engelmannii* at treeline.

Sitka spruce does not sprout from stumps or roots. Assisted asexual propagation can be accomplished by air-layering or by rooting of stem cuttings (Harris 1990). Sitka spruce cuttings have been successfully rooted in the greenhouse (Ruth 1958), and the many references on this subject by Mason and co-workers (Mason 1984; Mason and Gill 1986; John and Mason 1987; Mason and Harper 1987; Rook 1992) indicate that in the United Kingdom, cuttings are now a standard method for propagating improved stock obtained from tree-breeding programs. Clonal propagation of Sitka spruce has revealed that clones differ in their ability to root or graft. Clones that graft easily do not necessarily root easily and vice versa. Cuttings from shoots of the current year root more easily than cuttings from older branches (Harris 1990).

Based on information from many published sources, Minore (1979) indicated that Sitka spruce was not as well suited for successful establishment on organic seedbeds as are western hemlock and western redcedar. Organic matter, as a seedbed, was not defined by Minore; we assume that it refers to humus because in many sites Sitka spruce seedlings are abundant on certain forms of organic matter, such as rotten logs. On mineral soil seedbeds, red alder is the most successfully regenerating west coast species, followed by Sitka spruce. Western redcedar and western hemlock are less suited than Sitka spruce to germination on mineral soil.

Seedbed influences on Sitka spruce germination are described in more detail in the following subsection. Daniel and Schmidt (1972) suggested that seeds of several conifers are killed by fungal pathogens on humus seedbeds. B. van der Kamp (pers. comm., Oct. 1996) has shown that this phenomenon does not occur on mineral soil or rotten logs; he also confirmed the presence of fungal seed pathogens in humus of interior spruce–subalpine fir forest stands. Although humus seedbeds have not been tested for Sitka spruce seedling germination, this suggests that there may be fungal pathogens that are always present in the forest floor. If so, they may often infest and kill spruce seed before it has a chance to germinate.

Regeneration after Clearcutting

This subsection summarizes some observations on Sitka spruce regeneration in relation to harvest methods. Revegetation of a clearcut 100-year-old stand in coastal Oregon, one-half of which was burned, indicated that rotted tree stems on the forest floor provided the best conditions for Sitka spruce seedling establishment. The second best germination medium was mineral soil, followed by slash made up of mixtures of coarse woody debris, humus, and soil (Berntsen 1955).

Studies on Vancouver Island revealed that four years after high-lead clearcutting near Port McNeill, logging residues were considerably decayed and natural Sitka spruce regeneration averaged 1,976 seedlings per hectare compared with 7,412 seedlings per hectare for western hemlock and amabilis fir combined (Roff and Eades 1959). Clearcutting results in higher soil temperature than in uncut or partially cut stands, which favours seed germination and humus decomposition (Anderson 1954, 1956; Gregory 1956). The combined effect of a disturbed humus layer, exposed mineral soil, and increased soil temperature is beneficial to seed germination and results in an increased proportion of Sitka spruce in mixed spruce-hemlock stands. In Alaska, moss-covered seedbeds were less desirable than disturbed soil for Sitka spruce germination (Gregory 1956), but once seedling roots were established in moist layers below the moss, the chances for survival were better. Log surfaces that retain litter best also retain seeds best, and have the highest rates of Sitka spruce and western hemlock seedling recruitment (Harmon 1985; Harmon and Franklin 1989).

Regeneration success is related not only to the presence of seedbeds conducive to Sitka spruce seedling establishment but also to the availability of a seed source. For Alaskan clearcuts, James and Gregory (1959) suggested as a rule of thumb that no part of a clearcut should be more than 400 m from a Sitka spruce seed source. On the Queen Charlotte Islands, Warrack (1957) recommended that continuous clearcuts not exceed 360 ha, and that encircling seed sources within 800 m should be left for a period up to 10 years. These early silvicultural suggestions on ways to encourage natural regeneration of Sitka spruce on clearcuts are now supplemented by guidelines (B.C. Ministry of Forests and B.C. Ministry of Environment, Lands and Parks 1995k) prepared under the Forest Practices Code of British Columbia Act.

Regeneration Dynamics in Relation to Age Structure of Sitka Spruce Stands

Recent data from the Carmanah Creek valley in British Columbia provide insights into Sitka spruce regeneration in relation to stand age-class structure. Field observations by Beese (1989), together with his review of existing inventory data, showed that a considerable range of size and age classes are represented in the Carmanah alluvial spruce forests. The most common diameters of large spruce at breast height were 1.5-2.5 m, with scattered veterans occasionally exceeding 3 m dbh. There are also younger mature stands in which spruce range in dbh from 25 to 125 cm, accompanied by scattered larger trees. Age sequences representing a range of immature stand conditions of spruce were observed on aggrading portions of Carmanah Creek. Thus, there are younger spruce that could replace the large specimens over the long term.

In this example of age structure in the Carmanah Creek valley, Sitka spruce regeneration takes place primarily on gravel bars, old stream channels, and debris fans that have formed from slope failures in major gully systems on the steep slopes adjacent to the floodplain. Gravel bars typically have red alder, with spruce and hemlock underneath. Old stream channels have alder, dense salmonberry, and only a few scattered spruce. Small canopy openings from individual tree mortality regenerate almost exclusively to western hemlock. Only 10% of the MacMillan Bloedel Ltd. inventory plots in the Carmanah Creek valley recorded any significant spruce regeneration. Most of the regeneration of hemlock, and minor amounts of amabilis fir and Sitka spruce, occur on decaying stems (Beese 1989).

McKee et al. (1982) also found that decaying stems on the forest floor were the predominant microsites for spruce regeneration on alluvial sites in the Hoh River valley in the Olympic Peninsula. Unlike in the Carmanah Creek valley, they found that Sitka spruce was more abundant in understorey regeneration than western hemlock. Possible reasons for this abundance of spruce in the South Fork Hoh River were the open, parklike stand conditions and the preferential grazing of western hemlock regeneration by Roosevelt elk (see Appendix 3 for the scientific names of animal species).

Without disturbance, the regeneration patterns in the Carmanah Creek valley suggest that western hemlock and amabilis fir would eventually replace Sitka spruce in climax forests. Spruce lives to greater ages than western hemlock (Table 4), so surviving veterans persist in hemlock–amabilis fir stands. The presence of spruce in a variety of age classes on the Carmanah

Creek floodplain suggests that repeated disturbance from stream channel changes, flooding, and windthrow is a natural part of these alluvial ecosystems, and that these processes influence natural renewal of Sitka spruce stands. For long-term management, the important point is that on alluvial sites periodic disturbances create a continuous supply of old trees in stands that are not scheduled for harvesting, but on non-alluvial sites seedbeds suitable for spruce regeneration are needed now if the desire is to have 300-year-old trees three centuries from now. Those who advocate long-term maintenance of today's old-growth forests may overlook the fact that a disturbance that removes the present forest canopy and provides a mineral soil surface suitable for Sitka spruce germination is required if future generations are to experience old-growth Sitka spruce forests three or more centuries from now.

Sitka spruce is naturally adapted to alluvial ecosystems and their natural geomorphic processes. In a study of fluvial processes in the Hoh River valley, Swanson and Lienkaemper (1982) found that floodplain surfaces as much as 2.5 m above the low water level with 250-year-old trees were still affected by flooding. These dynamic processes create periodic new opportunities for Sitka spruce regeneration on alluvial mineral soil, with resulting development of uneven-aged stands. The ability of spruce to regenerate on newly exposed alluvial parent material suggests that in areas such as British Columbia's Carmanah Creek valley, the young spruce now on the floodplain will likely eventually replace the present giant Sitka spruce as they are lost to natural causes. However, the infrequent occurrence of the largest spruce suggests that disturbance is more the rule than the exception in the history of areas such as the Carmanah Creek valley (Beese 1989).

Growth and Productivity

An overview of Sitka spruce's productivity in its natural range in British Columbia is provided by Tables 14, 15, and 16. Site series where Sitka spruce is a naturally prominent species or where it has silvicultural potential are listed in these tables in order of decreasing productivity. For the Vancouver Forest Region (Table 14), productivity is expressed as site index, defined as median height of dominant trees (m) at breast-height age of 50 years, based on data from Green and Klinka (1994). For the Prince Rupert Forest Region (Tables 15 and 16), median values are not involved because site index estimates in these tables are based on estimates from site index bar graphs published in Green and Klinka (1994). The highest productivities shown in Tables 14, 15, and 16 are 34-37 m for height of dominant Sitka spruce at breast-height age 50. These high productivities can potentially occur on the following site series: Ss–Salmonberry, BaCw–Salmonberry, Cw–Devil's club, Ss–Lily-of-the valley, CwSs–Foamflower, CwSs–Skunk cabbage, CwSs–Devil's club, and PlYc–Sphagnum.

Table 19

Average diameters, average heights, and numbers of stems per hectare for Sitka spruce damaged and undamaged by white pine weevil attack at Port Renfrew, Franklin River, and Mooyah Bay, British Columbia

	North aspect				South aspect						Valley bottom (high elev.)						Valley bottom (low elev.)					
Initial spacing (m)	2.7		3.7		2.7		3.7		4.6		2.7		3.7		4.6		2.7		3.7		4.6	
	D[a]	U	D	U	D	U	D	U	D	U	D	U	D	U	D	U	D	U	D	U	D	U
Port Renfrew																						
(age: 24 yrs from seed)																						
Stem count (per ha)	434	841	458	267	773	258	229	309	108	327	990	95	–	–	–	–	–	–	–	–	–	–
Average diameter (cm)	19.0	16.1	19.8	18.1	16.9	13.4	9.3	8.6	11.3	9.6	16.7	9.2	–	–	–	–	–	–	–	–	–	–
Average height (m)	11.3	10.8	11.9	11.0	8.9	8.2	4.5	4.4	5.7	4.9	8.8	7.5	–	–	–	–	–	–	–	–	–	–
Franklin River																						
(age: 25 yrs from seed)																						
Stem count (per ha)	597	733	–	–	618	359	313	88	396	44	482	515	298	427	249	225	1,180	82	709	29	469	0
Average diameter (cm)	17.7	14.7	–	–	15.8	11.7	17.4	13.2	20.4	21.2	19.6	16.0	23.7	20.9	19.4	17.5	18.4	13.1	22.0	14.9	24.3	–
Average height (m)	11.0	10.2	–	–	8.3	7.0	8.2	6.8	8.3	8.0	9.9	8.3	7.3	10.7	9.3	9.0	9.0	8.5	9.7	8.9	9.1	–
Mooyah Bay																						
(age: 26 yrs from seed)																						
Stem count (per ha)	–	–	–	–	165	1,004	252	466	186	283	–	–	–	–	–	–	–	–	–	–	–	–
Average diameter (cm)	–	–	–	–	9.9	8.4	14.1	10.5	10.5	7.5	–	–	–	–	–	–	–	–	–	–	–	–
Average height (m)	–	–	–	–	6.5	5.5	8.1	7.1	6.5	5.2	–	–	–	–	–	–	–	–	–	–	–	–

Source: Omule (1988)

[a] D = damaged; U = undamaged

The relatively high growth potential of Sitka spruce on the moist and nutrient-rich sites where it grows best has been confirmed by measurements of its early growth rates within its natural range in British Columbia, as summarized in the subsections that follow. It should be emphasized that there are two aspects to estimates of productivity of forest ecosystems. One is the site component that influences ultimate productivity, as reviewed by Weetman (1982). The other involves managerial influences, including genetic improvement of yield, stand conversions that emphasize species with the greatest potential for rapid growth, full stocking to maximize the use of each site's biogeoclimatic potential to grow trees, and silvicultural steps such as fertilization and thinning.

Early Growth and Comparisons with Other Species

Sitka spruce was one of four coastal plantation species studied in the CWHvm1 variant of western Vancouver Island for survival and early growth in relation to initial spacing (Omule 1988). The evaluations involved spacings that ranged from 2.7 × 2.7 m, to 3.7 × 3.7 m, to 4.6 × 4.6 m, and the 1988 report summarized results up to age 26.

At the test sites studied by Omule, the white pine weevil repeatedly attacked Sitka spruce trees after age 8. This influenced height growth and caused forking, crooked stems, and poor tops. For this reason, crop tree yields could not be computed, as was done for Douglas-fir and western redcedar. Sitka spruce trees were grouped into damaged and undamaged trees, and the number of stems per hectare, average diameter, and average height at ages 24-26 were recorded for each initial spacing that was sampled. The results are reproduced in Table 19.

Omule (1988) noted that damaged spruce trees at a given spacing were generally taller and larger than the undamaged ones, suggesting that weevils preferred large, vigorous trees. As recorded in more detail by Alfaro and Omule (1990), the three spacings had similar average numbers of stem defects per tree, but the close spacing showed a significantly higher frequency of trees of good form relative to the more open plantations. It was concluded that trees planted at close spacing had lower levels of weevil damage. For a site with Sitka spruce reaching top heights of 12 m at age 25 (site index 24 at breast-height age 50), Alfaro and Omule (1990) made the management recommendation that Sitka spruce plantations be started at close spacing (2.7 × 2.7 m) and then be pre-commercially thinned at age 25. By this age, trees will average about 19 cm dbh and 12 m in height, thus ensuring a first log of good quality.

Also in the CWHvm1 variant on the west coast of Vancouver Island, Sitka spruce height growth to age 26 was evaluated in relation to four different sites (Omule and Krumlik 1987). In general, Douglas-fir was the tallest tree species on all sites at age 26, followed by western hemlock, Sitka spruce, and

Table 20

Characteristics of four site units in the CWHvm1 variant on which Sitka spruce juvenile height growth was evaluated by Omule and Krumlik (1987)

CWHvm1 site unit	Soil moisture regime	Soil nutrient regime
SU2	slightly dry to fresh	very poor to medium
SU3	fresh to moist	very poor to medium
SU4	fresh to moist	medium to very rich
SU6	moist to very moist	medium to very rich

western redcedar. Stem defects were more prominent in Sitka spruce and Douglas-fir than in western hemlock and western redcedar.

Omule and Krumlik (1987) compared Sitka spruce juvenile height growth on four different site units. Following the definition from Klinka et al. (1984), a site unit was considered to represent a grouping of closely related forest sites that can support forest stands of the same composition and structure, have similar production potential, and can be managed by the same silvicultural system. Characteristics of the four site units, defined from edatopic grid No. 9 of Klinka et al. (1984, p. 30), are summarized in Table 20.

Regression equations relating Sitka spruce top height to total age were depicted graphically by Omule and Krumlik (1987) for each of the four site units, and separately for trees damaged and undamaged by the white pine weevil. Height growth curves for spruce without weevil damage are reproduced in Figure 27. For comparison, Figure 27 also shows the height growth curve for Queen Charlotte Islands spruce (BA30) from Barker (1983), and a United Kingdom spruce plantation growth curve (UK24) from Hamilton and Christie (1971). Unlike in Douglas-fir, western hemlock, and western redcedar, the shapes of Sitka spruce growth curves from Vancouver Island differ substantially from the growth curves produced by Barker (1983) and by the U.K. Forestry Commission, no doubt because of significant weevil damage to the Vancouver Island sample trees. Based on the data from Omule and Krumlik (1987), Figure 28 compares early growth of coastal coniferous species sampled on the poorest site (SU2, shown as A in Figure 28) and on a more productive site (SU4, shown as B in Figure 28).

Additional data on early height growth are available from stem analyses carried out by Smith et al. (1986) on red alder, Sitka spruce, and western hemlock on Queen Charlotte Islands landslides and surrounding logged areas. The cumulative heights (in centimetres) of these three species at age 10 are shown in Table 21. Approximate values were read from height-age regressions shown in Figure 23 of Smith et al. (1986).

The Queen Charlotte Islands data indicated that red alder height growth was at least twice that of conifers on logged areas and 5-6 times that of conifers on landslide areas. For western hemlock and Sitka spruce, height growth in logged areas was about double the growth rate on landslides for the same species. Height growth of red alder was also greater in logged areas than on landslides, but the differences were much less marked than for hemlock and spruce. For both hemlock and spruce, height growth on landslides was significantly less higher in the slope than further down.

When the growth of Sitka spruce and western hemlock are compared over a longer period than shown in Figure 28, there is evidence from coastal

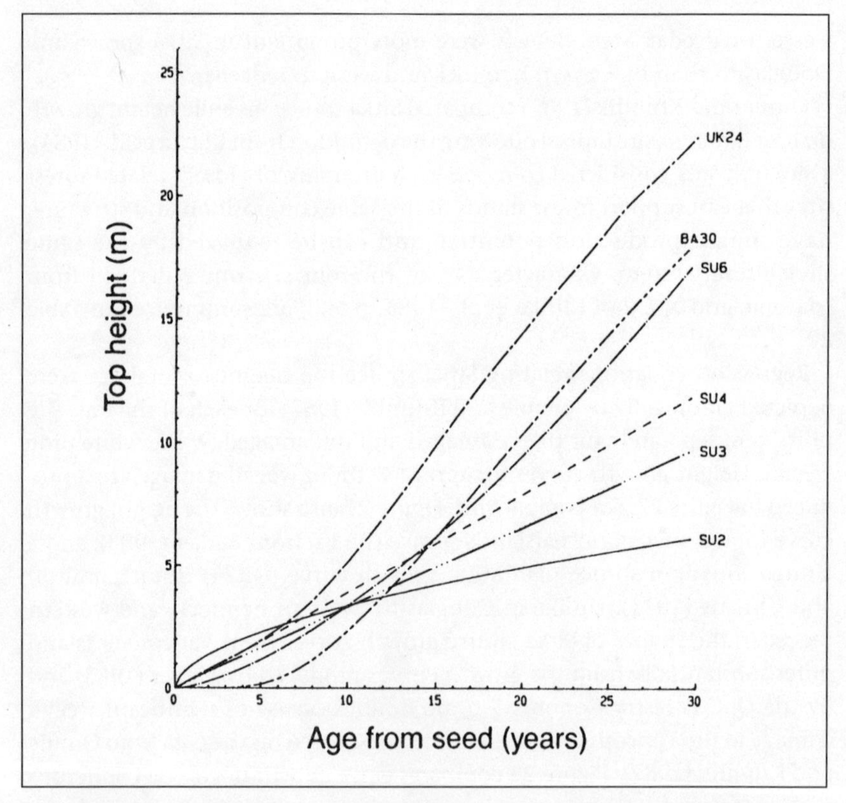

Figure 27. Height growth curves from the CWHvm1 variant for young Sitka spruce plantations without weevil damage on four Vancouver Island site units (SU2, least productive, to SU6, most productive) (Omule and Krumlik 1987) in relation to curves from Barker (1983) for Queen Charlotte Islands spruce (BA30) and Hamilton and Christie (1971) for United Kingdom spruce plantations (UK24). The BA30 curve is the Queen Charlotte Islands site index curve for 30 m at total tree age of 50 years. The UK24 curve is for Hamilton and Christie's yield class 24, which refers to a site that can attain a maximum mean annual volume increment of 24 $m^3/(ha \cdot yr)$.

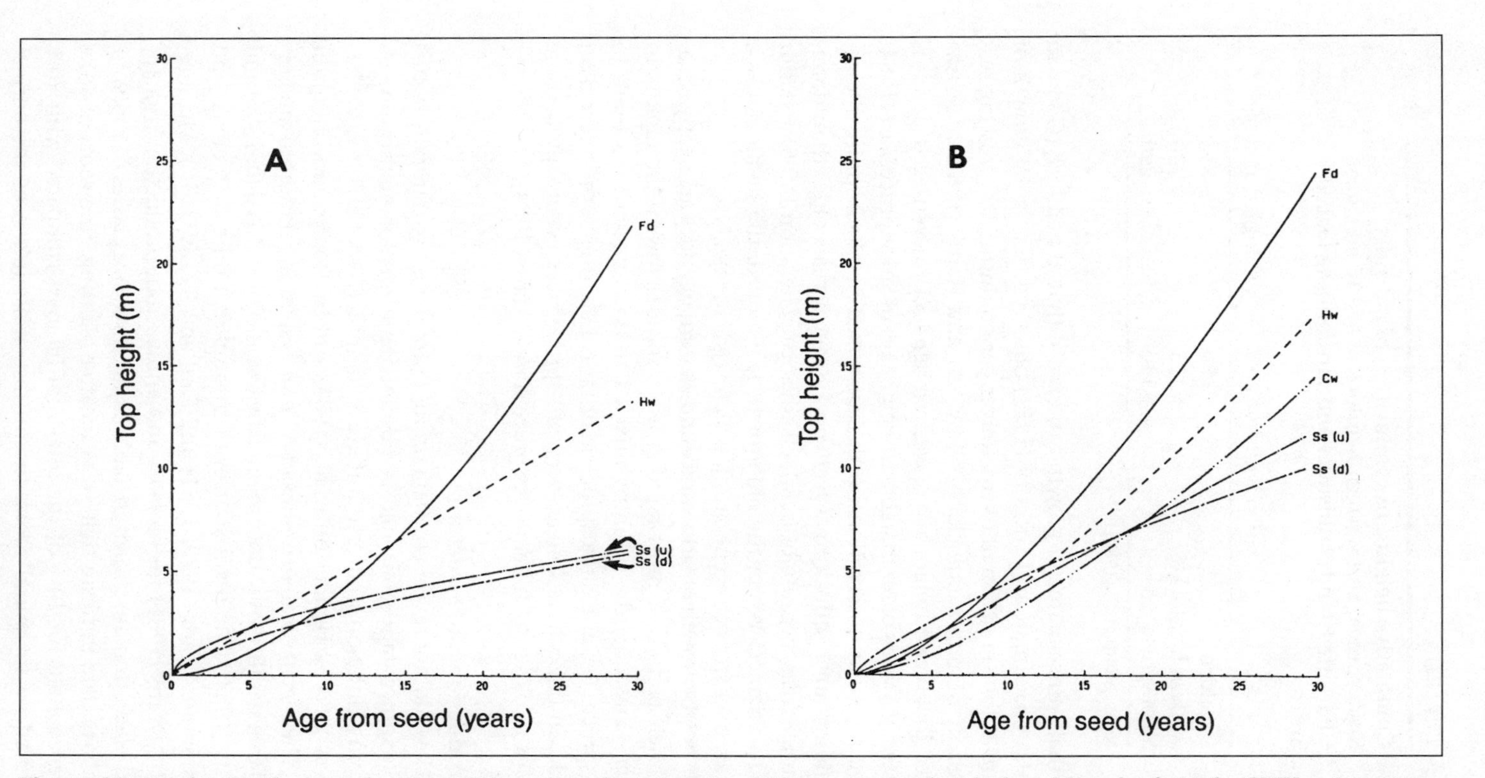

Figure 28. Height growth curves for young Sitka spruce plantations in relation to western hemlock and Douglas-fir in the CWHvm1 variant on slightly dry to fresh, nutrient-poor sites (A) and on fresh to moist, nutrient-medium to very rich sites (B) (Omule and Krumlik 1987). Two curves are shown for Sitka spruce: (d) for weevil-damaged trees and (u) for undamaged trees.

Table 21

Cumulative heights, in centimetres, of red alder, Sitka spruce, and western hemlock at age 10 on Queen Charlotte Islands landslides and surrounding logged areas

	On landslide area	On logged area
red alder	680	815
western hemlock	125	320
Sitka spruce	130	290

Source: Smith et al. (1986)

Alaska that spruce diameter growth can exceed that of hemlock (Godman 1949). Also, Godman and Gregory (1953) observed that radial growth of spruce near Juneau occurred over a longer period during the growing season than in western hemlock. Work by Farr and Harris (1983) in Alaska indicated that breast-height age of western hemlock in even-aged stands is an average of 5 years older than breast-height age of Sitka spruce. Farr (1984) recorded a similar observation in Alaskan mixed hemlock-spruce stands, where dominant hemlock trees were on average 5 years older than dominant spruce trees, because of hemlock advance generation in older stands before disturbance. When the old growth is removed, the hemlock has a head start over spruce regeneration, which seeds in later.

While height growth of Sitka spruce and western hemlock are nearly equal during their periods of most rapid growth, spruce grows more rapidly in diameter. Consequently, thinning mixed stands of smaller trees tends to favour spruce. Because Sitka spruce maintains height growth longer than hemlock, and also lives longer, very old spruce trees eventually become dominant in old-growth hemlock-spruce stands (Harris 1984).

Root Production and Structure

This review did not reveal any data on root production in naturally occurring Sitka spruce stands, but there is substantial information on this spruce's roots from plantation studies in Britain. Sitka spruce's root architecture has been described by Henderson et al. (1983) and by Coutts and Philipson (1987). The mechanical attributes of its root system in relation to anchorage and tree stability were documented by Coutts (1983a, 1983b, 1986) and by Deans (1983). The main sources of information on root biomass and production are Deans (1979, 1981) and Ford and Deans (1977). Measurements by Deans (1981) for biomass and annual root production in a 16-year-old plantation are shown in Table 22 for three size classes of roots.

The above data indicate that 16 years after planting, fine roots of Sitka spruce accounted for 14% of the total root biomass compared with 80% contributed by structural roots. In contrast, most (62%) of the annual

Table 22

Measurements for biomass and annual root production for three size classes of roots in a 16-year-old Sitka spruce plantation

	Biomass		Annual production	
Root diameter (cm)	kg/ha	kg/tree	kg/ha	kg/tree
0-0.2 cm fine roots	3,534	0.93	5,244	1.38
0.2-0.5 cm	1,368	0.36	38	0.01
> 0.5 cm structural roots	20,102	5.29	3,154	0.83
Total	**25,004**	**6.58**	**8,436**	**2.22**

Source: Deans (1981)

below-ground production was directed to the formation of fine roots (1.38 of 2.22 kg per tree), compared with only 37% to structural roots. Last et al. (1983) noted that because annual production of fine roots exceeded the contribution of fine roots to the total biomass (1.38 compared with 0.93 kg per tree in the example above), the high fine-root turnover rate may profoundly influence the development of associated mycorrhizae.

Two prominent aspects of Sitka spruce root form that relate to tree stability are (1) shallow rooting on many upland sites where it has been planted, and (2) restricted lateral root development, which gives an asymmetrical root form in plan view. There is also a tendency for major lateral roots to branch near the stem base (Deans 1983), a feature that reduces this tree's mechanical strength under bending loads (Coutts 1983a). Sitka spruce shows a remarkably uniform production of main woody branches along the length of lateral roots (Henderson et al. 1983). An important feature is its abundant production of adventitious roots, a feature shared with other spruces and with larch (Coutts and Philipson 1987). The role of adventitious roots in the development of the final root form has not been studied in detail in Sitka spruce.

It should be emphasized that the root information summarized here is based exclusively on studies in British spruce plantations. It is not known to what degree these aspects of root structure and growth apply to Sitka spruce in ecosystems in its natural range in western North America. Researchers who have given the most attention to this spruce's root system, such as Coutts and Philipson (1987), believe that the growth rates and development of Sitka spruce roots generally lie within the ranges of variation reported for other conifers of northern temperate forests, but detailed comparative work has not been done.

Nutrient Relationships and Comparisons with Other Species

In its natural range, Sitka spruce occurs on a narrower range of sites than western hemlock, western redcedar, or Douglas-fir. This may be partly a result of other conifers having a competitive advantage over Sitka spruce in terms of regeneration biology, shade tolerance, and growth rates, but it may also be a result of Sitka spruce's comparatively high nutrient demands. For example, Miller and Miller (1987) noted that, compared with pines, Sitka spruce demands a relatively rich soil – a circumstance that has led to substantial research into fertilizers that can widen the range of sites upon which Sitka spruce can be planted. Fertilization as a part of basic silviculture is discussed in the section 'Successful Basic Silviculture' (page 184) in Chapter 3. The intent of this section is to highlight Sitka spruce's key nutritional features.

This tree has a relatively heavy crown; consequently nutrient demands are high prior to canopy closure. This is one reason why Sitka spruce's growth responses to fertilizers are most pronounced when stands are young and the crown canopy is being established. After crown closure, annual production of foliage by Sitka spruce is no greater than that of other species with comparable growth rate. At crown closure and thereafter, establishment of nutrient cycles within Sitka spruce trees, as well as through litter, allows efficient reuse of nutrients. Hence, nutrient demands placed upon the soil are dramatically less than demands in the young stages preceding canopy closure (Miller and Miller 1987).

The relatively high nutrient demands made by young Sitka spruce trees are thought to be related to the rapid buildup of a green crown that has a large biomass compared with other young conifers of the same age. After spruce develops a closed canopy, however, the maintenance demand for nutrients is reduced because it is then mainly a function of the amount of foliage replaced each year. Miller and Miller (1987) posed the still-unanswered question of why Sitka spruce should retain so much foliage compared with other conifer genera, with no apparent advantage in growth rate. There is speculation that long retention of older foliage provides a reservoir of nutrients and stored energy, thereby making spruce better buffered against possible exceptional nutrient demands or reductions in site nutrient supply than species with shorter needle-retention periods. Research to test this hypothesis is still under way.

There is another aspect of Sitka spruce nutrition that is of silvicultural importance. In British upland plantations, there is good evidence that Sitka spruce grown in mixture with other conifers (often lodgepole pine or larch) is less likely to show symptoms of nitrogen deficiency than pure spruce on the same sites (Weatherell 1957; O'Carroll 1978; McIntosh 1983). Miller and Miller (1987) suggested that the greater nutrient demands made by spruce relative to other conifers during canopy formation may be counteracted by planting spruce in mixture with pine or larch. Such mixtures have the advantage of lessening the nutrient demands placed upon the soil per unit area of plantation. This is likely not the only beneficial effect of mixed planting, but it appears to be an important way to reduce nutrient loss in sites with poor soils.

It is well recognized that internal nutrient cycling can contribute a large proportion of the annual nutrient supply required for new growth in trees. For Sitka spruce, internal cycling has been best demonstrated by recent studies of phosphorus (P) supply in relation to seasonal growth of seedlings (Proe and Millard 1995a), which found evidence of translocation of P from old shoots to support new shoot growth. This research is part of a program to address the problem that in many British forest soils, P deficiency limits the growth of Sitka spruce. Besides P retranslocation, greenhouse studies of 3-year-old clonal cuttings of Sitka spruce have confirmed internal cycling of nitrogen (N) (Millard and Proe 1992, 1993; Proe and Millard 1994). This is not surprising, as there appears to be a strong correlation between N and P retranslocation (Proe and Millard 1995b), although the timing of retranslocation of these two nutrients may differ. Millard and Proe (1992) have shown that nitrogen is stored over winter in roots and current-year needles. During the first few weeks of spring growth, it is remobilized for new foliage growth, but subsequent growth is dependent on root uptake of N. In the autumn, the cycle is completed by rapid uptake of N into roots and current-year needles.

It is well known that photosynthetic capacity of plants is influenced by their N nutrition, but only in recent research (Brown et al. 1996a) are details of the relationships between foliar N concentrations and photosynthesis being worked out for Sitka spruce. As described by Brown et al. (1996b), the design of appropriate N fertilization programs for coniferous seedlings requires an understanding of how seedling growth and survival are influenced by tissue N concentrations. This recent work showed that seedlings of Sitka spruce, western redcedar, and western hemlock had similar patterns in their growth responses to whole-plant N concentrations. Despite these similarities, the three species differed in their growth and mineral nutrition responses to N additions in ways consistent with field observations of their natural habitats. Sitka spruce, which is normally associated with fertile sites, required the highest whole-plant N concentrations to achieve a given relative growth rate.

Poor growth of plantation Sitka spruce on northern Vancouver Island has been studied in detail with regard to nutritional processes (Weetman et al. 1989, 1990; Messier and Kimmins 1991; Messier 1993; DeMontigny et al. 1993; Keenan 1993; Keenan et al. 1993, 1995; Prescott et al. 1993, 1995; Prescott and Weetman 1994; Chang et al. 1995). A key conclusion from these studies is that the chlorosis and near-cessation of growth of young Sitka spruce, western hemlock, western redcedar, and amabilis fir following clearcutting and slashburning of old-growth cedar-hemlock forests is at least partly a consequence of the low capacity of cedar-hemlock forests to provide nutrients to the regenerating forest (Prescott et al. 1995). The northern Vancouver Island research also confirmed that soil fauna play a key role in site nutritional relationships based on comparisons of two distinct forest types: cedar-hemlock (CH) and hemlock–amabilis fir (HA). The HA is considered to be a more productive forest type than the CH, possibly because the CH is N- and P-deficient for good tree growth (Germain 1985; Prescott et al. 1993). Battigelli et al. (1994) tested the hypothesis that better tree growth in the HA could be the result of a larger soil fauna community in this forest type. HA forests did, in fact, maintain a higher abundance and biomass of soil fauna than CH forests, indicating that forest site productivity is positively correlated with soil fauna biomass.

This decade-long interdisciplinary research project, under the Salal Cedar Hemlock Integrated Research Program (SCHIRP), did not result in Sitka spruce being recommended as a plantation species for regenerating northern Vancouver Island areas that have been clearcut and slashburned. The most effective way to regenerate these cedar-hemlock forests is to plant hemlock and cedar, and to fertilize at five years with at least 200 kg/ha N and 50 kg/ha P, or organic wastes (Prescott and Weetman 1994). Aside from being one of the second-growth species observed and measured in the SCHIRP nutrition-related studies, Sitka spruce has received little attention compared with other conifers in the Pacific Northwest (Perry et al. 1989).

The visible symptoms of various nutrient deficiencies result from physiological responses that are an inherent part of the genetics of a given tree species. Therefore, the illustrated guide to nutrient deficiencies in British coniferous forests (Binns et al. 1976), which includes many photographs of Sitka spruce, is a good reference for managers anywhere that this spruce is grown. Combined with foliar analyses and a knowledge of nutrient regimes in site series where Sitka spruce is being managed, deficiency symptoms are a practical basis for prescribing fertilizer treatments. In general, the best time of year to observe deficiency symptoms is during autumn, when height growth is complete, foliar nutrient concentrations and colours have stabilized, and winter winds have not yet caused foliar colour changes.

Mycorrhizal Relationships

Sitka spruce has ectomycorrhizal associations with a large number of fungi (Trappe 1962), but none of these appears to be host-specific (Molina and Trappe 1982). Overall, about 110 fungi are likely mycorrhizal associates with conifers in the Pacific Northwest. In contrast to western redcedar, which has vesicular-arbuscular mycorrhizae, Sitka spruce and western hemlock are associated with ectomycorrhizae and ectendomycorrhizae (Minore 1979). Mycorrhizal fungal diversity is maximum around the time of canopy closure, when forest floor litter depth is highest and when trees reach a peak of nutrient deficiency (Dighton 1987; Dighton et al. 1986, 1987). What is not known is how long these influences persist. In his review of Sitka spruce mycorrhizae based on the British experience, Walker (1987) concluded that there are few instances where lasting benefit has been derived from using mycorrhizal fungi in commercial forestry, and none relating to success with Sitka spruce.

This lack of success is likely due to insufficient knowledge of factors affecting natural populations of mycorrhizae, and of how manipulations in artificial conditions further influence populations. It is known that factors such as changes in soil conditions, nutrient availability, additions of fertilizers, clonal differences in natural populations, and site variability can affect mycorrhizae (Walker et al. 1986). Both strains and species of mycorrhizal fungi are known to influence Sitka spruce seedling growth differently in a nursery, effects that are further influenced by soil type (Holden et al. 1983). Using fungal species from field studies in northern Britain, Lehto (1992a) showed that differences in mycorrhizal colonization of Sitka spruce seedlings occurred during well-watered, moderate, and severe drying cycles, but the seedlings with high proportions of mycorrhizae tended to have the highest nutrient concentrations in all water regimes. Studies of mycorrhizae and drought resistance in conditions of adequate versus deficient nutrients (Lehto 1992b, 1992c) indicated that during drought the improved performance of the mycorrhizal seedlings could be attributed to their better phosphorus (P) and potassium (K) nutrition and their more extensive root systems with mycelial strands. In the case of seedlings with adequate nutrition, mycorrhizal plants had significantly lower water potentials than non-mycorrhizal plants at any given substrate moisture content; mycorrhizal infection had no effect on stomatal conductance or net photosynthetic rate in either well-watered or drought-stricken Sitka spruce seedlings in this glasshouse study. Similarly, Last et al. (1990) found that 2-year-old spruce seedlings with similar numbers of mycorrhizae tended to be the same height irrespective of inoculation treatment.

Sitka spruce may also have shared symbioses. Studies in the United Kingdom have found evidence that Sitka spruce has mycorrhizal relationships with fungi associated with both birch and Scots pine. Plantings of Sitka spruce in mixture with pine or larch alleviate a problem of poor growth without addition of nitrogenous fertilizer. The 'mixture effect' is poorly understood but may be related to mycorrhizae and improved nitrogen and phosphate status resulting from increased mineralization from mixed litter rather than pure spruce litter (Miller et al. 1986; Dighton 1987). Different isolates of the same fungal species can have different effects on Sitka spruce seedlings, and some isolates respond to nursery conditions but not to forest conditions. Walker (1987) stressed the urgent need to develop reliable criteria for selecting isolates to test in forest conditions.

Further, Mason et al. (1983) suggested that the concept of fungal succession for mycorrhizal species necessitates selection of the correct stage of fungi relative to plant age for outplanting. For example, planting trials of Sitka spruce in Alaska used three species (*Hebeloma crustuliniforme, Laccaria laccata,* and *Cenococcum geophilum*) for inoculation (Shaw et al. 1982, 1987; Sidle and Shaw 1987). The fungi survived for two years, but the mycorrhizae remained confined to the original planting plug, and inoculation did not result in any improvement in survival or growth. Research in both Alaska (Shaw et al. 1984) and Scotland (Taylor and Alexander 1990) show that fertilizer treatments appear to have no influence on root colonization, or at most only slightly reduce the level of infection of mycorrhizae and the proportion of the dominant mycorrhizal type.

Despite Walker's (1987) conclusion that there are few instances where lasting benefit has resulted from the encouragement of mycorrhizal fungi in forestry, several British reviews based on forest trials and nursery studies (Mason et al 1983; Thomas and Jackson 1983; Thomas et al. 1983; Walker 1987; Wilson et al. 1987) have shown that inoculation with selected mycorrhizal fungi can positively influence the early growth of Sitka spruce. Ectomycorrhizal inoculation of tree seedlings has frequently increased the survival and growth of seedlings, with improvements reported in nutrient uptake, increased tolerance to low water availability, and increased resistance to some fungal pathogens (Ingelby et al. 1994). Regardless of the types of mycorrhizae that developed on Sitka spruce seedlings, their height growth was enhanced and was exponentially related to the total numbers of mycorrhizae present (Last et al. 1990).

Nursery inoculation as an economic practice requires the inoculum to be applied at a time when it is capable of surviving in the soil until seedling root systems develop to the stage where they can form mycorrhizae, a period of several weeks. Tests in two Scottish nurseries where seedbeds were inoculated with the ectomycorrhizal fungus *Laccaria proxima* and seeded to Sitka spruce indicated that the numbers of emergent seedlings were affected by both inoculum and fertilizer treatments. Up to 114% more seedlings were found in plots inoculated with live fungus in a vermiculite-peat carrier than in control plots. Application of fertilizer to the seedbed before sowing, and later as a top dressing, reduced the numbers of emergent seedlings (Ingelby et al. 1994). These results offer encouragement for the use of mycorrhizal inocula in Sitka spruce nurseries, but they also indicate the need for discretion in applying fertilizer to nursery beds, whether or not it is done in conjunction with mycorrhizal inoculation.

Problems with inoculating nursery seedlings indicate that it is important to maintain the inherent inoculum potential in coastal forest sites. In general, inoculum on recently logged sites in the Pacific Northwest appears adequate. However, clearcuts invaded by weeds that do not form mycorrhizae, and clearcuts where reforestation is delayed, may have low site inoculum potential. The main agents disseminating mycorrhizae in the Pacific Northwest are small mammals, such as the red-backed vole (*Clethrionomys gapperi*), which eat truffles formed by vesicular-arbuscular and ectomycorrhizal fungi and then leave droppings of fungal inoculum wherever they travel (Russell 1989).

Additional general information on coniferous mycorrhizae, applicable but not specific to Sitka spruce, is provided by Russell (1978), DeYoe and Cromack (1983), Roth (1989), and Trofymow and van den Driessche (1991).

Soil and Soil-Water Relationships

Day (1957) suggested that Sitka spruce and western hemlock require less moisture than western redcedar. Wood (1955) reported that Sitka spruce dominated western hemlock on moist sites in southeastern Alaska, but not on dry sites. On the basis of available information, Minore (1979) listed major western tree species in the following order, beginning with those that have optimum development on moist sites and ending with those that have optimum development on dry sites:

mountain hemlock, amabilis fir, black cottonwood
grand fir
western redcedar
Sitka spruce
western hemlock
bigleaf maple
red alder
Douglas-fir
ponderosa pine
arbutus
Garry oak

Sitka spruce typically occurs on moist seepage sites in the coastal zone. The critical level of well-aerated soil moisture required for this species has not been defined, but it is known that if rooting space or seepage supply is restricted, or if vertical drainage is impeded, there is a tendency for Sitka spruce crowns to thin out or for trees to die. In all Sitka spruce sites examined by Stanek and Krajina (1964), soil moisture was abundant. Each site was characterized either by ground water 50-100 cm below the forest floor or by a permanent supply of seepage water. Even where Sitka spruce occurs on coastal dunes, there is abundant soil moisture in the rooting zone. Also near its natural eastern limits, Sitka spruce is confined to sites in valley bottoms that have reliable soil moisture, as in the Mount Baker region of the Cascade Mountains or in moist *Oplopanax* sites, as in the Malcolm Knapp Research Forest near Haney, British Columbia.

There are indications that differences in water regimes of various soil horizons are responsible for growth responses of Sitka spruce, probably through nutrients in seepage water. For example, the absence of a cemented hardpan beneath the root system is often accompanied by greater height growth of Sitka spruce as well as an increase in its abundance in the species composition of the site (Day 1957). Soil compaction as a result of harvesting or silvicultural operations has received much less attention for Sitka spruce than for other managed conifers, but Miller et al. (1996) addressed this subject in three coastal clearcuts in Washington following ground-based yarding on sites with fine-textured soils. On primary skid trails, soil bulk density initially increased by 40% or more, and by eight years following harvesting, it was 20% greater than on areas adjacent to skid trails. Changes in soil bulk density did not affect the eight-year survival and growth of planted Sitka spruce and Douglas-fir, but reduced western hemlock survival by 50%. Tilling of skid trails improved survival and early height growth, especially in hemlock, but average tree height and volume of Sitka spruce, Douglas-fir, and hemlock showed only minor differences between treatments.

Drought Tolerance

There was insufficient information from studies within Sitka spruce's natural range to permit Minore (1979) to include Sitka spruce in a ranking of drought tolerance among western tree species. However, it is known that Sitka spruce and western hemlock are more susceptible to moisture stress than Douglas-fir (Fraser and Cordes 1967). Although stomatal closure is only an indirect indicator of drought tolerance, Hodges (1967) found that Douglas-fir stomata closed more completely at night than those of Sitka spruce or western hemlock.

Based on many years of field experience in Great Britain, Sitka spruce managers there generally follow the guideline that Sitka spruce should not be planted on sites where mean annual precipitation is less than 1,000 mm or where the soil is shallow and freely drained (Fourt 1968; MacDonald 1979; Jarvis and Mullins 1987). The most detailed information on the effects of drought on growth of Sitka spruce comes from the modelling work by Jarvis and Mullins (1987), based on data collected from spruce plantations in Scotland. Their comparisons between measured annual spruce stem growth on a freely drained podzol and nearby peaty gley sites confirmed that soil water deficit limited tree growth in dry years. This modelling work indicates that for sites studied in Scotland, reasonably high yield classes can be achieved even when annual rainfall is as low as 700-800 mm.

Response to Flooding

Sitka spruce's natural occurrence on floodplains of river valleys in British Columbia and the Pacific Northwest is an indication of a relatively high tolerance to flooding. This tolerance is also evident in upland areas where Sitka spruce seedlings are able to survive in microsites that experience prolonged periods of standing water (Plate 12). Reviews by Minore (1979) and Walters et al. (1980) revealed examples of Sitka spruce showing 84% survival after four weeks of flooding. Survival of flooding is greater during the dormant season than during the growing season. Sitka spruce was also listed by Oliver and Hinckley (1987) as an important contributor to biodiversity of riparian areas because of its tolerance to high water tables.

During the 1948 spring flood on the Fraser River delta, Sitka spruce, lodgepole pine, and western redcedar all experienced lower mortality from flooding than Douglas-fir (Brink 1954). Sitka spruce is among the tree species that Krajina (1969) listed as a survivor on floodplains where Douglas-fir cannot grow. In wet sites and along streams in the Olympic Peninsula, Minore and Smith (1971) ranked red alder and western redcedar as the most tolerant to shallow water tables, followed by Sitka spruce; western hemlock was the least tolerant. Minore (1979) listed major western tree species in the following order of decreasing tolerance to excess moisture:

lodgepole pine, western redcedar, mountain hemlock, yellow-cedar, subalpine fir, black cottonwood
red alder
Sitka spruce, ponderosa pine, grand fir
western hemlock
Douglas-fir

Compared with other west coast conifers, Sitka spruce exhibits substantial tolerance to flooding, but forest managers interested in promoting the species on alluvial floodplains face a dilemma. Floodplains are among the most productive ecosystems in British Columbia, and they have the potential to support valuable stands of Sitka spruce, western hemlock, western redcedar, or black cottonwood. Because of the high productivity, a forest manager has incentives to reforest these ecosystems after harvesting. As indicated by research in the floodplains of the lower Skeena River (Beaudry et al. 1990), if there are no special silvicultural efforts, the post-harvesting vegetation is commonly red alder of low merchantability together with unmerchantable Sitka willow. The choice of whether to replace alder and willow with Sitka spruce and black cottonwood is made difficult by the fact that the stability and longevity of different parts of a floodplain are so variable that there is considerable uncertainty about whether particular areas will be eroded or permanently flooded before planted trees become harvestable. Fortunately, large areas of flat valley floors and terraces are stable over the time scales used for forest management.

The same, however, is not true for channel margins and islands. As Hogan and Schwab (1990) emphasized, the fact that a particular channel island has been stable long enough to produce a mature timber stand does not guarantee its future stability. Furthermore, citing Church (1983) and unpublished data gathered by T.P. Rollerson on northern Vancouver Island, these investigators point out that some islands become more unstable after logging. Thus forest managers need to carefully evaluate island longevity if such ecosystems are to be considered as investment options for Sitka spruce plantations.

For the reasons outlined above – and also because alluvial floodplain sites are characterized by severe vegetation competition, have high value as fish and wildlife habitat, and also have high potential for land-use conflicts involving recreation – particular attention needs to be given to decisions on where to locate plantings of alluvial-site Sitka spruce. Recent studies in the lower Skeena River valley of British Columbia were designed to address these management questions by developing biophysical criteria for identifying sites favourable for the establishment and growth of Sitka spruce (Beaudry et al. 1990). In general, survival and growth of spruce on alluvial sites is influenced by the frequency, duration, and depth of flooding. Ground elevation is the most direct measure for estimating flooding frequency and depth, whereas terrain features control flood duration. Observations by Beaudry and Hogan (1990) in the Skeena River floodplain confirm that the greatest growth problems for spruce seedlings are in the depressions and lower areas of the floodplain. Therefore, important questions for forest managers in floodplain areas typified by the Skeena River valley include the following:

- Is flood hazard solely a function of site elevation?
- Is there a critical elevation below which Sitka spruce cannot survive?
- Is it possible to identify the differences between good and unacceptable growing sites for Sitka spruce?
- Can the flooding potential of a site be described and classified, and then related to the chance of success of various silvicultural activities?

Beaudry and Hogan (1990) found that, in general, where flooding from spring peak flows was deep (greater than 1.2 m), the growth of Sitka spruce was poor. Small gains or losses in elevation substantially influenced the recorded heights of Sitka spruce seedlings. These researchers' linear regression of Sitka spruce tree height over ground elevation for one sample transect is reproduced in Figure 29. Ground elevation does not explain all of the

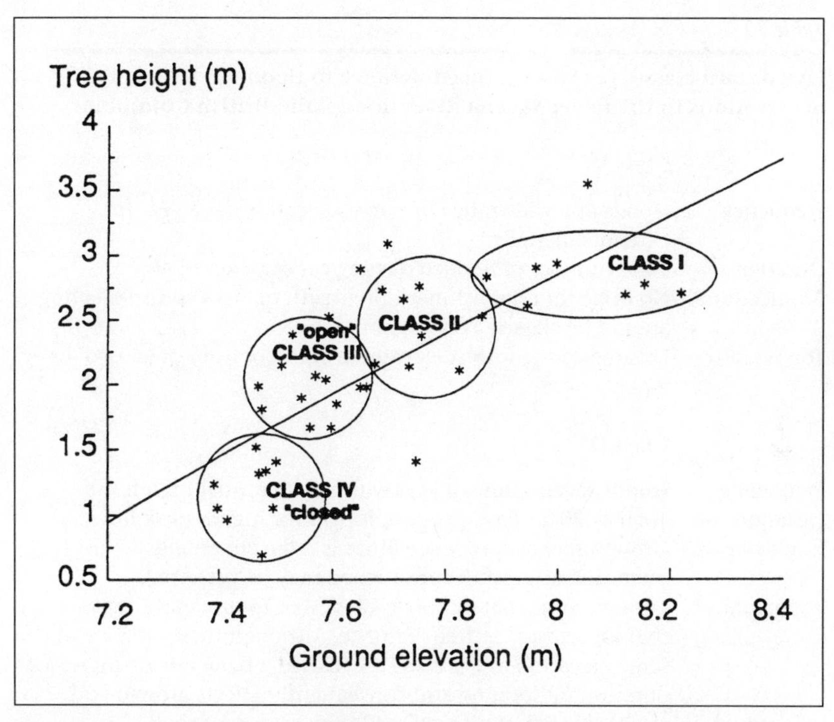

Figure 29. Flood hazard classes (Class I, most suitable for Sitka spruce seedlings, to Class IV, least suitable for spruce) superimposed upon a regression of Sitka spruce tree height over ground elevation for a floodplain site on the lower Skeena River, British Columbia (from Beaudry and Hogan 1990). The flood hazard classes shown on this figure are explained in Table 23.

variation in height of Sitka spruce; variation in growth is best explained by a combination of both depth and duration of flooding. In Figure 29, the flood hazard classification developed by Beaudry and Hogan (1990) is superimposed upon the tree-height/ground elevation regression. The five classes of the flood hazard rating and the implications of each class for establishment and growth of Sitka spruce seedlings are explained in Table 23. Class V, the most severe flooding class, did not occur in the sample transect shown in Figure 29.

In this example from the Skeena River valley, as river banks are overtopped the most dramatic effects on young Sitka spruce seedlings occur if there are extended periods of high flows in June, which is only a few weeks after seedlings would normally be planted in April or May. Two landscape features, both easily identifiable on aerial photographs, are important indicators of differences in flood duration that can be significant for Sitka spruce seedling growth. Low areas that are open at one end, such as abandoned back channels, allow free drainage of floodwaters. In contrast, flooded sites

Table 23

Five hazard classes for Sitka spruce tolerance to flooding, based on observations in the lower Skeena River floodplain, British Columbia

	Class V
Frequency	Floods at a wide range of flows, several times a year for extended periods.
Duration	Total number of flooded days a year can exceed 40.
Implications	No hope for establishment or growth of Sitka spruce seedlings. Should be classed as non-commercial.
Topography	Located in the lowest elevations at the bottom of 'closed' or 'open' depressions.
	Class IV
Frequency	Floods several times a year with mean annual high flows.
Duration	Total of 20-25 days per year, longer for higher peak flows.
Implications	Growth rate of spruce seedlings is reduced; establishment is often not successful; cottonwood may be preferred.
Topography	Located immediately above Class V or in moderate depressions that are 'closed' to free drainage. Although these sites are at the same elevation as Class III, the 'closed' characteristic increases duration of flooding and consequently affects growth and establishment of planted seedlings.
	Class III
Frequency	Floods several times a year with mean annual high flows.
Duration	Total of 10-15 days per year, longer for higher peak flows; the shorter duration (10-15 days versus 20-25 days) differentiates Class III from Class IV.
Implications	Growth of Sitka spruce is less than ideal, but can be acceptable. Spruce survival can vary, depending on extent of flooding the year the seedlings are planted.
Topography	Same elevation as Class IV, but sites are not confined. The most important characteristic is the 'open' configuration.
	Class II
Frequency	Floods annually, usually once in the spring and once in the fall to a depth of 30-90 cm.
Duration	Short; several hours to 1-2 days.
Implications	Growth is good.
Topography	High ground located between Classes III and I. On aerial photographs, it is difficult to distinguish between Classes I and II. However, because both are good sites for establishment of Sitka spruce, this has little consequence.

▶

◄ *Table 23*

	Class I
Frequency	Flooded not more than once a year.
Duration	If flooded at all, duration is short – a few hours to half a day.
Implications	The short-duration flooding has no detrimental effect on growth of Sitka spruce.
Topography	High ground.

Source: Beaudry et al. (1990)

also contain areas that are closed to drainage back to the river. These two distinct ecosystems are identified in Figure 29 and Table 23 as 'open' and 'closed.'

Open and closed areas can be at the same elevation but can vary greatly in their duration of flooding. That is why ground elevation alone is not a totally satisfactory predictor of Sitka spruce seedling success. The growth of Sitka spruce is much better in an open ecosystem than in a closed one at the same elevation, not only because of shorter flood duration in open systems but also because such systems have greater input and flow-through of nutrients than closed areas. For these reasons Beaudry et al. (1990) and Beaudry and Hogan (1990) distinguished between open and closed depressions in the classification system used in Table 23 and Figure 29. In summary, the Skeena River valley studies indicated that the landscape features of greatest importance to forest managers planning plantations of alluvial-site Sitka spruce are as follows: distinction between linear depressions that are open or closed to the main channel; distance of the proposed planting site from river flow in the mainstem stream or in back channels; and presence of hummocks and other kinds of elevated areas.

When the flood hazard classification developed by Beaudry and Hogan (1990) is overlain on tree-height/ground-elevation data in Figure 29, the best spruce heights are achieved in Class I (as defined in Table 23), but the height growth of trees in Classes I and II tend to be influenced by factors other than flooding. Trees in Class III are taller because they are in more open, better-drained terrain than the Class IV areas. Sitka spruce planted in Class V normally do not survive, and such areas should not be considered for silvicultural investment.

Physical and Chemical Damaging Agents

The following subsections focus on Sitka spruce responses to wind and frost, which are potentially damaging physical agents, and to chemical agents such as airborne pollutants and herbicides.

Wind Influences and Blowdown/Mass-Wasting Relationships

Western North America is spared the frequent typhoons and hurricanes experienced by the coastlines of eastern Asia and eastern North America. The infrequency of hurricane-strength winds along the west coast of North America is one reason why conifers in this region can grow to large sizes (Franklin and Waring 1979). However, ridgetops and areas directly on the coast are exposed to high winds. Hurricane-force gales do occur often enough within the natural range of Sitka spruce to be an important, if infrequent, source of forest disturbance. Even in the absence of hurricane-force weather systems, Sitka spruce is subject to windthrow losses of standing trees, especially at the edges of clearcuts (Figure 30). For these reasons, wind damage in hemlock–Sitka spruce stands is recognized as a factor in planning for stand tending and harvesting in British Columbia, mainly because occasional but rare extreme winds can result in large areas of blowdown in certain topographic situations. Scattered blowdown can occur during years when there are no catastrophic storms, and this process cumulatively results in large timber losses.

Sitka spruce was ranked as having medium resistance to windthrow in Minore's (1979) comparison of five coastal tree species. Douglas-fir and western redcedar were considered to be more resistant, whereas western

Figure 30. In its natural range Sitka spruce is usually not subjected to major stand losses from windthrow, but wind damage is a potential concern at stand edges, as in this case of 1993 post-logging blowdown of spruce adjacent to a small experimental patch clearcut near Rennell Sound, Queen Charlotte Islands, British Columbia. (Photograph by F. Pendl)

Table 24

Windthrow hazard evaluation for British Columbia forests, with potential application to stands containing Sitka spruce

	High hazard	Moderate hazard	Lower hazard
Wind force factors	Topographically exposed locations	–	Topographically protected locations
	Boundaries on the windward edge of a stand	Boundaries parallel to the storm wind direction	Boundaries on the lee edge of a stand
	Tall trees	Trees of intermediate height	Short trees
	Large, dense crowns	Moderately dense crowns	Small, open crowns
Resistance to overturning	Trees with low taper and no butt flare	Trees with moderate taper and moderate butt flare	Trees with high taper and large butt flare
	Shallow rooting (< 0.4 m)	Moderately deep rooting (0.4-0.8 m)	Deep rooting (> 0.8 m)
	Root rot areas	–	No evidence of root rot
	Shallow soils (< 0.4 m)	Moderately deep soils (0.4-0.8 m)	Deep soils (> 0.8 m)
	Poorly drained soils	Imperfectly to moderately well drained soils	Well-drained soils
Other indicators	Moderate to extensive natural windthrow present	Minor natural windthrow	No natural windthrow
	Extensive windthrow present on similar adjacent cutting boundaries	Minor to moderate windthrow present on similar adjacent cutting boundaries	No windthrow on similar adjacent cutting boundaries
	Pit and mound microtopography	–	No evidence of pit and mound topography

Source: Stathers et al. (1994)

hemlock and amabilis fir were ranked as less windfirm than Sitka spruce. More recently, Stathers et al. (1994) provided a windthrow hazard evaluation for British Columbia forests that has potential application to stands containing Sitka spruce (Table 24).

Blowdown is classified into two main groups for management purposes: (1) uprooted trees on poorly drained soils and on shallow soils that prevent deep root development, and (2) stem breakage on well-drained soils that support deep rooting. Knowledge of local wind directions, which winds generally cause blowdown, the time of year these occur, local topography, timber types and age classes, soil conditions, and cutblock boundaries are all factors that must be evaluated to reduce losses from wind. Meteorological records for the return periods of stand-destroying winds are not much help to forest managers because of major differences in intensity and frequency of high winds in different topographic settings. The best that Sitka spruce managers can do with currently available data is estimate the probability of a stand-destroying windstorm (Sanders and Wilford 1986) during an interval of forest management that spans several decades.

Mass wasting has been linked to blowdown in the Queen Charlotte Islands (Sanders and Wilford 1986). The related processes are storms, which bring high-intensity rainfall that can trigger mass wasting, and high winds that cause blowdown. In addition, blowdown itself is a process that in some circumstances may encourage mass wasting.

Mass wasting is common on steep slopes throughout the range of Sitka spruce. Typically it involves initial failure of a relatively shallow, cohesionless soil mass that overlies an impermeable layer on steep slopes. Considerable work has been done to facilitate the recognition of potential mass-wasting problems and to minimize human-caused mass wasting during road construction and logging (Swanston 1974; Rollerson 1980; Wilford and Schwab 1982; Carr 1985; Krag et al. 1986; Sanders and Wilford 1986; Sauder and Wellburn 1987, 1989; Carr and Wright 1992; Chatwin and Smith 1992; Chatwin et al. 1994). Suggestions have included modifications to clearcut harvest methods, such as staggering the setting boundary in a configuration that resembles the cutting edge of a rip saw, and removing all the tallest trees within half a tree length of a clearcut boundary. Other steps for improved management of landslide-prone terrain include: slope stability mapping; reductions in timber supply areas in locations where there is concern about slope stability; implementation of detailed logging prescriptions such as those prepared under the coastal fisheries/forestry guidelines (B.C. Ministry of Forests et al. 1993); use of alternative harvesting systems, such as skylines and helicopters, for removing harvested trees from steep slopes; and improved road-building practices (Chatwin and Smith 1992). Sanders and Wilford (1986) stressed that regardless of the action taken, little can be done to counter major climatic events. Such events are of such magnitude on the Queen Charlotte Islands that any forest ecosystem, no matter how well managed, is vulnerable to mass wasting and blowdown.

In the Coast Ranges of Oregon, Ruth and Yoder (1953) found no correlation between wind damage and size of clearcuts in Sitka spruce–western hemlock stands. Losses from windthrow averaged 0.64 $m^3/(ha \cdot year)$ where 16-24% of the volume had been removed by thinnings. Losses were also heavy where dominant and codominant trees had been removed. Windthrow is generally more severe in areas with a high water table or where the soil is very shallow. Wind-caused mortality increases the proportion of shade-tolerant species such as western hemlock, because understorey trees are released after the canopy dominants are removed. Thus, when trees in the released stand die, the contribution of dead wood is from smaller-diameter, more rapidly decomposed western hemlock stems.

Recently Greene et al. (1992) documented the impact of wind disturbance on growth, mortality, and biomass change in the Sitka spruce–western hemlock forests of Oregon's Coast Ranges. They suggested that biomass accumulation may be limited by wind disturbance. Wind was shown to have a direct effect on the location and extent of regeneration by creating disturbance patches ranging from discrete gaps where one or a few trees are windthrown to large blowdown areas. Initial small canopy openings may expand and join nearby small openings, eventually forming larger, indistinct disturbance patches. This process is characteristic of forests exposed to periodic windstorms. A successional consequence is an early peak in forest stand biomass after disturbance, followed by the eventual decline of biomass per unit area of landscape.

Bormann (1989) and McClennan et al. (1990) reported that in the Tongass National Forest in Alaska, soil turnover caused by windthrow is an important factor in reducing the soil degradation effects of podzolization. When trees are uprooted by windthrow, humus is mixed with podzolized mineral soil, thus aiding in nutrient cycling and maintenance of ecosystem fertility (Ugolini et al. 1989). On Prince of Wales Island in southeastern Alaska, Harris (1989) recorded 1,010 patches of partial or complete blowdown that had occurred up to 1971-72, ranging from 0.8 to 70 ha. These blowdowns made up 1.6% of the area of productive forest land and 1% of the net sawtimber volume in the study area. Blowdown was most common in the 250-360 m^3/ha volume class. Sitka spruce–western hemlock stands were the most susceptible to blowdown, whereas western redcedar stands were the most windfirm. Nearly all blowdowns occurred in uneven-aged stands in advanced successional condition or in stands previously damaged in some way. About 14% of blowdowns were next to the edges of clearcuts, and about 66% of these were leeward of cutover units. Trees adjacent to salt water were quite windfirm in the face of onshore winds, but less so in the face of offshore winds.

In southeastern Alaska and in the wettest biogeoclimatic subzones of coastal British Columbia, where disturbances from wildfires are very uncommon compared with interior forest regions, uprooting is a predominant disturbance influencing stand development in Sitka spruce–western hemlock forests. McClellan et al. (1990) recorded that in southeastern Alaska, spruce and hemlock site index is positively correlated with the degree of uprooting disturbance. They report unpublished data indicating that basal area per hectare for spruce and hemlock is up to four times greater on sites where mounds from uprooting are common than in undisturbed sites with little or no uprooting.

Management to reduce the risks of blowdown is not addressed by a specific guidebook under the Forest Practices Code of British Columbia Act. However, the relationships between mass wasting and storms of high wind and heavy precipitation, as outlined above, are part of the rationale behind at least five guidebooks that deal with erosion-related topics: *Hazard Assessment Keys for Evaluating Site Sensitivity to Soil Degrading Processes; Coastal Watershed Assessment Procedures; Gully Assessment Procedures; Mapping and Assessing Terrain Stability;* and *Soil Conservation* (B.C. Ministry of Forests and B.C. Ministry of Environment, Lands and Parks 1995b, 1995d, 1995e, 1995f, 1995l).

The most detailed guides on management to minimize wind damage include the following: for British Columbia, the handbook by Stathers et al. (1994); for Alaska, the review by Harris (1989); and for Sitka spruce in Great Britain, the text by Coutts and Grace (1995), the Forestry Commission Bulletin by Quine et al. (1995), and the earlier wind hazard classification by Miller (1985). The handbook by Stathers et al. (1994) recommends three alternative strategies for regenerating windfirm stands on sites of high to moderate windthrow hazard:

- Grow trees at wide spacing to develop the natural windfirmness of each tree (suitable for sites where any type of forest cover is acceptable, such as at high elevations).
- Grow trees at a medium spacing, either by planting or by using early thinning, to promote windfirmness (these stands should not be thinned after the height exceeds about 15 m, to allow the canopy to close and to increase the windfirmness of the stand during its more vulnerable older stage).
- Grow trees at close spacing and harvest at the onset of windthrow or at a specified height, such as 20 m, when the risk of windthrow approaches a critical threshold (these stands should not be thinned at any age because of the increased risk of windthrow).

The British experience is helpful in defining the critical threshold referred to above. Recent evidence from windthrow monitoring in Britain has shown that wind climate is very variable in space and time, and that adaptive growth by trees can ameliorate the effects of strong winds (Quine 1994). British foresters now prefer to assess the likelihood (risk) rather than inevitability of damage (hazard) from wind (Coutts and Grace 1995; Quine 1995).

Frost Tolerance

Among British Columbia's highly shade-tolerant tree species, the group that includes Sitka spruce was listed by Stathers (1989) as having moderate resistance to frost; others in this group were subalpine fir, amabilis fir, and grand fir. Mountain hemlock was the only highly shade-tolerant species ranked with a high resistance to frost; western hemlock and western redcedar have low frost resistance. Minore's (1979) summary indicated that a November 1955 cold wave in western Washington damaged true firs less than Sitka spruce, and Sitka spruce less than Douglas-fir (Duffield 1956). Day (1957) reported no incidence of frost injury on any Sitka spruce on Graham Island, but there was evidence of frost damage on Sitka spruce and western hemlock around Terrace.

For reforestation with Sitka spruce in Scotland and northern England, Queen Charlotte Islands provenances have traditionally been favoured because of their superior frost tolerance compared with Washington and Oregon provenances. Sitka spruce provenances from the southerly latitudes of Washington and Oregon grow faster in Britain than more northerly provenances, but this advantage must be balanced with provenance differences in susceptibility to spring and autumn frosts (Lines 1987; Nicoll et al. 1996).

Experience in Great Britain has shown that Sitka spruce shoots are particularly susceptible to frost damage before hardening in the autumn, when even a slight frost can cause damage. The frost-hardiness of spruce needles increases during autumn as temperatures fall and day length shortens. It is before this hardening, or when needles become de-hardened during a warm period in autumn, that needles are most vulnerable to frost damage. Symptoms are usually evident within two weeks of a damaging frost, visible as reddish-brown patches on the undersides of current-year needles, with older needles remaining green. In cases of severe frost damage, all current-year needles become completely reddish-brown; with slight frost damage, the tips and bases of affected needles remain green, and needles around the apical bud escape injury (Redfern and Cannell 1982; Nicoll et al. 1996). Research by Nicoll et al. (1996) indicates that during a frost of –5°C in October 1993 in Scotland, frost damage was positively correlated with lateness of root dormancy. Provenances from Alaska, the Queen Charlotte Islands, and Washington show substantial clonal variation in the onset of root dormancy. The relationship between susceptibility to frost damage and lateness of root dormancy suggests a criterion for screening clones from southern provenances, some of which may be less susceptible to frost damage because of earlier onset of root dormancy (Nicoll et al. 1996).

A review of frost heaving of forest tree seedlings (Goulet 1995) identified several effects on Sitka spruce seedlings: high mortality; reduced seedling growth; and damage to seedlings from mechanical breakage of roots or from lifting of roots from the soil, which exposes them to desiccation by wind and sun (Low 1975; Shaw et al. 1987). Frost heaving of Sitka spruce seedlings may cause trees to be less stable later in life because of malformed roots.

Response to Herbicides, Salt, and Airborne Pollutants

In British Columbia five herbicides are registered for forestry use: glyphosate, hexazinone, simazine, 2,4-D (as 2,4-D amine, 2,4-D ester, and 2,4-D ester + 2,4-DP ester), and triclopyr (Otchere-Boateng and Herring 1990; Biring et al. 1996). These are used as foliar, soil, cut-stump, stem injection, and basal bark applications. In relation to Sitka spruce management, the use of herbicides can be summarized as follows, based on Biring et al. (1996):

- Foliar applications of 2,4-D are used to release Sitka spruce from broadleaf tree species and shrubs; they are applied in late summer or early fall after dormant buds have formed.
- Foliar applications of glyphosate are similarly applied and used to control competing red alder, salmonberry, thimbleberry, and willows.
- Hexazinone can be applied as foliar treatment for perennial grasses, black cottonwood, and willow, but conifers should not be planted until at least one growing season after application.
- Foliar applications of triclopyr can be used in spring or late summer for site preparation to control bigleaf maple, red alder, salmonberry, and thimbleberry; they are applied only when conifers are dormant.

Early spring is an ideal time to control red alder, before flushing commences in conifers. Salmonberry, one of the main competitors of Sitka spruce in coastal British Columbia, can be herbicidally controlled by foliar applications of glyphosate at 1.5-2.1 kg active ingredient (ai) per hectare in early foliar, late foliar, and late summer stages of salmonberry seasonal growth. Soil applications of hexazinone in all seasons can also inflict severe injury to salmonberry at 2.0-4.0 kg ai/ha, and very severe injury at 4.0-8.0 kg ai/ha.

Deliberate application of herbicides to control or eliminate Sitka spruce is an unlikely management goal. However, if herbicides do come into contact with Sitka spruce, several types of potential damage can result, as listed by Biring et al. (1996):

- Very severe reactions (90-100% injury) occur with: cut-stump applications of glyphosate (30-100% rate) in all seasons; triclopyr ester (20-30% rate) in all seasons; stem injection of glyphosate (1 EZJECT® capsule for each 5 cm of dbh) in all seasons; and basal bark application of triclopyr ester (3-5% rate) in all seasons.
- Less severe injury (25-60%) to spruce can result from late foliar application of 2,4-D ester (2-3 kg active ester [ae] per hectare in the late foliar stage, when more than 66% of leaves are fully expanded and actively growing), and when triclopyr ester is applied at 2-4 kg ae/ha in late summer or dormant stages.
- Light (< 25%) injury to spruce results from foliar application of triclopyr ester at 1-2 kg ae/ha during early foliar or late summer to dormant stages.

The review of Carnation Creek herbicide research on Vancouver Island (Reynolds et al. 1989, 1993) recorded no herbicide-induced mortality of conifer crop trees when glyphosate was aerially applied to target species (red alder, salmonberry, and salal) at the rate of 2 kg ai/ha. The leaders of some western redcedar and western hemlock were damaged by the glyphosate treatment, but recovered within one year. No leader damage was observed in Sitka spruce, the major planted crop tree. Spruce released by the glyphosate treatment showed a significant increase in stem diameter compared with spruce in areas not treated with herbicide. Karakatsoulis et al. (1989), in an appendix to their Vedder Mountain study, recorded physiological responses in the 5-year-old Sitka spruce planted at Carnation Creek after glyphosate treatment to reduce vegetative competition. There were no significant differences between treated and untreated areas in Sitka spruce seedling stomatal conductance, plant moisture stress, soil temperature, or air temperature.

However, carbon dioxide uptake was significantly higher in herbicide-released seedlings than in control seedlings for both 1-year-old and current foliage. Light levels were twice as high in the herbicide-treated areas than in control plots, and this may account for the higher photosynthetic rates for spruce seedlings in the glyphosate-treated plots.

In a greenhouse and nursery study at Surrey, British Columbia, Prasad (1989) evaluated tolerance of Sitka spruce, Douglas-fir, and western redcedar to glyphosate sprayed at 2.1 kg ai in 380 litres of water per hectare. Of the three species, Douglas-fir was the most sensitive to glyphosate after 4 weeks of needle development. Sitka spruce hardened at a faster rate than the other two species, and by 8 weeks was reasonably tolerant to glyphosate. Because hardening of needles is essential for establishing resistance and tolerance to the action of glyphosate, Prasad recommended that at least 12 weeks elapse from the onset of bud flush before glyphosate is applied to conifers for successful weed control without damage to the crop trees. In general, there is evidence that Sitka spruce is less damaged than western hemlock when it comes in contact with herbicides used for brush control (Krygier and Ruth 1961).

Silvicultural herbicides have the potential to significantly reduce food sources and alter available wildlife habitat. In British Columbia herbicide use has increased in recent years, with glyphosate and 2,4-D formulations most commonly used for brushing and weeding, as well as for conifer release. Aerial spraying has also increased, and is the method of greatest concern to wildlife resource managers. Within the province, the Coastal Western Hemlock Zone, which encompasses the natural range of Sitka spruce, is one of three biogeoclimatic zones that have experienced the greatest herbicide use. Based on the assessment by Biggs and Walmsley (1988), the likelihood of herbicide-induced habitat alteration is of greatest concern for the management of the following wildlife species: grizzly bear (very high concern); mule deer, Roosevelt elk, forest-dwelling passerine birds, aquatic and terrestrial furbearers (high to very high concern); small mammals, raptorial birds, upland game birds, and amphibians and reptiles (moderate concern).

In addition to the herbicide information outlined above, there is some information, although limited, about Sitka spruce's responses to airborne compounds that may influence its growth. Sitka spruce is less sensitive to the cations and anions borne by ocean spray than either western hemlock or western redcedar (Cordes 1972), and was ranked by Krajina (1969) as the British Columbia tree species most tolerant to ocean spray. The intensity of ocean spray tolerated by Sitka spruce was considered by Cordes to be the main factor influencing the replacement of western hemlock–western redcedar forest by Sitka spruce in a narrow band along the west coast of Vancouver Island.

The responses of western North American forests to air pollution have been summarized by Olson et al. (1992) and Peterson et al. (1992). Earlier research had indicated that, in coastal British Columbia, western hemlock is more tolerant to sulfur dioxide than Sitka spruce or amabilis fir, and that among west coast tree species, western redcedar is the least tolerant to sulfur dioxide exposure (Miller and McBride 1975). Until the 1980s, air pollution in the Pacific Northwest came primarily from point sources such as smelters and power plants. The expansion of metropolitan areas, particularly in the Puget Sound area and in the Fraser River Lowland in British Columbia, has resulted in increased levels of phytotoxic gases from nonpoint sources. Olson et al. (1992) concluded that there was little evidence of deleterious effects of air pollution in most forests of the western United States. Most forests in the Pacific Northwest fall within the lowest risk rating proposed by Bormann (1985), in which air pollution is low and any effects are barely detectable. There is little information specific to Sitka spruce, however.

Elevated levels of ozone reported to occur during the growing season from May to August in western Washington have often coincided with periods of drought. When episodes of elevated ozone levels are followed by fog, there is a potential for formation of hydrogen peroxide, a possible damaging agent in coastal forests. Fog is efficient at scavenging gaseous pollutants from the lower atmosphere, and the prevalence of fog in summer on the Pacific Northwest coast (Basabe et al. 1987, 1989) may significantly influence the concentration and distribution of airborne pollutants. The main concentration of Sitka spruce on the outer coast of Washington (Figure 17) and the western part of Vancouver island, in the path of eastward-moving weather systems off the Pacific Ocean, suggests that in its area of best development this species is relatively free of the potentially damaging airborne pollutants that originate in Puget Sound and in the Lower Mainland of British Columbia.

Fire

Within the natural range of Sitka spruce, fire is not a prominent cause of forest disturbance in southeastern Alaska (McClellan et al. 1990), on the mainland of northern British Columbia, or on the Queen Charlotte Islands. Fire is more common on Vancouver Island, on the mainland of southern British Columbia, and in Oregon and Washington, where fire hazards can build up quickly in summer, particularly in logging residue. For this reason hazard abatement, in the form of broadcast burning, is more common in the southern parts of the range of Sitka spruce than further north. The risk of damage from escaped broadcast burns is correspondingly greater along the south coast than on the north coast.

Wind is a more persistent disturbance factor than fire in coastal forests (Agee 1993). Minore (1979) ranked Sitka spruce and western redcedar as the least fire-resistant tree species in coastal areas, whereas Douglas-fir and grand fir were the most fire-resistant. Sitka spruce is characterized by a very thin bark, shallow rooting habit, moderately dense branch habit, and medium foliage flammability, all of which contribute to a low degree of fire resistance (Spalt and Reifsnyder 1962; Lotan et al. 1981). Parminter (1983) indicated that lightning-caused fires are rare on British Columbia's north coast, based on records from the *Provincial Fire Atlas;* only four such fires were identified in this atlas for the Queen Charlotte Islands for the period 1940-82. At the heads of marine inlets and in the vicinity of Sitka spruce's inland limit, there are more lightning strikes than along the coast. Within Sitka spruce's natural range in British Columbia, the greatest chances of locally abundant lightning strikes are in the area between Terrace and Kitimat and just to the east of Bella Coola.

Both Parminter (1992) and Bunnell (1995) suggest fire return intervals of greater than 250 years for most of the CWH and MH zones of British

Table 25

Minimum, average, and maximum fire return intervals and fire sizes in western redcedar–Sitka spruce stands of the Coastal Western Hemlock Zone

	Minimum	Average	Maximum
Mean fire return interval (yr)	100-150	150-350	350-500
Mean fire size (ha)	0.1-5.0	50-500	> 500

Source: Parminter (1992)

Columbia. This figure is comparable to the average return interval of 242 years estimated by Ripple (1994) for unharvested areas near the coast in northern Oregon. Agee (1990) estimated a fire return cycle of 400 years for cedar-spruce-hemlock forest stands in Oregon. Parminter's more detailed breakdown indicated that in western redcedar–Sitka spruce stands of the CWH zone, minimum, average, and maximum fire return intervals and fire sizes had the ranges shown in Table 25.

In general, average conifer longevities (Table 4) tend to be significantly longer than estimated fire return intervals because some trees are located in relatively fireproof areas and the degree of tree mortality from fires varies according to site, topographic position, fire intensity, and type of fire (surface versus crown fires). In coastal temperate forests, there is also a characteristic pattern of more small fires and fewer large ones, compared with inland forest areas (Hansen et al. 1991). The wet climate of the CWH zone ensures that large, stand-destroying fires are uncommon. In this context Larson (1992) stressed that the substantial downed woody biomass that accumulates on the forest floor of the Western Hemlock–Sitka Spruce forest type in southeastern Alaska does not present a fire hazard through fuel build-up.

In a review of likely regeneration and early-growth response to slashburning in major tree species in Vancouver, Otchere-Boateng and Herring (1990) rated Sitka spruce and its companion species according to how they were affected by slashburning:

1 Sitka spruce – generally beneficial
2 western hemlock – generally adverse except in perhumid climates (CWHvh and CWHwh subzones)
3 western redcedar – adverse on all but the wettest sites
4 Douglas-fir – generally adverse on drier, nutrient-poor sites; beneficial on other sites except possibly for severe burns

Among the species most likely to compete with regenerating Sitka spruce, the response to fire was summarized by Otchere-Boateng and Herring (1990) as follows:

Alnus rubra. Fire can reduce stump sprouting and seedling establishment if mineral soil exposure is low; after fires, red alder invades primarily by seed dispersal; seedlings can be abundant if fire has created excessive mineral soil exposure.

Rubus parviflorus. After fire this shrub sprouts and suckers; cover may increase; can be set back on dry sites.

Rubus spectabilis. After fire this shrub sprouts, suckers, and seeds in, often achieving maximum cover 3-5 years after fire; cover increases, but can decline relatively early on dry sites.

Sambucus spp. Response to fire variable; sprouts after fire; may seed-in; severity of burn affects response, but buried seeds on subhygric sites are usually stimulated by fire to germinate.

Acer circinatum. Severe fire can kill this species, but with less intense burning there is vigorous sprouting from basal portions of the stem and from the roots.

Cornus sericea. Set back by severe fires; favoured by light fires; usually increases after burns on subhygric sites.

Ribes bracteosum. Sprouts after fire but can take 2 or 3 years to recover.

Gaultheria shallon. Generally slow to recover to pre-burn levels; fire can successfully control this species on dry sites but usually not on wet sites; fire stimulates resprouting from roots and stem bases.

Polystichum munitum. Burning has variable results depending on fire severity and soil moisture.

Except that slashburning is a traditional step after clearcutting in many coastal areas where Sitka spruce is a candidate for post-harvest planting, prescribed fire is not a widely used silvicultural approach for Sitka spruce regeneration. Although not specifically directed to Sitka spruce management, the use and benefits of prescribed fire are reviewed for the Pacific Northwest region by Lotan et al. (1981) and Walstad et al. (1990), and by Hawkes et al. (1990) and Haeussler (1991) for British Columbia. The response of Sitka spruce to slashburning for regeneration and early growth was rated generally beneficial (Hawkes et al. 1990).

Insects and Diseases

This section does not enumerate all the insects and diseases that use Sitka spruce as a host for some portion of their life cycle. Instead the emphasis is on insects and diseases of greatest concern to foresters responsible for forest renewal involving Sitka spruce planting stock and for protecting the growing stock. Table 26 provides a summary of these pests.

Among the main insect and disease pests of Sitka spruce, the most attention by far has been given to the white pine weevil (*Pissodes strobi*), sometimes referred to as Sitka spruce weevil, spruce weevil, or tip weevil. So great is the risk of weevil damage to Sitka spruce in coastal British Columbia,

Table 26

Summary of pests and diseases of Sitka spruce in managed forests

Pest or disease	Potential damage[a]	Actual damage[b]
Defoliators		
Black army cutworm (*Actebia fennica*)	X	
Bud moths (*Zeiraphera* spp.)	X	
Hemlock sawfly (*Neodiprion tsugae*)		X
Spruce budworm (*Choristoneura* spp.)	X	
Western black-headed budworm (*Acleris gloverana*)		X
Western hemlock looper (*Lambdina fiscellaria lugubrosa*)		X
Sucking insects		
Cooley spruce gall adelgid (*Adelges cooleyi*)		X
Giant conifer aphid (*Cinara* spp.)	X	
Green spruce aphid (*Elatobium abietinum*)		X
Woody tissue feeders		
Conifer seedling weevil (*Steremnius carinatus*)		X
White pine weevil (*Pissodes strobi*)		X
Broom rusts		
Spruce broom rust (*Chrysomyxa arctostaphyli*)		X
Dwarf mistletoes		
Hemlock dwarf mistletoe (*Arceuthobium tsugense*)		X
Needle casts, blights, and rusts		
Large-spored spruce–Labrador tea rust (*Chrysomyxa ledicola*)		X
Siroccocus tip blight (*Siroccocus strobilinus*)		X
Spruce needle cast (*Lirula macrospora*)		X

►

◀ *Table 26*

Pest or disease	Potential damage[a]	Actual damage[b]
Root diseases		
Annosus root disease (*Heterobasidion annosum*)		X
Armillaria root disease (*Armillaria ostoyae*)	X	
Black stain root disease (*Leptographium wageneri*)	X	
Laminated root rot (*Phellinus weirii*)	X	
Rhizina root disease (*Rhizina undulata*)	X	
Tomentosus root rot (*Inonotus tomentosus*)	X	
Mammals		
American porcupine (*Erethizon dorsatum*)		X
Black bear (*Ursus americanus*)	X	
Deer (*Odocoileus* spp.)	X	

Source: Finck et al. (1989)

[a] Sitka spruce possibly involved because all conifers can be host.

[b] Sitka spruce specifically identified as a host by Finck et al. (1989).

except for the Queen Charlotte Islands and coastal areas further north, that this spruce has been excluded from use in most reforestation plans, and interior species of spruce are similarly threatened.

In the past five years, an impressive amount of information has been assembled on the biology, damage, and management of white pine weevil, the most comprehensive of which is the symposium proceedings edited by Alfaro et al. (1994). Other advances in our knowledge of this weevil have been made by the researchers cited at the beginning of the subsection 'White Pine Weevil Damage,' on page 120. At the forefront is research at the Pacific Forestry Centre, Victoria, on biological control of the white pine weevil using insect predators and parasites, and on the natural resistance to weevil attack exhibited by some spruce trees. Evidence suggests that such resistance can be inherited. Research into chemicals present in needles and bark, which can provide information on the genetic structure of the trees, may lead to a way of reliably identifying resistant trees, thereby providing forest managers with a strain that could be used in replanting. Further aspects of this research are described in the subsections 'White Pine Weevil Damage' and 'White Pine Weevil Control.'

There are several key sources of information for the various stages of Sitka spruce plantation establishment. Diseases and insects of possible concern for nursery production of Sitka spruce are identified by Sutherland et al. (1989). For young established stands, Allen (1994) provided a comprehensive summary of insects and diseases important for damage appraisal. An additional source of information on pests of managed forests in British Columbia is the guide by Finck et al. (1989), which includes the insects and diseases of potential concern for managed stands of Sitka spruce in the province.

An important point for Sitka spruce managers is that ecosystem changes brought on by forest pathogens are not necessarily 'bad.' Pathogens make a major contribution to ecosystem diversity by breaking up what would otherwise be uniform forest landscapes. Using Armillaria root disease as an example, van der Kamp (1991) described the possible benefits in terms of spatial and temporal diversity created by root disease centres. Thies and Sturrock (1995) also suggested that tree losses from laminated root rot may be biologically beneficial, by creating forest gaps or stands of lower density that are more attractive to certain wildlife species, and by contributing to large woody debris on the forest floor that can enhance small-mammal and invertebrate biodiversity.

Hennon (1995) made the same point in an analysis of heart-rot fungi as a major source of disturbance in gap-dynamic forests, using as an example southeastern Alaska's Sitka spruce–hemlock forests, which contain large quantities of heart rot. Combined with the observation by Harris (1989) that catastrophic windthrow in southeastern Alaska is greatest in stands with heart rot, it appears that Sitka spruce at that latitude is influenced by important relationships between stem decay, stand-destroying storms, and stand development sequences related to gap dynamics. An unanswered question about the disturbance and gap-creation role of these fungi is the degree to which the presence or absence of heart-rot fungi actually influences the incidence of uprooted trees, trees with boles broken off some distance above ground level, and trees that die but remain standing.

The following subsections provide more detail on the main insects and diseases associated with Sitka spruce.

Insects

This subsection focuses first on the insects that can significantly affect the plantation establishment phases of Sitka spruce management. As reviewed by Finck et al. (1990), information is presented for seed orchard, nursery, and plantation aspects of forest renewal. Insect and disease control are especially important in high-value trees in seed orchards because scion collection, propagation, planting, and orchard maintenance represent substantial investments. There is some evidence that tolerance thresholds for pests in orchards are lower than in forest plantations (Finck et al. 1990). The review by Finck and co-workers identified the following common insect pests for Sitka spruce cones and for Sitka spruce in seed orchards.

- **Spruce seedworm** (*Cydia strobilella*), north-coastal B.C. (Hedlin 1974; Hedlin et al. 1980; Ruth et al. 1982). One larva can destroy 40% of the seed in a cone; more than two can destroy 100%. Usually no more than three larvae are present in one cone. They mine through the cone but feed only on seeds.
- **Spruce spiral-cone borer** (*Strobilomyia neanthracinum*). One larva can destroy up to 50% of the total seeds in a cone. Where infestations are severe, the total seed crop can be destroyed.
- **Gall aphids** (*Adelges cooleyi*). On spruce, this aphid damages new shoots and reduces the number of potential cone-producing sites. A heavy attack in a year conducive to cone initiation could limit the subsequent crop; if cones themselves are attacked, spruce seed yield can be greatly reduced.
- **Green spruce aphid** (*Elatobium abietinum*). This aphid can cause serious defoliation in coastal spruce seed orchards.
- **Spruce spider mite** (*Oligonychus ununguis*). This mite can cause serious defoliation in coastal spruce seed orchards.

Seed orchards are costly to establish and maintain. Thus it is important for seed orchard managers to have criteria for deciding when pest management steps are needed and whether treatments are effective. For example, the green spruce aphid and the spruce spider mite can cause serious defoliation in coastal spruce seed orchards (Finck et al. 1990). Both must be constantly monitored during their active periods because their numbers can increase dramatically within a few weeks.

Conifer seedling nurseries have a unique complex of insect pests because succulent seedlings can be hosts to insects that would not normally feed on mature trees. Many of these insects are general feeders, and they directly affect the quantity and quality of nursery stock. Nursery managers can find further information in Sutherland et al. (1989) about insects that damage nursery seedlings. Control measures, including timing and application rates of insecticides, can be found in the *Nursery Production Guide for Commercial Growers,* published annually by the British Columbia Ministry of Agriculture, Fisheries and Food. Although Finck et al. (1990) identified several insects that attack the species of spruce grown in nurseries, only those species that appear to be of risk to nursery-grown Sitka spruce are listed below. Spruce-feeding insects that Finck and co-workers identified as a problem at

Table 27

Defoliators, sucking insects, and inner-bark feeders that are potentially damaging to plantation Sitka spruce

Group	Potentially damaging insects	Potential for damage to Sitka spruce plantations
Defoliators	None	–
Sucking insects	Giant conifer aphid (*Cinara* spp.) (Johnson 1965)	Throughout British Columbia this aphid can cause growth reduction in all young conifers except western redcedar; the aphid may also predispose a plantation tree to secondary insects, fungi, or stress from drought.
	Cooley spruce gall aphid (*Adelges cooleyi*) (Wood 1977; Duncan 1996)	On Sitka spruce, repeated attacks by this species can produce stunted and deformed trees, but normally it is not a concern for Sitka spruce plantations. Conelike galls on branches are the most obvious sign of adelgid infestation.
Inner-bark feeders on terminal shoots	White pine weevil (*Pissodes strobi*) (Duncan 1982)	All spruce south of lat. 56°N, except on the Queen Charlotte Islands, especially trees 1.5-10.0 m tall, are attacked; these weevils prefer vigorous, open-grown trees and can attack more than 50% of a stand in a year; white pine weevil causes terminal die-back of 2 years' growth or more. In addition to growth loss, they reduce wood quality by causing forked and crooked stems.
	Plantation weevil (*Steremnius carinatus*) (Condrashoff 1968)	Seedlings of Sitka spruce and Douglas-fir are preferred foods; partially girdled seedlings may recover but complete girdling kills stem; damage is most severe in first-year plantations.

Source: Finck et al. (1990)

only one or a few nurseries, or at nurseries in the British Columbia interior, are excluded from the following list.

- **Cutworms**, especially variegated cutworm (*Peridroma saucia*) (Palmer and Nichols 1981). All species of conifers, especially 1+0 stock (either bareroot or container seedlings), can be destroyed by a single larva each night; there are probably populations of cutworms at all nurseries in every growing season.
- **Needle tiers** (*Archips rosana* and *Choristoneura rosaceana*), the latter sometimes referred to as budworm (Furniss and Carolin 1977). Infestations often occur in spruce, but most of them are incidental and spotty in their distribution.
- **Cooley spruce gall aphid** (*Adelges cooleyi*) (Wood 1977; Sutherland et al. 1989). The woolly stage can be a serious nursery pest year round; the gall stage seldom kills Sitka spruce seedlings, but heavy infestations can reduce vigour and cause deformation of small trees.
- **Green spruce aphid** (*Elatobium abietinum*) (Koot 1992). This aphid is particularly attracted to spruce; initial feeding results in mottled needles followed by chlorosis and needle drop; severe infestations can lead to spruce seedling death.
- **Woolly aphid** (*Mindarus obliquus*). This aphid has killed the leaders of spruce seedlings and caused stem deformation in several nurseries throughout B.C., but Finck et al. (1990) did not specify whether Sitka spruce was involved.

The most damaging insects for young plantations can be placed in three main groups: defoliators, sucking insects, and inner-bark feeders. Those potentially damaging to plantation Sitka spruce are summarized in Table 27.

For the Alaska portion of Sitka spruce's natural range, Holsten et al. (1985) listed the following insects for which Sitka spruce is a host (scientific nomenclature follows Holsten and co-workers):

Defoliators
- Black-headed budworm (*Acleris gloverana*)
- Spruce budworm (*Choristoneura* spp.)
- Western hemlock looper (*Lambdina fiscellaria lugubrosa*)
- Spruce bud moth (*Zeiraphera* sp.)

Sap-sucking insects
- Green spruce aphid (*Elatobium abietinum*)
- Giant conifer aphid (*Cinara* sp.)
- Spruce gall aphid (*Adelges* sp.)
- Woolly aphid (*Adelges* sp.)

- Spruce bud scale (*Physokermes piceae*)
- Spittlebug (*Aphrophora* spp.)

Bark beetles

- Spruce beetle (*Dendroctonus rufipennis*)
- Engravers (*Ips* sp.)
- Four-eyed bark beetles (*Polygraphus* sp., *Dryocoetes* sp.)

Wood borers

- Ambrosia beetles (*Trypodendron lineatum*)
- Shipworms (*Bankia setacea*)

Seed and cone insects

- Spruce seed cone moth (*Cydia youngana*)
- Spruce cone maggot (*Hylemya* sp.)

Bud and shoot insects

- Spruce bud midge (*Dasineura* sp.)

White Pine Weevil Damage

White pine weevil was recognized as an important insect for Sitka spruce managers in British Columbia in an early Forest Pest Leaflet (Holms 1967). A current example of why this weevil remains a concern is provided by the three-year review of forest health in young managed stands in British Columbia (Nevill et al. 1995), which indicated that in 1994 the highest level of cumulative attack by white pine weevil in the province was in the Cecil Creek area of the CWHws1 variant in the Kitimat River valley, where Sitka spruce made up 40% of the stand.

Detailed scientific study of this weevil in the province began in earnest in the 1970s (McMullen 1976a, 1976b; VanderSar and Borden 1977a, 1977b; Alfaro et al. 1980) and continued through the 1980s and early 1990s (for example, Alfaro and Borden 1985; McMullen et al. 1987; Mitchell et al. 1990), reaching its highest intensity in the past five years (Van Sickle 1992; Warkentin et al. 1992; Alfaro et al. 1993, 1994, 1995; Fraser and Heppner 1993; Sahota et al. 1994a, 1994b; Tomlin et al. 1996; Turnquist and Alfaro 1996).

The biology of white pine weevil, the dominant pest of Sitka spruce in British Columbia south of the Queen Charlotte Islands, is documented in much detail. The synopsis by Alfaro (1994) is an up-to-date account of this weevil's life-cycle phases, epidemiology, intra-stand dynamics, and natural enemies, as well as the sources of Sitka spruce's resistance to weevil attack. Details of the life cycle are readily available in other publications and are not summarized here.

Generalized features of the weevil's life cycle are shown in Figure 31, based on research summarized by Sahota (1993). Differences in Sitka spruce leaders with and without white pine weevil damage are shown in Figure 32, and a close-up view of typical weevil damage, involving a broken-over leader with a lateral branch that has become a new leader, is shown in Figure 33.

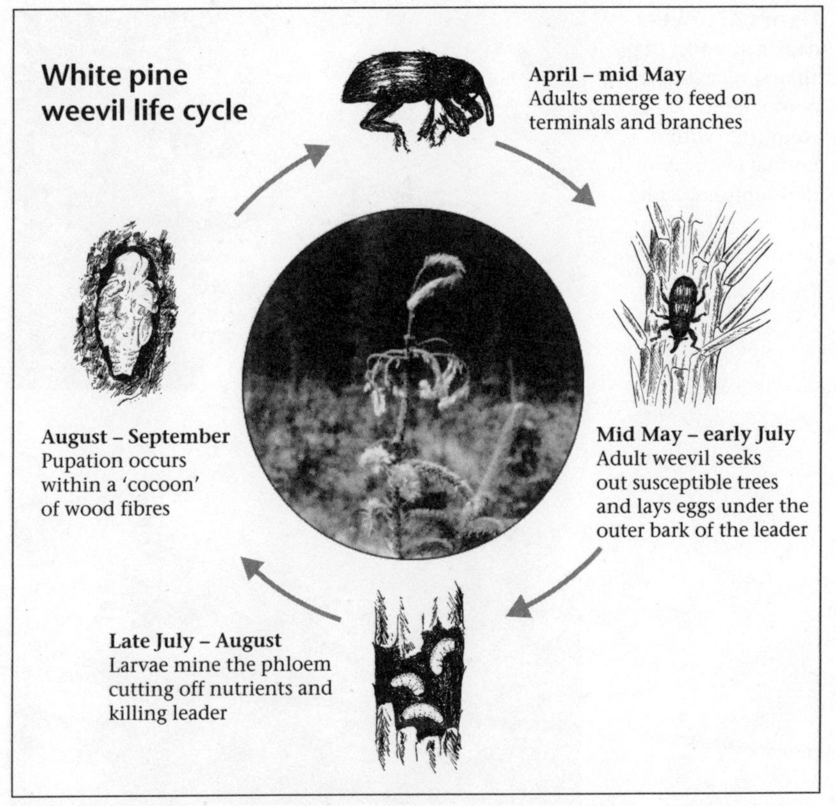

Figure 31. April-to-September stages in the white pine weevil life cycle (from Sahota 1993).

The sequential stages of weevil outbreak in open-grown Sitka spruce in an area of high weevil hazard are shown in Figure 34. These stages involve a phase of rapid increase in the weevil population, a phase of insect-host equilibrium, and then a phase of decline in the percentage of Sitka spruce trees attacked (Turnquist and Alfaro 1996). The white pine weevil repeatedly destroys the terminal leader of Sitka spruce trees, causing reduced height growth and deformed stems (Figure 33), a process that hampers the development of young plantations (Alfaro 1982). Serious damage is most common on Vancouver Island, at least as far north as Sayward (Heppner and Wood 1984).

It is known that a complex combination of volatile and non-volatile chemicals determines the weevil's acceptance or rejection of Sitka spruce as a host (Harris et al. 1983; Alfaro and Borden 1985). Factors that influence the weevil's feeding behaviour have been studied in detail (VanderSar and Borden 1977a, 1977b; VanderSar 1978; Alfaro 1980; Alfaro et al. 1980, 1981, 1984; Alfaro and Borden 1982). Substantial research over several decades

Figure 32. Weevil-damaged leader of Sitka spruce (upper photograph) contrasted with normal leader growth (lower photograph).

Figure 33. Typical weevil damage in Sitka spruce, involving a broken-over leader with a lateral branch that has become the new leader (from Turnquist and Alfaro 1996).

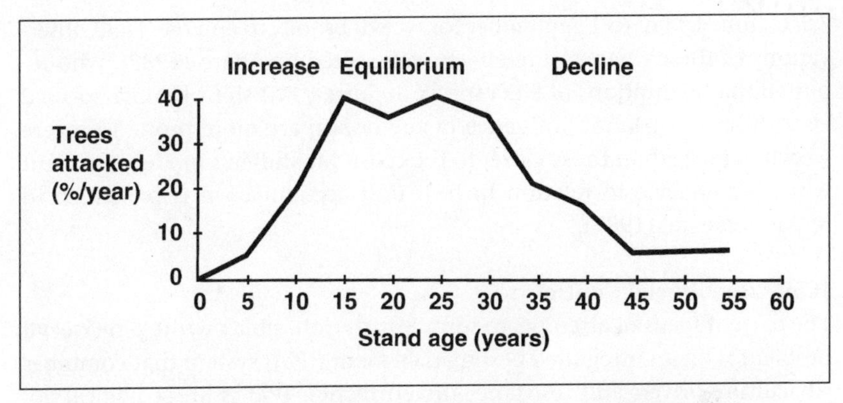

Figure 34. Typical stages of weevil outbreak in open-grown Sitka spruce in a locality with high weevil hazard, demonstrating a phase of rapid increase in the weevil population, a phase of insect-host equilibrium, and then a phase of decline in the percentage of trees attacked (from Turnquist and Alfaro 1996).

has attempted to identify indicators of weevil resistance (Kriebel 1954; Wilkinson et al. 1971; Squillace 1976; Harris et al. 1983; Brooks et al. 1987; Brooks and Borden 1992; Alfaro et al. 1993; King 1994; Sahota et al. 1994a, 1994b; Ying and Ebata 1994). Early field observations indicated that some individual Sitka spruce trees are resistant to the weevil (Silver 1968; Wright 1970; Mitchell et al. 1974), and there is now optimism that plantations of resistant trees from seed orchard seed can not only achieve increases in growth rates of about 10% but also maintain high levels of genetic diversity. In addition, evidence that spruce stands around the Strait of Georgia show resistance to weevil damage despite high weevil numbers has led to an interest in screening material from gene archives and other natural populations in the Strait of Georgia lowlands in order to breed the resistance trait into Sitka spruce from other areas (Parrish and Lester 1993). Also, provenances from the Skeena River valley near Terrace and from two other locations on the mainland seem to be less susceptible than others (Alfaro and Ying 1990; Ying 1991). This is consistent with the statement by Ruth and Harris (1979) that Lutz spruce, the hybrid between Sitka and white spruce, appears to be resistant to the weevil.

How soil moisture and soil nutrient supply influence the incidence of weevil attack is not clear. Stanek (pers. comm., Dec. 1988) observed that, when fertilized with N and P, a plantation of Sitka spruce that was chlorotic and growing poorly near Port McNeill, British Columbia, grew long, vigorous leaders that attracted a high incidence of weevil attack. However, local differences in spruce susceptibility to weevil damage involve factors other than nutrient supply and spruce leader length. For example, McMullen (1976a) recorded the minimum numbers of accumulated degree-days above

7.2°C from 1 May to 1 September for weevil broods to emerge. Field observations of these temperature effects were noted by Alfaro (1982), who reported that plantations of Sitka spruce at Surrey, British Columbia, located away from the cooling influence of the ocean, are more prone to severe weevil damage than those closer to the coast. McMullen's model of percent weevil emergence in relation to heat unit accumulation is published in McMullen et al. (1987).

White Pine Weevil Control

The current focus of attempts to solve British Columbia's white pine weevil problem is on an integrated pest management (IPM) system that combines silviculture-driven and resistance-driven tactics. IPM is an ecological approach to pest control that relies on a combination of varying tactics to *reduce damage* rather than to eliminate the pest. An important goal of IPM is to minimize environmental impact through a unified program that specifies strategies and procedures in the context of the crop productivity cycle (Alfaro et al. 1995). Such an approach is now technically possible because of significant progress in weevil research in the areas of genetic resistance, silvicultural control, and chemical control (Alfaro et al. 1994; see also the 1992-96 sources cited in the previous subsection).

As described by Alfaro et al. (1995), an IPM system for weevil management requires an accurate hazard rating of plantation sites, continuous monitoring of attack levels, and forecasting of plantation productivity under various IPM tactics through the use of a decision support system. One element of these information needs is the ability to forecast impacts of the weevil upon plantation volume growth. To address this need, a weevil ATTACK (SWAT) module has been developed for use in the Tree and Stand Simulator (TASS) model of the British Columbia Ministry of Forests (R. Alfaro, pers. comm., March 1994). The preliminary IPM system proposed by Alfaro and co-workers suggests two broad strategies and numerous specific tactics for existing and new spruce plantations. Other tactics can be added as needed, when economically feasible and when proven to be effective. As of 1995, the weevil management tactics proposed by Alfaro et al. (1995), grouped under two different strategies, were as follows:

1 Manage existing plantations of mainly susceptible trees to reduce weevil incidence or impact
   - Favour non-host species in spacing and thinning.
   - Interplant with non-host coniferous species in young plantations.
   - Interplant with non-host deciduous (shade) trees in young plantations, or conserve shade trees that have regenerated naturally until spruces reach height at which susceptibility diminishes.
   - Delay spacing and thinning to maintain high stand density.

- Interplant with resistant trees, if available, in young plantations.
- Monitor weevil populations.
- Model weevil impacts and effects of control actions.
- Apply stem injections of insecticide to crop trees.
- Clip leaders to remove weevils and enhance stem form.
- Conserve parasitoids in clipped leaders.

2 Develop and manage new plantations to minimize susceptibility to weevils and to reduce weevil incidence or impact
  - Rate sites according to weevil hazard, prior to harvest, to guide selection of management tactic.
  - Model weevil impacts and effects of control actions.
  - If ecologically and economically sound, and effective in reducing weevil incidence, implement alternative silvicultural systems, e.g., conventional and group selection, shelterwood.
  - Practise high-density planting.
  - Mix spruce with non-host conifers.
  - Mix spruce with non-host deciduous (shade) trees.
  - Plant resistant stock mixed with susceptible stock.
  - Implement brush control and intensive silvicultural treatments to minimize weevil incidence or impact.
  - Monitor weevil populations.
  - Apply stem injections of insecticide to crop trees where necessary.
  - Clip leaders to enhance stem form, remove weevils, and conserve parasitoids.
  - Mass-trap weevils with semiochemicals (if/when available).
  - Augment parasitoid populations (if/when possible).

The control of white pine weevil is obviously a difficult task. The best control is prevention, which implies that Sitka spruce should not be planted in high-risk areas. Besides the tactics suggested above, continuous monitoring of existing and new plantations is very important to help define criteria for recognizing high-risk areas.

Other Insects Associated with Sitka Spruce

Except for the white pine weevil, there are few serious insect problems related to Sitka spruce. However, there are periodic outbreaks of the black-headed budworm (*Acleris gloverana*), often in conjunction with the hemlock sawfly (*Neodiprion tsugae*). These outbreaks commonly spread northward from the Olympic Peninsula. Western hemlock is the preferred host of the budworm, but Sitka spruce is also attacked, particularly during epidemic levels (Harris and Farr 1974; Ruth and Harris 1979; Harris and Johnson 1983). In British Columbia, the most serious budworm defoliation occurs on the

Queen Charlotte Islands and northern Vancouver Island (B.C. Ministry of Forests and B.C. Ministry of Environment, Lands and Parks 1995c), where outbreaks can last up to four years. In 1985, 28,000 hectares were attacked by the black-headed budworm on the Queen Charlotte Islands, including substantial areas of spruce established after logging in 1940 (Sanders and Wilford 1986).

Of minor commercial importance but of significant aesthetic or urban significance is the green spruce aphid (*Elatobium abietinum*), a sap-sucking insect that defoliates and can kill Sitka spruce (Phelps 1973; Ruth and Harris 1979). Sanders and Wilford (1986) speculated that aphid infestations may be enhanced with open canopy conditions, as indicated by its presence in recently thinned stands.

The Cooley spruce gall aphid (*Adelges cooleyi*) frequently attacks young Sitka spruce seedlings, causing the formation of cone-shaped galls on terminal shoots (Phelps 1973). Phelps (1973) also described a weevil, *Steremnius carinatus*, that has caused damage to natural regeneration on the Queen Charlotte Islands. Occurring at endemic levels, the Sitka spruce bark beetle (*Dendroctonus* sp.) rarely becomes epidemic and is not considered a serious problem (Harris and Farr 1974). Other seed-destroying insects that feed on Sitka spruce cones and seeds include *Megastigmus piceae, Torymus* spp., *Laspeyresia youngana,* and *Heinrichia fuscodorsana* (Fowells 1965). Insects that influence cone and seed production of Sitka spruce include the spiral spruce cone borer (*Lasiomma anthracina*), the spruce seedworm (*Cydia strobilella*), and the spruce cone axis midge (*Dasineura rachiphaga*) (Eremko et al. 1989).

The green spruce aphid may cause severe defoliation of Sitka spruce. Severe infestations are associated with mild weather, and population declines are linked to the effects of colder winters (Koot 1992). Infestations occur on trees of all sizes, from saplings to mature stands in natural stands and plantations (Wood and Van Sickle 1993). The first indications of infestation are yellow patches on needles in winter or spring, and by summer needles turn yellow to brown and ultimately drop off. Severe attacks may cause complete defoliation. Weather is the main natural control agent. The aphid prefers needles older than current growth, and most aphids are found on the lower side of needles and in the lower, shadier portion of the crown. This aphid has been a chronic pest of spruce on the Queen Charlotte Islands and the adjacent mainland since 1960. On the Queen Charlotte Islands, populations have continued at damaging levels for three years on average, unless reduced by colder than normal winter temperatures. The green spruce aphid is also known to severely attack Sitka spruce in Great Britain (Seaby and Mowat 1993; Straw 1995), where it has caused complete defoliation, death of buds, and subsequent development of forked main stems (Carter 1989). Research in Scotland has shown that applications of N and P fertilizers can hasten the recovery of losses in diameter growth following defoliation by the green spruce aphid (Thomas and Miller 1994).

The biology and control of spruce beetles (*Dendroctonus rufipennis*) has been studied in Lutz spruce (hybrids between Sitka and white spruce) in Alaska (Hard 1992; Holsten and Werner 1993). This research indicated that pruning to remove the lower live branches of Lutz spruce reduced the incidence of attack by spruce beetles.

Diseases

Like any other tree species, Sitka spruce can be a host to many fungi and microorganisms. Only the main pathogens are reviewed here, relying almost exclusively on Sutherland and Hunt (1990), who restricted their review to those species most likely to be of concern to foresters responsible for regeneration, protection, and management of forests stands. The main diseases important at various stages of the life cycle of Sitka spruce and at various stages of stand renewal and management are summarized in the subsections that follow.

Cone Diseases

Inland spruce cone rust, *Chrysomyxa pirolata,* is the only disease that consistently causes serious losses of cones in British Columbia. Although this rust occurs throughout the province, Sitka spruce is usually not damaged by it, because the rust is relatively rare on the coast and is most pronounced in the interior, where white spruce, Engelmann spruce, and the alternate hosts required for spread from tree to tree, *Orthillia secunda* and *Moneses uniflora,* more commonly occur. A related species, coastal spruce cone rust (*Chrysomyxa monesis*), for which *Moneses uniflora* is the only alternate host, has damaged cones of Sitka spruce on the Queen Charlotte Islands (Sutherland et al. 1987). Occurrence of *Chrysomyxa monesis* on spruce is restricted to areas where the telial host (*Moneses uniflora*) occurs, but occurrence of this rust on the telial host is not restricted to areas where spruce occurs. Up to the time that Ziller (1974) reported this rust, its damage to Sitka spruce cones had been confined to the Queen Charlotte Islands. This somewhat restricted area of *Chrysomyxa* damage was significant, however, because most of the Sitka spruce seed exported to Britain and other European countries came from the Queen Charlotte Islands. To guard against introduction of this rust to other regions of the world, Ziller (1974) recommended that west coast Sitka spruce seed be surface-sterilized before export.

For both species of *Chrysomyxa,* diseased cones first become noticeable in mid to late summer, when they dry out, open prematurely, and shed massive amounts of yellow-orange spores. They should not be included in cone collections because they yield few seeds or seeds that germinate abnormally. No practical methods have been developed for managing cone rust in the forest. However, the disease is less prevalent in areas where the alternate hosts are absent, either naturally or as a result of widespread forest fires. Seed orchard cones can be protected by applying one or two sprays or fungicides (such as ferbam) to the cones at pollination, the time when the pathogen enters cones (Sutherland and Hunt 1990). For a forest manager seeking more information, summaries of cone diseases can be found in Sutherland et al. (1987), and detailed life-cycle information on tree rusts are provided in the publication by Ziller (1974).

Seed Diseases

The two diseases that currently affect seeds in British Columbia reforestation species are the seed or cold fungus, *Caloscypha fulgens,* and Sirococcus blight, caused by *Sirococcus strobilinus*. The Sitka spruce manager needs to be concerned with the latter only. In British Columbia this blight is confined mainly to western hemlock, Sitka spruce, and lodgepole pine. On Sitka spruce and western hemlock, Sirococcus blight is most likely to occur in dense stands of trees before they reach pole size. In container nurseries, Sirococcus blight is known to be seed-borne, mainly on all species of spruce. Spruce seedlots become contaminated with diseased seeds when old cones, on which the fungus grows and sporulates, are included in cone collections. In nurseries, disease centres develop from these seeds and act as a source of secondary inoculum. Small, dark fruiting bodies of *S. strobilinus* are fairly common on old spruce cones. However, since other fungi produce similar structures, identification of *S. strobilinus* must be confirmed by microscopic examination of the spores.

As more seed is obtained from seed orchards, Sirococcus blight resulting from seed-borne inoculum should eventually disappear, because orchard collections will not contain old cones. In bareroot nurseries where spruce, Douglas-fir, and lodgepole pine seedlings are commonly affected, some losses may still occur due to wind- and rain-borne inoculum from nearby diseased trees. Managing Sirococcus blight in nurseries includes removing diseased windbreaks and other nearby trees, applying fungicides, and reducing watering where practical, as moisture favours *Sirococcus* infection and spread. Sutherland et al. (1987) provide more detail on this blight.

Nursery Diseases

Although many diseases affect seedlings in nurseries (Sutherland et al. 1989), grey mould (*Botrytis cinerea*) on container-grown seedlings and storage moulds on bareroot seedlings are the only nursery-induced diseases that field foresters in British Columbia are likely to encounter (Sutherland and Hunt 1990). The latter authors indicated that grey mould is the most serious disease on container-grown seedlings in the province. It is most severe on Douglas-fir, western hemlock, and all the spruces. Symptoms are most conspicuous following seedling canopy closure and after the onset of cool, wet autumn weather. Masses of light grey mould develop on the lowermost needles, spreading upward on the shoot. If not controlled, the fungus can kill the needles on the lower half of the shoot and eventually penetrate the bark, killing the woody lateral branches and the seedling's main stem. In certain cases, root regeneration capacity tests may be justified to determine seedling condition. If the foliage of seedlings is severely damaged or defoliated, and the vascular tissues are rotted, the seedlings should be culled (Sutherland and Hunt 1990).

Nursery practices to help control grey mould include: reducing or withholding irrigation; rearranging styroblocks to improve ventilation and reduce humidity; and applying fungicides before seedling canopy closure and again before the stock is lifted for storage. Where possible, removing the polyethylene/fiberglass cover of covered growing houses during summer also alleviates grey mould by exposing seedlings to full sunlight.

Grey mould causes direct damage to growing seedlings; if uncontrolled or undetected, it can also lead to further damage during or after storage. In fact, some of the most serious losses to grey mould have occurred on stored stock. Symptoms on stored seedlings are usually the same as those on growing seedlings, except that the disease may originate and spread from any portion of the shoot. It has not been seen to damage seedling roots.

Managing grey mould on stored seedlings involves the following: (1) applying a pre-storage fungicide to seedlings; (2) storing stock with a nursery history of the disease (even incipient disease) for the shortest possible time; (3) storing stock that will withstand frozen storage at –1° to –2°C; and (4) frequently inspecting stored seedlings for the disease and outplanting the stock as quickly as possible when grey mould is first noticed. After reaching the planting site, seedlings must be planted as soon as possible. Interim measures that should be taken at the planting site include keeping seedlings cool by storing them in shade under thermal blankets and by improving ventilation around the seedling shoots (Sutherland and Hunt 1990).

Various species of *Penicillium, Fusarium,* and *Rhizoctonia* are responsible for moulding of bareroot stock. Sitka spruce is not immune to this problem as virtually all species and age classes of seedlings can be affected, except for lodgepole pine and ponderosa pine, which seem to be less prone than other conifers to storage moulds. The symptoms, which appear after the seedlings have been in cold storage, include watery or dry moulding of the foliage in localized areas or in patches over the entire shoot. A mouldy odour may also be noticed when boxes of seedlings are first opened. If string is used to tie seedlings together in bundles, moulding may start and spread from around the string. Transplanted seedlings are particularly vulnerable to storage moulding, especially if they are stored for a long time before being transplanted. The procedures for determining the severity and prevention of storage moulds on bareroot seedlings are the same as those for assessing grey mould on container-grown stock (Sutherland and Hunt 1990). Sutherland and Hunt (1990) caution that some fungi seen on roots of stored seedlings may be beneficial mycorrhizae that have proliferated during storage.

Root Diseases of Young Sitka Spruce Stands

Sutherland and Hunt (1990) indicated that the six most serious root diseases in British Columbia forest tree species are the following:

- Armillaria root disease (*Armillaria ostoyae*), to which all conifers, including Sitka spruce, are highly susceptible for about the first 30 years
- Tomentosus root rot (*Inonotus tomentosus*), to which all spruces and lodgepole pine are the most susceptible hosts
- Laminated root rot (*Phellinus weirii*), to which all spruces, all true firs, all hemlocks, Douglas-fir, and western larch are the most susceptible species in British Columbia
- Annosus root rot (strain 'S' of *Heterobasidion annosum*), to which Sitka spruce and all true firs are the most susceptible
- Black stain root disease (*Leptographium* [= *Verticicladiella*] *wageneri* var. *pseudotsugae*), which is a disease mainly of young Douglas-fir and to which the spruces, true firs, broadleaf species, and most pines (except lodgepole pine) are relatively resistant
- Rhizina root rot (*Rhizina undulata*), to which all conifers in British Columbia are susceptible and all hardwoods are resistant.

The early review of Sitka spruce by Phelps (1973) emphasized infection of spruce by *Armillaria, Phellinus,* and *Heterobasidion,* but he also indicated that the most important root rot is the brown cubical butt rot caused by *Phaeolus schweinitzii.* British Columbia's current *Root Disease Management Guidebook* (B.C. Ministry of Forests and B.C. Ministry of Environment, Lands and Parks 1995h) specifically names Sitka spruce as a host only for *Heterobasidion,* to which it is highly susceptible, and *Phellinus,* to which it is rated as moderately susceptible. The 1995 guidebook lists spruces in general as being highly susceptible to Armillaria root disease and tomentosus root rot, and tolerant or immune to *Leptographium.* The latter is therefore not described in the summaries below. The recent pest leaflet on tomentosus root rot by Hunt and Unger (1994) did not single out Sitka spruce, hence this disease is not addressed here either. Also, brown cubical butt rot, referred to by Phelps (1973), does not appear to be a concern in the current *Root Disease Management Guidebook.* Based on current information, therefore, *Heterobasidion,*

Phellinus, Armillaria, and *Rhizina* are the main root rot diseases of concern to Sitka spruce managers, and the following summaries are limited to them.

Annosus Root Rot In the *Root Disease Management Guidebook* (B.C. Ministry of Forests and B.C. Ministry of Environment, Lands and Parks 1995h), Sitka spruce is listed as being highly susceptible to Annosus root rot (*Heterobasidion annosum*). Like other root diseases, Annosus root rot is very persistent in natural ecosystems. In an Annosus root rot survey in the Prince Rupert Forest Region, Morrison et al. (1986) determined that host susceptibility, in terms of percentage of stumps colonized, was in the following decreasing order: amabilis fir, Sitka spruce, western hemlock, and Douglas-fir, with none detected on lodgepole pine stumps. In this study, 23.6% of all sampled Sitka spruce stumps were colonized by Annosus root rot.

Despite two decades of detailed study of the biology of *H. annosum* in relation to cut stumps of Sitka spruce and other conifers (Morrison and Johnson 1970; Morrison et al. 1985), the forestry role of this fungus remains an active subject of research. Recent population studies (Morrison et al. 1994) have confirmed that many of the roots connected to stumps infected with *H. annosum* are dead ends for fungal spread. Morrison and Redfern (1994) noted that only about 20% of the contacts between Sitka spruce stumps infected with *H. annosum* and the roots of other trees resulted in transfer of the fungus to adjacent trees.

Experiments by Redfern (1982) indicated that living stumps are more susceptible to infection than those that die after felling. Although live stumps were infected more frequently (92%) than dead ones (80%), the area colonized by the fungus was similar, possibly because drying of sapwood in these stumps was sufficiently extreme to inhibit development of *H. annosum.* Although the overall difference in susceptibility between live and dead stumps seems to be only minor, live stumps may play a special part in long-term disease development. Redfern (1989) noted that living stumps that were infected remained alive for at least 10 years and were relatively undecayed compared with dead infected stumps of the same age.

Research in the past 15 years suggests the following relationships between amount of precipitation and intensity of Annosus infection. Redfern (1993) noted that in southern Scotland, Sitka spruce stumps exposed to high rainfall had increased Annosus infection in the heartwood but reduced infection of the sapwood compared with covered stumps. Redfern did not think that the reduced sapwood infection in uncovered stumps was due to the loss of spores in heavy rain, because some of the exposed stumps that had no *H. annosum* in the sapwood did have extensive heartwood infection. Furthermore, exposure to rain increased heartwood infection. Earlier, Redfern (1982) had suggested that stump infection may be reduced by high rainfall.

Also, near Terrace, British Columbia, Morrison et al. (1986) hypothesized that the low incidence of stump surfaces colonized during winter may be due to low spore inoculum levels caused by low temperatures and heavy precipitation. Redfern (1993) placed these observations from the 1980s in context by suggesting that overall, in wet upland forests, there is unlikely to be a general reduction in the amount of woody inoculum containing *H. annosum* compared with drier forests. Redfern's assessment is that overall susceptibility of the stump is reduced by exposure to high rainfall, despite the difference in response between heartwood and sapwood, which was possibly a reflection of the relative proportions of the two wood types (sapwood constitutes about 75% of total stump area). The effect of high rainfall on stump infection appears to change with tree age in relation to spruce's age changes in heartwood/sapwood ratios. The exclusion of *H. annosum* from sapwood by very heavy rainfall on cut stumps may help reduce infection or prevent it from spreading to surrounding trees, regardless of the amount of inoculum in the heartwood. Endogenous factors may have a greater influence on infection than environmental factors or the availability of inoculum (Redfern 1993).

Laminated Root Rot Among western North American conifers, Sitka spruce is intermediate in susceptibility to laminated root rot caused by *Phellinus weirii* (Thies and Sturrock 1995). Test conifers that were planted around infected stumps in Oregon and Washington indicated the following susceptibilities based on mortality over 17-20 growing seasons: grand fir experienced nearly 30% mortality from *P. weirii;* Douglas-fir mortality exceeded 20%; Sitka spruce, noble fir, giant sequoia, western hemlock, and ponderosa pine averaged less than 10% mortality; western white pine and lodgepole pine mortality was less than 1%; and western redcedar and redwood showed no mortality from *P. weirii* (Nelson and Sturrock 1993). This order of tree susceptibility to *P. weirii* is similar to that reported by Hadfield (1985).

Based on the review by Thies and Sturrock (1995), the highly susceptible trees native to the zones where Sitka spruce can occur in British Columbia are Douglas-fir, grand fir, amabilis fir, and mountain hemlock. Among Sitka spruce's other potential tree associates, western hemlock, Pacific yew, and Engelmann spruce share with Sitka spruce an intermediate susceptibility to *P. weirii* – they are readily infected by *P. weirii* but are usually not killed by it, although development of butt decay is common. Both yellow-cedar and western redcedar are listed as being resistant to *P. weirii,* and bigleaf maple, red alder, and vine maple are immune to this root rot. Detailed recommendations for management of *P. weirii,* not specific to Sitka spruce stands, are provided by Thies and Sturrock (1995). The *Management Guidelines for Laminated Root Rot in the Vancouver Forest Region* (Beale 1989) suggest that one strategy is to use less susceptible species in establishing plantations. Sitka spruce is listed among the less susceptible species and is recommended for use specifically in the Pl–Sphagnum site series of the CWHdm subzone and in the CwSs–Skunk cabbage site series of the CWHmm1 variant.

Armillaria Root Disease Although Armillaria root disease is prominent in the current forest pathology literature, it has not been emphasized as an important source of disease centres in Sitka spruce stands. A guide to the diagnosis and management of this disease appears in Morrison (1981). To the extent that Sitka spruce is a host for this fungus, the key biological features of this decay organism were described by Allen (1994) as follows.

Armillaria is a wood decay organism in roots, lower boles, and stumps of dead and living trees. It spreads to living trees through rhizomorphs growing from infected stumps or debris, or when the roots of an infected tree come into contact with an old infected root or stump. Signs and symptoms of *Armillaria* infection first appear on trees at about age 5, with a first major kill from contact with infected debris and stumps within about 10 years. Resistance to infection increases as trees age. However, if trees become stressed because of moisture or nutrient deficiency, attack from insects or other pathogens, or physical damage, *Armillaria* growth can increase, leading to reduced growth or mortality of the host tree. This fungus attacks the roots of trees of all ages, killing the cambium and inner bark and causing decay of both sapwood and heartwood. On trees up to 10 years old, *Armillaria* usually kills within a year or two of infection. Older infected trees are not killed, but show reduced growth associated with chlorotic foliage and stunted shoot growth. However, mortality can occur in older trees that become damaged or stressed.

Rhizina Root Rot Rhizina root rot, caused by the fungus *Rhizina undulata,* is associated with seedling damage or death in several coniferous species, including Sitka spruce, especially in slashburned clearcuts (Watts 1983; Callan 1993). A widespread outbreak discovered in 1988 in the Prince Rupert Forest Region continued at a high level in 1989. Surveys in 1989 revealed that 22% of Sitka spruce seedlings had been killed by *Rhizina* at seven plantation sites. *Rhizina* is present in all regions of British Columbia and the Pacific Northwest, where early reports of this fungus were most often associated with Douglas-fir and western hemlock (Ginns 1974; Thies et al. 1977, 1979). In British Columbia, the most favourable conditions for *Rhizina* reproduction are in wetter regions such as the Coastal Western Hemlock and Interior Cedar Hemlock biogeoclimatic zones.

Conditions created by wildfires or prescribed burns stimulate the development of *Rhizina.* The incidence of fruiting bodies is highest in areas previously covered by old-growth stands and on acid soils (Callan 1993). These fruiting bodies, or apothecia, develop above dying roots or burnt wood, emerging approximately 15 weeks after a burn. They can release ascospores that are borne by wind throughout the growing season. Ascospores can remain dormant but viable for up to two years. Heat is required for germination, usually 37°C for three days based on laboratory tests. Three conditions are necessary for *Rhizina* to become established: presence prior to burning, acidic soil, and presence in the soil of live conifer roots.

Severely infected seedlings have extensive white mycelia attached to their roots (Callan 1993). Infected seedlings appear girdled at or below the soil line, often with the appearance of beetle galleries, and frequently have discoloured foliage, a symptom indistinguishable from drought or other root rots. Seedling mortality is most severe in the first year after a fire. Definitive diagnosis requires fruiting bodies to be found within 0.5 m of the dead seedling, but often there is little correlation between occurrence of fruiting bodies and seedling loss (Wood and Van Sickle 1989).

Where Sitka spruce managers use broadcast burning as part of site preparation, it is important to consider the potential for increased incidence of Rhizina root rot after burning (Callan 1993). Significant losses of spruce seedlings as a result of *Rhizina* infection have been recorded in plantation sites in the Prince Rupert Forest region. *Rhizina* incidence is greatest in the first two years after a burn. In wetter regions, therefore, where *Rhizina* fruits more abundantly than in dry interior regions, one possible control measure is to delay planting spruce for at least two years. Surveys for fruiting bodies from 10 to 16 months after a burn can indicate whether further delay is warranted. A disadvantage of such planting delays is increased weed competition. When planting commences, avoiding sites adjacent to stumps and large pieces of burnt wood may prevent the further spread of this fungus.

No chemical or biological controls for *Rhizina* are known at present (Callan 1993). In Europe control measures have included banning bonfires and digging trenches (0.3 m deep and 0.3 m wide) around burnt areas to reduce radial spread of the fungus. In western Great Britain, the greatest *Rhizina* damage to Sitka spruce occurs around fires that are lit during thinning operations.

Stem Rusts

Rusts affecting branches or stems are among the most destructive forest diseases on other conifers in British Columbia (Finck et al. 1989); fortunately Sitka spruce is free of attack by these fungi. Finck and co-workers did not list Sitka spruce as a host for any species of *Cronartium* or *Endocronartium* that typically cause severe damage to pines in the province. This agrees with Sutherland and Hunt (1990), who emphasized that stem rusts are of concern mainly in pines.

The detailed compilation of tree rusts on forest species in western Canada (Ziller 1974) lists the following species potentially present on Sitka spruce needles and cones: cone rusts – *Chrysomyxa monesis, C. pirolata,* and *C. ledicola;* needle rusts – *Chrysomyxa arctostaphyli, C. piperiana, C. weirii, C. ledicola, Melamspora medusae,* and *M. occidentalis.*

Stem Decay

The earliest study of stem decay involving Sitka spruce (Boyce 1929) documented the deterioration of windthrown trees in the valleys of the Olympic Mountains region in Washington state. Boyce found that downed Sitka spruce contained less decay than amabilis fir and western hemlock but more than Douglas-fir. More recent studies indicate that the rate of sapwood decay in slash on clearcut areas is about the same in Sitka spruce, Douglas-fir, and western hemlock, but heartwood decay is slowest in Douglas-fir, intermediate in Sitka spruce, and most rapid in western hemlock (Minore 1979). There is evidence from southeastern Alaska that live Sitka spruce as old as 400 years have less stem decay than western hemlock of comparable age (Farr et al. 1976).

In British Columbia the first detailed investigation of Sitka spruce stem decay (Bier et al. 1946) was stimulated by the great demand for a rot-free supply of this wood for wartime airplane construction. Bier's early study confirmed that, compared with many other species, Sitka spruce suffers relatively low stem decay losses from disease. One estimate from southeastern Alaska indicated that only 9% of the gross volume in old-growth Sitka spruce stands was cull, compared with 22% for mature western hemlock and 52% for mature western redcedar (Harris and Farr 1974). In western Washington and Oregon, decay following logging resulted more from injuries to the root and butt than injuries above 1.4 m on the stem. Western hemlock and Sitka spruce showed similar decay trends, except that in Sitka spruce the rate of spread was initially slower than in hemlock and then faster once decay was established (Wright and Isaac 1956).

Dwarf Mistletoe

In mixed hemlock–Sitka spruce stands, hemlock is the principal host of hemlock dwarf mistletoe (*Arceuthobium tsugense*). As cited in Ruth and Harris (1979), the few reported infections of Sitka spruce are from Chichagof Island, Alaska (Laurent 1966); Kitimat, British Columbia (Molnar et al. 1968); and southern Vancouver Island (R.B. Smith, pers. comm., 1977). Smith (1974) did not successfully inoculate Sitka spruce with hemlock dwarf mistletoe in plantation trials. In these trials, he observed that in the 11 species tested, the greatest mistletoe seed retention occurred on the branches of the spruces (Sitka, white, and Engelmann). However, the best mistletoe seed germination was on western hemlock, Douglas-fir, ponderosa pine, and grand fir, whereas the poorest germination was on the spruces. Since Sitka spruce is only an occasional or rare host (Baranyay and Smith 1972; Mathiasen 1994), parasitism of it by dwarf mistletoe is not considered a management problem.

Some Fish, Mammalian, and Bird Habitat Features of Sitka Spruce Ecosystems

The emphasis in this section is on wildlife species whose occurrence or habitat requirements are dependent on coastal forests where Sitka spruce is a component. No species of wildlife is totally dependent on Sitka spruce by itself, but the communities where Sitka spruce occurs provide specific habitat needs, and several wildlife species are listed as endangered-threatened (Red List) or sensitive-vulnerable (Blue List) in the British Columbia portions of Sitka spruce's natural range (Harper et al. 1994). Emphasis is also given to wildlife species whose presence interacts in important ways with forest management – salmonids, Sitka black-tailed deer, Columbian black-tailed deer, Roosevelt elk, porcupine, grizzly bear, and bald eagle. Further suggestions on management to protect fish and wildlife values are provided in the section 'Fish and Wildlife Considerations in Sitka Spruce Ecosystems' (page 245 in Chapter 3.

Sitka spruce–western hemlock forests have figured prominently in research focusing on relationships between old-growth forest structure and wildlife habitat. A key caveat was made in the review by Alaback (1984), namely that forest and wildlife managers cannot reliably apply single, generalized forest structural statistics to the classification of old-growth forests. Because there is very wide variation in stand structure of old-growth spruce-hemlock forests, especially in stands over 200 years old, a multi-factor characterization is more appropriate for evaluating the relative value of such forests as wildlife habitat. One generalization that can be made is that old-growth spruce-hemlock stands have a greater variation in mean standard diameter, wider average tree spacing, increased dominance of western hemlock, and increased understorey productivity than second-growth stands. But beyond this, wildlife managers should expect very great structural variation when old-growth Sitka spruce–western hemlock forests are compared within a region.

In terms of maintenance or enhancement of habitat values for wildlife, a Sitka spruce manager is constrained less by available information than by techniques for applying basic information. There is substantial recent literature on wildlife relationships in ecosystems that contain Sitka spruce in Alaska, British Columbia, Washington, and Oregon (Bunnell and Jones 1982; Meehan et al. 1982; Hanley 1984; Brown 1985; McNay and Davies 1985; Brunt 1987; Hanley and Rose 1987; Sadoway 1988; Schoen et al. 1988; Carey

1989; Hanley and Rogers 1989; Hanley et al. 1989; Nyberg et al. 1987, 1989, 1990; Bunnell and Kremsater 1990; Nyberg and Janz 1990; Meidinger and Pojar 1991; Ruggiero et al. 1991; Hanley 1993; Stevens 1995a, 1995b). The challenge is how to apply this information on a site-specific basis in Sitka spruce ecosystems.

Fish Relationships

Interest in interactions between fish and forests has been particularly keen on the Queen Charlotte Islands. A mild climate, high rainfall, and lack of summer drought combine to give this archipelago some of Canada's most productive forest land. Based on data from over a decade ago, summarized by Poulin (1982), the Queen Charlotte Islands land base contributed over 3% of British Columbia's annual timber harvest, with Sitka spruce as one component of this harvest. This important forest land base contains nearly 200 streams that contribute to sport and commercial fish production. Even though the Queen Charlotte Islands lack the very large salmon runs typical of major mainland rivers, total salmon escapement from these islands has been estimated at about 1 million annually, amounting to about 13% of British Columbia's total annual returning salmon stocks (Poulin 1982).

Studies at Carnation Creek on western Vancouver Island indicated that seemingly small natural or logging-related increases in stream temperature had important effects on the life history of young coho salmon (Hartman et al. 1982; Hartman and Scrivener 1990). Changes in stream temperatures in winter may be more obscure than those in summer after logging, but they are significant to production of coho salmon. Studies by Tripp and Poulin (1992) on the effects of logging and mass wasting on juvenile coho salmon, steelhead trout, and Dolly Varden on the Queen Charlotte Islands have indicated that stream reaches in logged areas had less undercut bank cover for fish than in unlogged areas, but otherwise did not differ from unlogged areas in any other habitat variable measured. In contrast, stream reaches in areas of mass wasting had even less undercut bank cover than logged reaches, less large organic debris, fewer pools and glides, and more riffles (Tripp and Poulin 1992). These researchers presented evidence that in summer and fall stream reaches in logged areas had significantly higher coho fry densities than in unlogged or mass-wasted areas. However, in streams that contain input from mass wasting, a combination of poor egg-to-fry survival due to excess gravel scour, and poor overwinter survival of juvenile coho due to loss of overwintering habitat, nullified any gains in coho production attributable to logging. It also nullified the high growth rates and large sizes achieved by coho in their first year in streams running through areas of mass wasting. Impacts on Dolly Varden and steelhead trout did not appear to be as serious as in coho. Impacts on chum and pink salmon were not investigated by Tripp and Poulin (1992), but these researchers presumed that these species of salmon would also be negatively influenced by increased gravel scour.

On coastal floodplains, Sitka spruce is a common tree along drainage features such as ephemeral swamps and intermittent tributaries. Hartman and Brown (1988) stressed the importance of these geomorphic features, adjacent to the main channels of coastal rivers and creeks, as overwintering habitat for juvenile coho, cutthroat trout, and steelhead trout. These sites, referred to as *off-channel habitat,* have a fishery importance much greater than would be expected on the basis of area. For example, in the 10 km^2 Carnation Creek watershed on Vancouver Island, off-channel coho overwintering habitat covers less than 2% of the total floodplain but accounts for as much as 24% of total coho production (Brown 1987). As much as 15% of this watershed's coho production can come from fish that have overwintered on sites that are devoid of surface water in summer. It is not surprising that the *British Columbia Coastal Fisheries/Forestry Guidelines* (B.C. Ministry of Forests et al. 1993) recognize these off-channel habitats as 'fisheries sensitive zones.'

The range of fluvial landforms referred to as off-channel habitat by Hartman and Brown (1988) included riverine ponds, beaver ponds, connected riverside channels, percolation channels, abandoned stream channels, seepage-fed tributaries, and ephemeral swamps. Forestry activities that can influence these sensitive landforms include road construction (both on the floodplain and upslope); rotation length (frequency of successive disturbance from forest harvesting); locations of tree removal; yarding across channels and across off-channel habitats; silvicultural steps to change tree species composition; and silviculture involving site preparation, herbicide use, or prescribed burning. Sitka spruce managers can benefit from the measures proposed for protection of these sensitive fish overwintering habitats by the *British Columbia Coastal Fisheries/Forestry Guidelines,* and from habitat protection measures proposed by Hartman and Brown (1988), the *Riparian Management Area Guidebook* (B.C. Ministry of Forests and B.C. Ministry of Environment, Lands and Parks 1995g), and the *Coastal Watershed Assessment Procedure Guidebook* (B.C. Ministry of Forests and B.C. Ministry of Environment, Lands and Parks 1995b).

Besides the scientific studies cited above, background information is provided by Sedell et al. (1988) for west coast salmonid streams where Sitka spruce can be a prominent component of coarse woody debris in and adjacent to channels. Similarly detailed studies of juvenile and adult salmonids in stream ecosystems of former old-growth Sitka spruce–western hemlock forests were reported by Hartman et al. (1982), Murphy et al. (1982), Myren and Ellis (1982), Poulin (1982), and Thedinga and Koski (1982).

Mammal Relationships

Sometime between 1900 and 1916, Sitka black-tailed deer were introduced to the Queen Charlotte Islands (Banfield 1974). The species now thrives there, and many foresters have noted their influence upon the vegetation of the Queen Charlotte Islands. Old-growth western hemlock–Sitka spruce forests provide important habitat for Sitka black-tailed deer. Even-aged, second-growth forests produce very little forage for black-tailed deer. However, young open stands under 20 years of age produce greater amounts of forage than do older even-aged stands or old-growth stands. Habitats differ in their canopy characteristics and in the amount and kind of forage they produce (Alaback 1982a, 1982b, 1982c; Alaback and Herman 1988). An understanding of these dynamic relationships between deer and their habitat is essential for developing management objectives for deer habitat.

A report by Pojar et al. (1980) stressed that there are, unfortunately, no demographic data for the deer population of the Queen Charlotte Islands, and little documentation of pre- and post-deer vegetation structure and plant species abundances. However, Pojar and co-workers believed that 'severe browsing has drastically changed the plant life of the Islands.' Pojar and Banner (1982) concluded that the serious silvicultural impacts of continued overbrowsing include the probable elimination of western redcedar and yellow-cedar as commercial timber species on the Queen Charlotte Islands, and increasing damage to Sitka spruce and western hemlock, especially in recent plantations.

Deer overbrowsing has also compounded problems of slope stability. As part of a recent survey of rare and endemic vascular plants in Gwaii Haanas/South Moresby National Park Reserve, Ogilvie (1994) recorded heavy deer browsing of Sitka spruce saplings near an estuarine meadow area at Hutton Inlet. The report by Pojar et al. (1980) contains several photographs showing heavily browsed krummholz-like Sitka spruce and young Sitka spruce that show browse lines on their stems as high as deer can reach.

Pojar and co-workers ranked the browsing preferences of deer in relation to various species on the Queen Charlotte Islands as follows, in order of decreasing preference:

1 *Lysichiton americanum*
2 *Rubus spectabilis, Oplopanax horridus, Thuja plicata*
3 *Athyrium filix-femina, Dryopteris assimilis, Polystichum munitum*
4 *Vaccinium parvifolium, V. ovalifolium, V. alaskaense, Blechnum spicant, Gymnocarpium dryopteris*
5 *Menziesia ferruginea, Gaultheria shallon, Chamaecyparis nootkatensis*
6 *Picea sitchensis*
7 *Tsuga heterophylla*
8 *Alnus rubra*

On the islands within Prince William Sound, Alaska, near the northern limits of Sitka spruce's natural range, Eck (1982) noted the following relationships between forest structure and winter habitat for Sitka black-tailed deer. Understorey plant communities in this region are similar to the coastal forest communities of southeastern Alaska. In general, diversity of deer forage species decreased as site quality for tree growth increased. Forage species diversity and abundance were lowest in stands dominated by Sitka spruce. Production of the key winter forage species – *Vaccinium ovalifolium* and *V. alaskaense* (the major browse species), and *Cornus canadensis, Rubus pedatus, Coptis aspleniifolia,* and *C. trifolia* (the critical evergreen forbs) – was inversely related to percent overstorey canopy closure and net volume of timber. Forage species were most abundant in the more open hemlock stands, and decreased as canopies became more closed and as the Sitka spruce component of stands increased.

Old-growth coniferous forests have traditionally been considered relatively poor habitat for Sitka black-tailed deer, but Rose (1982) documented circumstances in the Sitka spruce–western hemlock forests of Annette Island, southeastern Alaska, where old-growth forest is optimal winter range for deer compared with recent clearcuts or second-growth forests. South-facing slopes were used by deer seven times more than north-facing slopes. Rose's information that spruce-hemlock old-growth stands can be preferred deer winter habitat is an important consideration on forest lands where management for deer is a major goal. It appears that for Sitka black-tailed deer, the net effect of clearcut logging is to change the environment from an energy-limiting one to a protein-limiting one during the summer. During winter, energy is probably the most limiting factor for deer in all environments. Snow accumulation increases the problem of winter energy deficits. Thus, the productivity of black-tailed deer in managed forests depends on their ability to find protein-rich foods during summer months and digestible energy-rich foods during winter months (Hanley et al. 1991). Important habitat features that need to be provided in managed second growth are a well-developed understorey of shrubs and herbaceous plants, and a coniferous canopy to intercept snow effectively (Jones and Bunnell 1982; Kessler 1982). For southeastern Alaska, where deer predation by gray wolf is an added feature of deer management, Van Ballenberghe and Hanley (1982) predicted that the long-term effects of removing old-growth timber would be an ultimate reduction in the rate of increase of deer, even in the absence of hunting and wolf predation. The decline of the deer population would accelerate if hunting and wolf predation were intensive.

It is significant that Sitka black-tailed deer has been suggested as a key ecological indicator for forest management decisions that seek to balance economic development, biological conservation, and human cultural interests (Hanley 1993). Hanley's reasons for focusing on this ecological indicator in the Sitka spruce–western hemlock forests of southeastern Alaska are as follows: (1) the biology and habitat requirements of this deer are well known; (2) the deer have relatively large, seasonally migratory home ranges, and so require management of landscapes rather than isolated patches of habitat; (3) their need for a productive and nutritious food supply all year makes them largely dependent on old-growth forest and a variety of habitats, differing seasonally and in response to snow; and (4) they are an important game species in the subsistence economy of southeastern Alaska. Key ecological relationships between deer and their food resources centre on bioenergetics, digestible protein, the role of forest overstorey in the carbon-nutrient balance, and the chemical composition of understorey plants.

Ecological relationships in deer and elk habitats in coastal forests of British Columbia are now well documented (Nyberg and Janz 1990). Although drawing from today's most experienced forest wildlife researchers, this excellent compilation still emphasizes that there are no simple guidelines or prescriptions for optimal wildlife habitat management in coastal forest ecosystems. There are usually no simple answers to wildlife/forestry questions, and this is as true for sites where Sitka spruce is prominent as it is for other west coast ecosystems.

For Sitka spruce managers, a comprehensive source of information is available from Bunnell (1990) for the habitat requirements of the Sitka black-tailed deer, which occurs on the Queen Charlotte Islands and on the north coastal mainland from about Rivers Inlet (latitude 51°30′N) north to the vicinity of Stewart (about 55°55′N), and for the Columbian black-tailed deer, which occurs on Vancouver Island and on the coastal mainland from Rivers Inlet southward.

A similarly detailed account is available for the ecology of Roosevelt elk (Brunt 1990), which occurs on Vancouver Island and in valleys at the head of Loughborough Inlet and Phillips Arm on the mainland east of Vancouver Island, and on the Sechelt Peninsula, where elk were introduced in the late 1980s. Brunt's summary of habitat features for Roosevelt elk is not specific to Sitka spruce–dominated stands; in general, however, on the southern British Columbia coast thermal cover for elk is provided by coniferous stands greater than 10 m tall, with canopy closure exceeding 70%. Elk concentrate near boundaries between forested and unforested ecosystems, a behavioural feature important where Sitka spruce may be under management within the range of the elk. Where elk occur on northern and western Vancouver island, their most heavily used winter ranges are often stands of old-growth western hemlock and Sitka spruce on river floodplains. Recommendations to defer logging in critical winter ranges are usually based on field surveys of elk use and habitat suitability, but quantitative criteria for the size and distribution of these ranges have not yet been developed.

Elk winter ranges usually occupy a very small proportion of the forested land base. However, because areas where elk occur often support high volumes of valuable timber, the potential contribution of such sites to the total timber harvest is higher than their total area suggests (Nyberg 1990). There is some evidence from the Olympic Peninsula that, in valley-bottom sites where western hemlock was dominant, intense levels of elk browsing may in the long term be impairing new establishment of hemlock in favour of Sitka spruce (Woodward et al. 1994). This research suggests that intense levels of ungulate herbivory may locally influence the direction of succession in sites where Sitka spruce is present, but such disturbances on small scales over short time periods are usually masked by larger landscape-scale disturbances.

The examples summarized above indicate that Sitka spruce is generally not heavily browsed when alternative food sources are available. For example, western redcedar is the preferred species for deer, and when redcedar is sufficiently common, little browsing occurs on Sitka spruce. Coates et al. (1985) studied the effect of deer browsing on the Queen Charlotte Islands. They found few browsing losses on Sitka spruce or western hemlock, but heavy losses on western redcedar. Smith et al. (1986) reported that deer damage to trees established on landslide scars in the Queen Charlotte Islands was most severe on yellow-cedar, less on western redcedar, and considerably less on Sitka spruce and western hemlock. Of the major tree species established on slides, red alder was least browsed. Damage to individual trees tended to be greatest on landslide areas, intermediate in logged areas, and least in old-growth stands.

Because deer prefer to browse on species other than Sitka spruce, Sullivan (1990) suggested that one way a forest manager can reduce seedling damage from browsing is to use less susceptible species, such as Sitka spruce or western hemlock. As western redcedar and yellow-cedar continue to be eliminated by deer browsing, damage to Sitka spruce and western hemlock regeneration is expected to increase. As a result, systems of silviculture that involve small clearcuts, shelterwood, or selection will probably be ruled out by the intense browsing that occurs in all conifer regeneration in small openings of the Queen Charlotte Islands forest.

In coastal British Columbia, the most widespread porcupine damage is in the Kalum and North Coast districts of the Prince Rupert Forest Region. Porcupines are particularly abundant in 15- to 35-year-old (10-30 cm dbh) stands of western hemlock–Sitka spruce in the CWHws, CWHvm, and CWHwm subzones. A survey of stands along Khutzeymateen Inlet showed that western hemlock, which made up 67% of the trees in sampled stands, comprised 52.7% of the trees damaged by porcupines. Sitka spruce was second with 7.8% of trees damaged. The less abundant western redcedar and amabilis fir suffered very little damage (Sullivan et al. 1986; Sullivan 1990).

The latter researchers found that porcupines preferred to feed on the larger-diameter stems of the dominant and codominant trees. Damage to trees occurs mainly as a result of winter feeding, because porcupines rely on early-successional, post-harvest herbaceous vegetation for summer feeding.

Grizzly bear habitat specialists have not suggested that Sitka spruce itself directly influences bear habitat preferences, but many key grizzly bear habitats in coastal British Columbia are in sites where Sitka spruce is part of very productive floodplain ecosystems (MacHutchon et al. 1993). In the Kimsquit River basin, at the head of Dean Channel (about 53°N, 127°W), Hamilton and Archibald (1986) recorded that 64% of 112 feeding activity records for bears were in sites dominated by spruce with an understorey of *Oplopanax horridus*. All seral stages and variations of floodplain Sitka spruce–devil's club sites (mainly Ss–Salmonberry site series as defined by Banner et al. [1993] for the CWHmm1 variant) were heavily used by radio-collared bears studied by Hamilton (1987) in the lower Kimsquit River valley. In the Khutzeymateen Valley grizzly bear study, about 45 km northeast of Prince Rupert, MacHutchon et al. (1993) identified skunk cabbage sites in old growth spruce-hemlock stands as another preferred site series for grizzlies. These floodplain sites are important for grizzlies because of both their nearness to salmon as a food source in spawning channels and a high diversity of floodplain plants that are important as grizzly bear forage.

Bird Relationships

The Coast and Mountains Ecoprovince within British Columbia, as defined by Campbell et al. (1990), is home to numerous seabirds and shorebirds and is a migratory corridor for many other bird species. Sitka spruce is significant as a component of coastal old-growth forests on which certain bird species depend for habitat and parts of their life cycle. Most notorious recently in the Pacific Northwest are the Northern Spotted Owl and the Marbled Murrelet, but old-growth ecosystems are also used by Bald Eagles, Pileated Woodpeckers and other cavity nesters, and a variety of songbirds (Kirk and Franklin 1992). The significance of wildlife trees, defined as dead snags that provide essential habitat for any animal species (Backhouse 1990, 1993), in old-growth forests is that they deteriorate slowly and meet the needs of many different wildlife species over a long period of time. Current guidelines from British Columbia's Wildlife Tree Committee recommend that 5-10 snags per hectare be left for wildlife (Klenner and Kremsater 1993).

Sitka spruce in British Columbia does not figure prominently in the concern in the Pacific Northwest over the Northern Spotted Owl because the latter's natural range extends northward to include only a very small area of southwestern British Columbia. As of 1995, the known range of the spotted owl in the province extended west to the Capilano and Squamish River valleys, north to Birkenhead Lake (about 30 km north of Pemberton), and east to the Anderson River, which is an east-side tributary valley that enters the Fraser River canyon about 3 km south of Boston Bar (Dunbar and Blackburn 1995). Sitka spruce is not a significant forest component within this part of the Northern Spotted Owl's documented range. The greatest chance of Sitka spruce occurring within the Canadian portion of this owl's range would be at the eastern and northern fringes of the Fraser River Lowland and in the Squamish and Cheakamus River valleys.

Northern Spotted Owls have been strongly linked with old-growth forests in the Pacific Northwest (Ripple et al. 1991). More recently Mills et al. (1993), working on the Olympic Peninsula, determined that spotted owls tended to roost or nest in stands with high vertical canopy layering. Two of six community types examined were preferred by spotted owls: Sitka spruce–western hemlock/Oxalis; and western hemlock–Sitka spruce–Douglas-fir–western redcedar/Oxalis. The multi-layered canopy of old-growth forests provides owls with needed manoeuvring space and options for feeding at all levels from the ground to the canopy, as well as cover and protection in inclement weather. Cavities in the trees provide nest sites, and large branches provide platforms for owls, Pileated Woodpeckers, bats, and ants (B.C. Wildlife Branch 1991).

The Marbled Murrelet is believed to breed along the entire coast of British Columbia, but nesting areas are poorly understood. Primary nesting areas for this species are coniferous old-growth forests within 75 km of the coastline (Campbell et al. 1990). It is known to nest in old-growth forest habitat from southern Alaska to California (Binford et al. 1975; Quinlan and Hughes 1990; Singer et al. 1991). Studies on the Queen Charlotte Islands in four habitats (alpine, high-elevation old-growth forest, low-elevation old-growth forest, and second-growth forests) confirmed a strong association with old-growth forest (Rodway et al. 1991, 1993). Detections of Marbled Murrelets were most numerous in valley bottoms in the vicinity of old-growth spruce and hemlock forest with mean tree sizes over 1.0 m dbh. The relationship between tree size and number of detections is comparable with other studies of this species in California and Oregon. Few detections were recorded for second-growth forest (40-60 years) compared with old growth. Nesting requirements across the range of this species seem to pertain more to the structural characteristics of old-growth stands (Hamer and Cummins 1991; Singer et al. 1991) than to specific species associations; known nests of Marbled Murrelets from southern Alaska to California have been located on large moss-covered limbs greater than 36 cm in diameter in large conifers (Singer et al. 1991). Because the population levels and breeding biology of this species are poorly understood, it has been designated as threatened on the B.C. Environment Red and Blue lists and by the Committee on the Status of Endangered Wildlife in Canada (Campbell et al. 1990).

Nesting habitats that have been documented for Ancient Murrelets on the Queen Charlotte Islands have been mostly in forests over 140 years of age. Nesting occurs in burrows on the forest floor. Of the 12 species of seabirds that nest in colonies on the coast of British Columbia, only the Ancient Murrelet nests mainly in forested sites. Aside from the preference for old-growth forest, murrelet nesting sites are characterized by open understorey dominated by bryophytes, moderate to steep slopes facing the ocean, and shoreline within 325 m horizontal distance (Blood and Anweiler 1982). On Lyell Island, where 60,000 breeding pairs make up the largest colony of the estimated 204,000 breeding pairs on the Queen Charlotte Islands, western hemlock, western redcedar, and Sitka spruce – in this decreasing order – made up the forest cover in the highest-density nesting areas. Most nesting burrows were beneath large, living trees, with western redcedar apparently being preferred. The buttressed growth form of trees in the old-growth Lyell Island forest provide many potential nest cavities. Although Ancient Murrelets are known to nest on treeless islands, the apparent importance of old-growth forest near the coast for their nesting suggests that logging in such areas would destroy nesting habitat for a long time. None of the Queen Charlotte Islands nesting sites examined had ever been logged; Blood and Anweiler (1982) did not present data to indicate the age at which post-logging successional stands might be reoccupied by Ancient Murrelets.

On coastlines of the Pacific Northwest and British Columbia, Sitka spruce is one of the tree species contributing coarse woody debris to estuaries, coastal marshes, and beaches. It not only provides drift trees that lodge along tidally influenced rivers, upper and lower estuaries, salt marshes, and coastal beaches but is also a frequent regeneration species on nurse logs that have lodged in such coastal wetland sites. The use of such drift trees by birds has been described by Gonor et al. (1988), involving mainly Bald Eagles, herons, cormorants, and gulls. Reconnaissance surveys of winter wildlife use of stream and riparian habitats on the Queen Charlotte Islands (vandenBrink 1992b) showed the riparian zone to be especially important for Winter Wren, Golden-crowned Kinglet, and Red-breasted Sapsucker. In general, winter wildlife diversity on the Queen Charlotte Islands was found to be about 1.5 times greater in old-growth riparian areas than in non-riparian second-growth areas (vandenBrink 1992a).

An increasing amount of information is available to Sitka spruce managers who want to document the impact of forest operations on bird populations. On western Vancouver Island, bird assemblages in old-growth forests consist largely of species that are year-round residents in British Columbia (Bryant et al. 1993), whereas bird communities in second-growth forests, a key area where Sitka spruce occurs, consist mostly of birds that winter outside Canada. Forest managers can be guided by the fact that western Vancouver Island bird communities in clearcuts and in young stands up to 20 years old are mostly ground or shrub-nesting species (Bryant et al. 1993); bird communities in stands 30-60 years old include ground, shrub, and tree-nesting species; and old-growth stands support mainly tree- and cavity-nesting species.

Because of these succession-related changes to bird populations, Bryant and co-workers concluded that clearcutting on western Vancouver Island led to reduced diversity of bird species for at least 20 years. This influence was most pronounced at inland sites higher than 500 m above sea level. Bird diversity and abundance increased in stands 30-35 years old, was reduced in stands 50-60 years old, and peaked in old-growth stands on western Vancouver Island. The Very Wet Hypermaritime CWH Subzone (CWHvh), where Sitka spruce is best represented on Vancouver Island, contains cedar-dominated old-growth ecosystems that are less diverse in bird species than stands dominated by mixtures of redcedar, western hemlock, and grand fir (Bryant et al. 1993).

The 1995-96 report on *The State of Canada's Forests* (Canadian Forest Service 1996) noted that the Queen Charlotte Goshawk, which is very rare, has now been assigned to the provincial Red List of Endangered or Threatened Species by the British Columbia Ministry of Environment, Lands and Parks. Three nests have been reported on the Queen Charlotte Islands and six on Vancouver Island. This bird prefers to nest in large, unfragmented stands of mature forests with closed canopy cover; on the Queen Charlotte Islands especially, such forests may include Sitka spruce. The main threat to this vulnerable species is human disturbance and loss of suitable nesting trees and foraging habitat as a result of timber harvesting.

Plate 1. Sitka spruce is a prominent component of the coniferous rain forest of western North America. This spruce, of unknown age, is found at Hangover Creek on the west coast of Graham Island, Queen Charlotte Islands, British Columbia. (Photograph by R.B. Smith)

Plate 2. Sitka spruce reaches its greatest abundance on the coastal fringe of the Very Wet Hypermaritime CWH Subzone (CWHvh), as in this area near Bonanza Creek and Rennell Sound, Graham Island, Queen Charlotte Islands, British Columbia. The high-lead yarding system used for harvesting this clearcut is commonly employed in coastal British Columbia. (Photograph by R.B. Smith)

Plate 3. Present-day conventional roadside forest harvesting (lower right) and experimental helicopter harvest area (centre) near Rennell Sound, Queen Charlotte Islands, British Columbia. The experimental clearcut in the centre, flanked on the left by patch clearcuts where trees were harvested from 50% of the area and on the right by 25% patch clearcuts, was designed to assess ways to reduce the mass-wasting effects of logging on slopes with high natural potential for erosion (see Figure 2). (Photograph courtesy of Image Library, Forestry Division Services Branch, Ministry of Forests, Victoria; slide 930246)

Plate 4. Deer browsing on this open landslide scar has resulted in extremely dense, dwarfed Sitka spruce on Burnaby Island, Queen Charlotte Islands, British Columbia. (Photograph by R.B. Smith)

Plate 5. Abnormal bluish colour of Sitka spruce foliage after single-tree fertilizer response trials, near branch logging road NW 130, in the Port McNeill Unit of Naka Tree Farm License 25, Western Forest Products Limited, near Port McNeill, British Columbia. (Photograph by R.B. Smith)

Plate 6. Epicormic branching is common in immature Sitka spruce.

60°
138°
0
100 miles
0
100 kilometres

Plate 7. Distribution of biogeoclimatic zones in British Columbia

Plate 8. 'Golden spruce' with unusually pale outer sun foliage, growing on the west side of the Yakoun River, about 5 km southwest of Port Clements, near Masset Inlet, Queen Charlotte Islands, British Columbia. This spruce, revered by the Haida people, was felled by a vandal in January 1997. (Photograph by B. Davies in 1962; courtesy of Image Library, Forestry Division Services Branch, Ministry of Forests, Victoria, slide M055)

Plate 9. Typical coexistence of large scattered Sitka spruce and more numerous, smaller western hemlock near Deena River, Moresby Island, Queen Charlotte Islands, British Columbia. (Photograph by R.B. Smith)

Plate 10. Typical young stage of succession dominated by red alder with Sitka spruce understorey on the bottom portion of an approximately 23-year-old landslide scar, near Sewell Inlet, Moresby Island, Queen Charlotte Islands, British Columbia. (Photograph by R.B. Smith)

Plate 11. Typical stage of declining red alder dominance in succession from alder to Sitka spruce. Note the boles of fallen alder in the foreground. The largest moss-laden trees in this stand are red alder with younger, smaller-diameter spruce occurring throughout the stand, which was found on the bottom portion of an approximately 87-year-old landslide scar near Riley Creek, Rennell Sound, Graham Island, Queen Charlotte Islands, British Columbia. (Photograph by R.B. Smith)

Plate 12. Sitka spruce has high flood resistance, comparable with that of western redcedar.

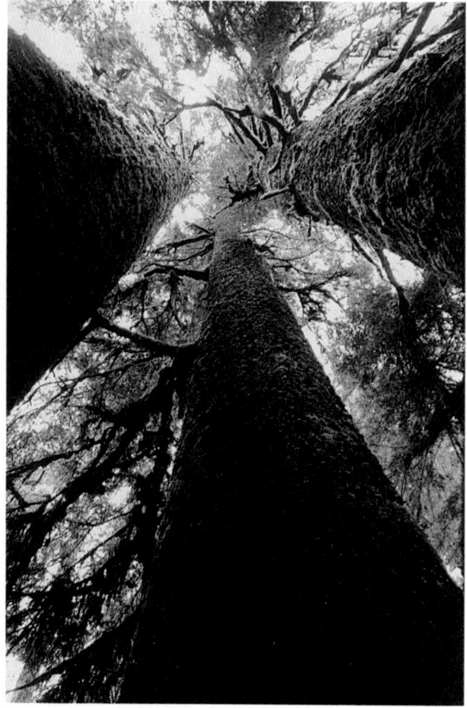

Plate 13. Recent research has shown the exceptional biological diversity of the upper canopy of coastal old-growth forests in British Columbia, as portrayed in this view of Sitka spruce from the forest floor, Carmanah Creek valley, British Columbia. (Photograph by A. Dorst)

3
Management of Sitka Spruce

Early in the history of its commercial use, the wood of Sitka spruce, with a high strength-to-weight ratio, provided an important basis for distinguishing it from its main coniferous companions – western hemlock, western redcedar, Douglas-fir, and grand fir. Yet from south coastal Oregon to southeastern Alaska, Sitka spruce has traditionally not been managed as a distinct species. For many decades it has simply been recognized as one component of the Western Hemlock–Sitka Spruce forest type (Ruth and Harris 1979; Harris 1990) that characterizes the outer coastal zone of western North America from latitude 39°N to 61°N.

This tradition is now changing. The development of guidelines for tree species selection and for stocking standards on a site-specific basis (Green et al. 1984; Silviculture Interpretations Working Group 1994), plus the current application of silvicultural steps specific to particular tree species on particular site series (Banner et al. 1993; Green and Klinka 1994) means that Sitka spruce can now be managed as a distinctive species that grows best on certain sites. This chapter describes today's concepts of Sitka spruce management in the context of the biogeoclimatic relationships outlined in the section 'Geographic Distribution and Ecological Characteristics of Ecosystems Supporting Sitka Spruce' (page 37) in Chapter 2.

Aside from the recent trend in British Columbia to base management and silvicultural decisions on the adaptability of a given tree species to a particular site series in the province's biogeoclimatic ecosystems classification system (MacKinnon et al. 1992; Britton et al. 1996), there are two reasons why Sitka spruce stands out from a management perspective:

- The white pine weevil (see 'Insects' on page 116 in Chapter 2) has had such a negative influence on the success of Sitka spruce plantations in British Columbia, except on the Queen Charlotte Islands, that the regeneration commitments given to other west coast conifers have had to be reassessed for Sitka spruce.

Figure 35. Sitka spruce, a species that is not very competitive in its natural environment, has proven to be an excellent plantation species beyond its natural range, as shown by this example from Northern Ireland.

- Sitka spruce has proven to be a very successful plantation species in the United Kingdom (Figure 35) and elsewhere in Europe. The enthusiasm for this spruce in European plantations has resulted in two things. First, the high yields achievable by this tree in Europe regularly rekindle interest in this species among forest managers who deal with it in its natural range; and second, much has been learned about this species from silviculturists who have successfully managed it outside its natural range. Sitka spruce management is clearly a part of the current interest in whether European silvicultural systems and precedents would be useful in British Columbia (Weetman 1996).

The natural forests in which Sitka spruce occurs are nearly all publicly owned, and are managed in very large sustained-yield units of hundreds of thousands of hectares, often with limited road access. Management of Sitka spruce in its natural range must focus on the role of this species in a very large forest area for which decisions must be made about levels of protection needed, harvesting alternatives (rate, timing, location, and size of harvest), and silvicultural steps needed for stand tending and stand renewal.

The present reality, which will persist for many decades, is that of a coastal British Columbia forest with very unbalanced age-class distributions. Large areas of very old timber are scheduled to be steadily harvested over many decades in an attempt to balance age-class distributions and to sustain timber yield and various non-timber values. Currently in British Columbia, apart from an attempt to regulate age classes, such objectives are not yet defined for Sitka spruce; there is no clear vision for Sitka spruce management. In recent years, however, British Columbia foresters have identified various stand management questions specific to Sitka spruce. For example, what site conditions on clearcuts are suitable for planting Sitka spruce? What steps would reduce or prevent damage of Sitka spruce leaders by white pine weevil? What soil, site, or vegetation competition factors are responsible for the frequently observed slowdown of Sitka spruce about a decade after planting?

These questions are part of the forest manager's concerns when deciding what must be done to achieve accepted standards of basic and incremental silviculture for west coast ecosystems that contain Sitka spruce. As defined by the British Columbia Ministry of Forests (1990b), 'basic silviculture' refers to the obligation of those who harvest old-growth forest stands to ensure successful regeneration. Aside from planning, surveys, and auditing, the four main field activities carried out in basic silviculture are site preparation, planting, brushing, and spacing. 'Incremental silviculture' is defined as those steps, beyond basic silvicultural activities, that are taken to improve growth rates and wood quality. At present, juvenile spacing, fertilization, pruning, and commercial thinning are the four main field activities in British Columbia's incremental silviculture program.

Activities that are classified under incremental silviculture are of particular interest to the administrators of the South Moresby Forest Replacement Account, which sponsored the preparation of this book. The intent of this program in the framework of forest estate planning is to achieve a sustainable increased allowable cut on the Queen Charlotte Islands through incremental silviculture, to offset the loss of harvestable forest land resulting from the creation of Gwaii Haanas/South Moresby National Park Reserve in 1988. This chapter summarizes current approaches to basic and incremental silviculture in British Columbia ecosystems where Sitka spruce is a conspicuous component.

Silvicultural Steps to Encourage Natural Regeneration

The silvicultural system best suited to the Western Hemlock–Sitka Spruce forest type and the Sitka Spruce forest type is clearcutting (Harris and Farr 1974; Ruth and Harris 1979; Harris 1990), followed by either natural or artificial regeneration. Both western hemlock and Sitka spruce are prolific seed producers, with small seeds that are widely dispersed over large areas. In some areas of southeastern Alaska, natural regeneration of hillside sites by spruce and hemlock is sometimes so dense that subsequent thinning is required. After clearcut logging, most areas are stocked with western hemlock advance regeneration, much of which dies within a few years to be replaced by seedlings that originate from windblown seed. Many of the latter are Sitka spruce (Harris and Farr 1974), so the proportion of this species in post-logging stands increases progressively during the stand's early years. In the southern part of Sitka spruce's range, organic seedbeds dry out faster than they do further north, and Sitka spruce seedling mortality is therefore relatively common in this part of the species' range.

Generally, competing vegetation is not a problem on sites that provide suitable germination media for Sitka spruce, except in some cases where control of red alder is required. Invasion of red alder is greatest where there is significant soil disturbance. Thus if a forest manager's goal is to reduce potential competition from alder, the approach should be to expose mineral soils as little as possible.

In mixed stands of western hemlock and Sitka spruce in southeastern Alaska, thinning had variable effects on conifer regeneration, depending mainly on stand age at the time of thinning. For example, thinning of dense, young (less than 30 years old) even-aged stands of hemlock and Sitka spruce allowed established understorey conifers to rapidly fill the available growing space, to the exclusion of new germinants. In contrast, thinning of older closed-canopy stands initiated germination of new coniferous regeneration, followed two to three years later by prolific establishment of hemlock regeneration, with 72-100% of all regeneration being hemlock (Deal and Farr 1994). The rest of the new regeneration was Sitka spruce. The amount of regeneration increased, and the percentage of hemlock regeneration decreased, with increasing thinning intensity.

For a manager interested in the broader aspects of understorey vegetation management, such as for wildlife habitat, the different regeneration responses to thinning have implications beyond the achievement of hemlock or spruce regeneration. In general, thinning of young hemlock-spruce stands on upland sites will benefit existing understorey conifers, which rapidly expand to occupy the available growing space. In contrast, heavy thinning in older stands promotes such dense germination of new understorey conifers (mainly hemlock) that it is difficult for other understorey plants to become established (Deal and Farr 1994). Before other understorey species can germinate and compete, they tend to be shaded out by dense hemlock regeneration. Even after thinning in younger stands, silviculturists may find it difficult to maintain a productive and diverse understorey for a long time because aggressive understorey conifers gradually outgrow and shade out other species.

In some areas, such as the Sewell Inlet area of the Queen Charlotte Islands, Sitka spruce regeneration on hillsides is hampered because of competition from grass (Sanders and Wilford 1986). As much as 75% of the area logged in the Sewell Inlet area is expected to be planted as a result (Bartlett 1977). To ensure a high spruce component, about 28% of the area logged on the Queen Charlotte Islands is now planted, particularly on brushy sites (Sanders and Wilford 1986).

Control of the Main Competing Species

Although the ecological characteristics of plant species that compete with coniferous regeneration in British Columbia are increasingly understood, little of the information is specific to Sitka spruce. For example, in the comprehensive compilation by Haeussler et al. (1990), red alder was the only competing species for which a relationship with Sitka spruce was specifically mentioned. A computerized bibliographic database (COMB) that summarizes the ecology, responses, and utilization of species that compete with conifers in British Columbia was developed at the Ministry of Forests Research Branch and the University of British Columbia (Comeau et al. 1990). The 23 species in this database as of October 1990 included several shrubs and herbs that occur in sites where Sitka spruce is a likely reforestation species. From the database, a forest manager can obtain ecological information for the following competitors of Sitka spruce: *Alnus rubra, Acer circinatum, Acer macrophyllum, Cornus sericea, Gaultheria shallon, Oplopanax horridus, Polystichum munitum, Rubus parviflorus, Rubus spectabilis,* and *Sambucus racemosa.* Additional recent information on the efficacy of forest vegetation management treatments currently used in British Columbia is available in Farnden (1992) and Biring et al. (1996).

The most practical way for a forester to approach the management of Sitka spruce's competitors is to assume that any prominent shrub or herb is a potential deterrent to spruce regeneration on fresh to wet sites and medium-rich to very rich sites, which is where Sitka spruce is most likely to be

Table 28

The main species that potentially compete with Sitka spruce regeneration in the Coastal Western Hemlock (CWH) and Coastal Douglas-Fir (CDF) zones

Sites	Vegetation type	Major competing species
Floodplains	Cottonwood–red alder	*Populus trichocarpa, Alnus rubra, Rubus spectabilis, Cornus sericea, Oplopanax horridus, Sambucus racemosa, Rubus parviflorus*
Moist or fresh to wet sites	Bigleaf maple–red alder–shrub	*Acer macrophyllum, Alnus rubra, Acer circinatum, Rubus parviflorus, Rubus spectabilis, Sambucus racemosa, Oplopanax horridus, Ribes* spp., *Polystichum munitum*
Fresh to wet sites	Salmonberry	*Rubus spectabilis, Rubus parviflorus, Acer circinatum, Ribes* spp.
Very dry sites	Salal	*Gaultheria shallon*

Source: Based on information from Conard (1984) as modified by Newton and Comeau (1990)

selected as a crop tree. This approach does not result in a large list. For example, the main species that potentially compete with Sitka spruce regeneration in the Coastal Western Hemlock (CWH) and Coastal Douglas-Fir (CDF) zones can be summarized as shown in Table 28.

This site stratification suggests that competition with spruce coming from some species will be localized, depending upon the local distribution of dry, fresh, moist, or wet site units. The manner in which competition occurs also varies by species. For example, densely sprouted bigleaf maple can produce substantial shade, and thick accumulations of fallen maple leaves can bury young Sitka spruce seedlings (Haeussler et al. 1990). In contrast, canopies of red alder allow considerable penetration of light, with the result that alder suppresses moderately shade-tolerant conifers such as Sitka spruce without actually causing shade-induced mortality.

Steps for Plantation Establishment

This section summarizes current practices for plantation establishment, drawing as much as possible from experience within the natural geographic range of Sitka spruce. Guidelines from experience with Sitka spruce plantations in other countries (Figure 35) are used if no information was available from western North America. The subsection 'An Overview of British Experience in Sitka Spruce Plantation Establishment' (page 170) is included because of over 50 years of experience in that country; it is instructive for those seeking to improve plantation procedures for this species in its natural range.

Stock Selection

Sitka spruce is now such a widely planted species internationally that genetic conservation techniques require both *in situ* (in the natural location) and *ex situ* (out of the natural location) efforts (Lester and Yanchuk 1996). As with other tree species, the quality of Sitka spruce seed influences reforestation success and the value of plantations (Woods et al. 1996). This point is recognized in the Silviculture Practices Regulations of the Forest Practices Code of British Columbia Act. These regulations require that available seed sources of the highest genetic quality be used to reforest Crown land in British Columbia. It is well established that the economic return from plantation forests increases with the genetic quality of the seed used. In British Columbia, this principle is the central goal in guidelines such as *Protocols for Rating Seed Orchard Seedlots in British Columbia* (Woods et al. 1996) and the *Seed and Vegetative Material Guidebook* (B.C. Ministry of Forests and B.C. Ministry of Environment, Lands and Parks 1995i). An obvious objective is to produce rapidly growing Sitka spruce, as shown in the Western Forest Products Limited example from northern Vancouver Island (Figure 36).

The potential increase in height growth in British Sitka spruce plantations as a result of planting genetically improved stock is shown in Figure 37.

Figure 36. Experimental Sitka spruce plantation established by Western Forest Products Limited in the Koprino River watershed, northern Vancouver Island, British Columbia. (Photograph by Western Forest Products Limited, Vancouver)

Analysis by Lee (1992) indicated that Sitka spruce yields in Britain could rise by the year 2000 to 25% above the yields of the control genetic sources from the Queen Charlotte Islands. This projected gain referred only to potential increases in annual rates of wood production by Sitka spruce; potential revenue gains from improved stem form and better wood quality were not documented by Lee (1992).

Silvical characteristics of Sitka spruce relevant to its use in regeneration were summarized by Klinka et al. (1990) as follows:

- frequent in the Coastal Western Hemlock Zone (wet mesothermal climate)
- infrequent in the Mountain Hemlock and Coastal Douglas-fir zones
- very frequent in moist sites

- frequent in fresh and wet sites
- infrequent in dry sites
- very frequent in very rich sites
- frequent in rich sites
- infrequent in medium sites
- low potential for natural regeneration in the shade
- high potential for natural regeneration in the open
- high spatial requirements
- intolerant to frost and snow
- tolerant of ocean spray, brackish water, and flooding.

In terms of Sitka spruce's potential for regenerating coastal forests in British Columbia, Klinka et al. (1990) identified the main hazards as aphids, weevils, frost, and wind. Sitka spruce should not be planted on moisture-deficient or nitrogen-deficient sites, whereas alluvial sites and sites that are winter-wet and summer-moist provide the greatest opportunity for assisted regeneration of this species. Many alluvial sites have severe brush competition, and Klinka et al. (1990) suggested that on such sites black cottonwood or red alder could be used as 'temporary nurse species,' to provide protection from frost and weevil attack, for example.

British Columbia has a very effective program of provenance testing of adaptive geographic variation. The purposes of this testing are to provide

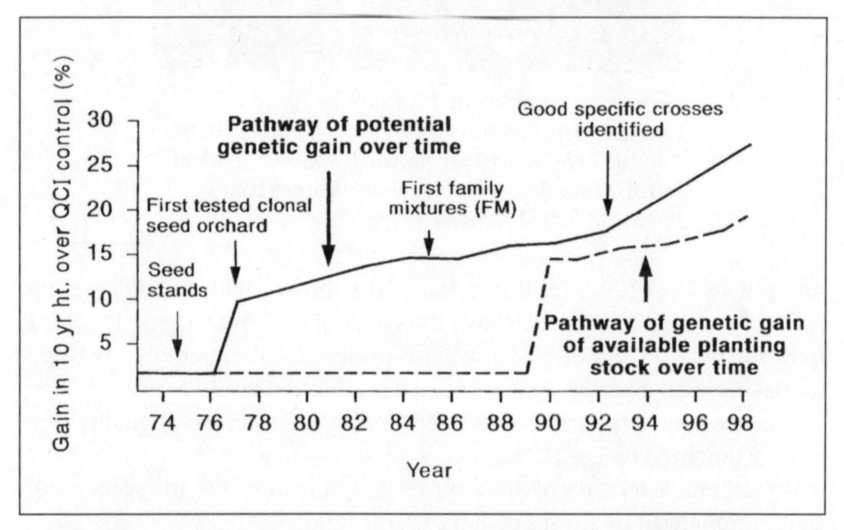

Figure 37. Projected increases in 10-year Sitka spruce height growth in Britain for genetically improved stock compared with growth in Queen Charlotte Islands controls, showing both the pathway for *potential* genetic gain and the pathway for gain possible from planting stock available in Britain as of the early 1990s (from Lee 1992).

guidelines for selection of seed sources, to identify geographic constraints between place of seed collection and place of seedling establishment, and to advise on ways to ensure plantation success. Sitka spruce is an important species in this provenance research (Ying 1990). Sitka spruce seed from British Columbia is also an integral part of work by the Canadian Forest Service, Victoria, to certify and register forest reproductive material moving in international trade (Portlock 1996a, 1996b).

Recent research by Ying (1990) indicates that Sitka spruce has a north-south pattern of differentiation in growth, winter-hardiness, lammas growth, and response to deer browsing. From north to south in British Columbia, there is a gradient of increasing growth potential. Higher growth is associated with decreasing winter-hardiness, high frequency of lammas growth, and quick recovery from deer browsing. This trend is most apparent in the wet and very wet hypermaritime variants of the Coastal Western Hemlock Zone (see Appendix 4 for a list of British Columbia's coastal biogeoclimatic zones). Based on results for 10 years, average volume increase was estimated at 1 m^3/ha per degree of latitude northward transfer. At sites free of the weevil problem, the most southern provenance, from Brookings, Oregon, grew twice the volume compared with the local source. Latitude alone explained about 90% of the among-provenance variation in growth, winter injury, and response to deer browsing. A few provenances exhibited a great tolerance to weevil attack, a subject still under investigation (Ying 1990).

Propagation and Nursery Practices

From the very beginning, Sitka spruce was involved in British Columbia's reforestation and forest nursery programs, which were initiated in 1927 when a small experimental forest nursery was established on Shelbourne Street on what was then the northern outskirts of Victoria. Sitka spruce and Douglas-fir were among the first 30 seedbeds planted in that nursery in 1928. These seedlings were the very first trees to be planted in British Columbia's first reforestation project, at the Green Timbers plantation in the Fraser River Valley near Vancouver in 1930 (Young 1989); many of them have since died.

Research by Chaisurisri et al. (1992) confirmed that stratification is essential for achieving germination uniformity in Sitka spruce, and that seed size has little operational importance. In the United Kingdom the usual way to overcome Sitka spruce seed dormancy is to use a moist pre-chill at 5°C (Jones et al. 1993). In Sweden seeds are invigorated at 15°C using a controlled moisture content. When combinations of the British pre-chill and the Swedish invigoration methods were tested, the best treatment combined the longest pre-chill and longest invigoration times studied (21 days at 5°C followed by 11 days at 15°C). This achieved a germination capacity of 92% and a mean germination time of 7.9 days. Moist seeds were also redried to storage moisture contents (5-7%) after treatment. Seeds were not harmed by redrying and the trends were similar, although the best treatment after redrying was 14 days at 5°C followed by 11 days at 15°C, which gave 89% germination and a mean germination time of 9.2 days.

Incubation at 5° or 15°C were both effective at overcoming dormancy in Sitka spruce, but the similar end results may be achieved via different metabolic routes. There are genetic differences in Sitka spruce seed dormancy and various other germination parameters (Chaisurisri and El-Kassaby 1993a, 1993b). El-Kassaby et al. (1993) reported significant differences among individual Sitka spruce seed parents with regard to three germination parameters (germination capacity, peak value, and germination value). This suggests that the common practice of harvesting cone crops from seed orchards, extracting the seed, and sowing them on a bulk basis needs to be re-evaluated.

In wind-pollinated Sitka spruce seeds from 18 seed orchard clones, Chaisurisri et al. (1994a) could not detect any statistically significant effects of seed size on seedling size at 8 months. This differs from white spruce, where Burgar (1964) confirmed a positive relationship between seed size and seedling size. In Sitka spruce, where the effects of clones on seedling attributes can be greater than the effects of seed size, the use of unsorted seeds is recommended for seedling production.

Recently the production of Sitka spruce stock has included the use of seedlings produced in styroblock containers. In 1987 the total British Columbia production of pure Sitka spruce was 2,939,000 seedlings, made up of 2,017,000 bareroot seedlings (2+0) and 922,000 containerized seedlings (1+0). In the same year, in addition to the pure stock, the Prince Rupert Forest Region produced 3,920,000 hybrid seedlings (Sitka spruce × white spruce) from the Terrace region, all in containers in nurseries in the Thornhill and Telkwa areas. To date, the main problem encountered in production of container stock has been difficulty in hardening off the stock.

Cuttings from young Sitka spruce trees will root relatively easily (Roulund 1971). Working with clones from 2- and 4-year old material, Van den Driessche (1983) found that cuttings from Sitka spruce that had been closely cropped at 1 m height (hedged) were able to root more freely than cuttings from the lower crown. Rooting occurred most readily when cuttings were taken from early January to early February. The concentrations of sugars in the stem and foliage showed little correlation with rooting. Dormant Sitka spruce cuttings require about 10 weeks of chilling at 2°C to obtain the most rapid and complete rooting (van den Driessche 1983).

In mainland and Vancouver Island nurseries, grey mould caused by the fungus *Botrytis cinerea* is a major disease in container-grown seedlings. It is worst on western hemlock and Douglas-fir seedlings but can also affect other species grown in nurseries, including Sitka spruce. Fungicide drenches were not recommended by Sutherland and Lock (1977) for spruce seedlings in coastal British Columbia nurseries because they failed to provide protection against damping-off. Control of grey mould by fungicides is difficult because *Botrytis* becomes tolerant to fungicides when there is long continued use of the same fungicides (Glover et al. 1987). For example, Glover and co-workers cited unpublished data indicating that 74% of the *Botrytis* isolates from Ministry of Forests and private container nurseries in British Columbia were tolerant to benomyl and 60% were tolerant to chlorothalonil. There is therefore a need for a wider variety of fungicides for use against *Botrytis* in this province.

In order to address this need, Sitka spruce was among the species tested for fungicidal control of grey mould. Anilazine, captan, dichloran, folpet, iprodione, and thiram were tested; all are currently registered in Canada for use on agricultural crops. Iprodione provided excellent *Botrytis* control in seedlings of Sitka spruce and other species; folpet provided good control in Sitka spruce; and anilazine provided adequate control in Sitka spruce, although it caused some needle browning. Glover et al. (1987) presented a composite ranking of fungicide efficiency and severity for the six tested herbicides, with 1 signifying the best control of *Botrytis* and 6 the poorest. The overall rankings for Sitka spruce were as follows: iprodione, 1; anilazine, 2; folpet, 3; dichloran, 4; thiram, 5; and captan, 6.

There was evidence of phytotoxic effects from some fungicides when they were applied to Sitka spruce container seedlings. The heights and diameters of these seedlings were significantly less than in untreated controls when iprodione, folpet, and anilazene were applied; spruce seedling heights, but not diameters, were significantly less when thiram, dichloran, and captan were applied. These results emphasize the fact that when selecting a fungicide, it is important to consider phytotoxic effects as well as the degree of control provided. For all coastal container-grown species tested by Glover et al. (1987), iprodione was recommended as the overall best fungicide for *Botrytis* control in terms of efficacy, phytotoxicity, and usefulness in pursuing registration for use in nurseries.

In tests of nursery stock performance at four planting sites in southeastern Alaska, Zasada et al. (1990) compared height growth over three years for seedlings produced in two different nurseries, one in Petersburg, Alaska, and one in Coeur d'Alene, Idaho. Although initial seedling height was greater for the Alaska-produced seedlings, third-year height growth of the Idaho-produced seedlings was equal to or greater than that of those produced in Alaska. Zasada and co-workers concluded that seedlings intended for outplanting in southeastern Alaska can be grown in nurseries further south in the Pacific Northwest if appropriate Alaskan seed sources are used.

For the interested reader, substantial information is available on root form and production of adventitious roots in nursery-produced Sitka spruce

Table 29

Site units, using the Vancouver Forest Region as an example, in which Sitka spruce has either a minor or major potential role in reforestation

	Subzone or variant[a]	Edatopic indicator group	Hygrotope	Trophotope	Site quality	Potential reforestation role of Sitka spruce
1	Windward Moist Maritime MH Variant (MHmm1)	*Vaccinium alaskaense* *Vaccinium ovalifolium* *Polystichum lonchitis* *Streptopus roseus*	Fresh	Rich	Poor-medium	An acceptable crop species at lower elevations in northern part of MHmm1 Variant in Port Hardy and Bella Coola forest districts
2	As above	*Rubus pedatus* *Tiarella unifoliata* *Streptopus roseus* *Veratrum viride*	Moist	Rich	Medium	As above
3	Wet Warm IDF Subzone (IDFww)	*Rhytidiadelphus triquetrus* *Rubus parviflorus* *Gymnocarpium dryopteris* *Athyrium filix-femina*	Fresh	Rich	Medium-good	An acceptable minor species
4	As above	As above but on alluvial floodplains	Moist	Rich	Good	Mixtures of western redcedar and Sitka spruce may be used as major species
5	As above	*Athyrium filix-femina* *Equisetum* spp.	Wet	Rich	Poor-medium	An acceptable minor species on raised mounds

►

◀ *Table 29*

	Subzone or variant[a]	Edatopic indicator group	Hygrotope	Trophotope	Site quality	Potential reforestation role of Sitka spruce
6	Very Dry Maritime CWH Subzone (CWHxm)	*Athyrium filix-femina* *Lysichiton americanum*	Wet	Rich	Poor-medium	An acceptable minor species
7	Dry Maritime CWH Subzone (CWHdm)	As above	Wet	Rich	Poor-medium	As above
8	Submontane Very Wet Maritime CWH Variant (CWHvm1)	*Vaccinium parvifolium* *Gaultheria shallon* *Rhytidiadelphus loreus* *Kindbergia oregana*	Fresh	Poor-medium	Poor	An acceptable minor species on nutrient-medium sites
9	As above	*Vaccinium parvifolium* *Hylocomium splendens* *Vaccinium alaskaense* *Blechnum spicant*	Fresh	Poor-medium	Medium-good	As above
10	As above	*Dryopteris expansa* *Kindbergia oregana* *Moneses uniflora* *Polystichum munitum*	Fresh	Rich	Medium-good	Sitka spruce is an acceptable alternative to western hemlock as a major species
11	As above	*Hylocomium splendens* *Rhytidiadelphus loreus* *Blechnum spicant* *Sphagnum girgensohnii*	Moist	Poor-medium	Medium-good	An acceptable minor species on nutrient-medium sites

▶

◀ *Table 29*

	Subzone or variant[a]	Edatopic indicator group	Hygrotope	Trophotope	Site quality	Potential reforestation role of Sitka spruce
12	As above	*Dryopteris expansa* *Polystichum munitum* *Tiarella trifoliata* *Rubus spectabilis*	Moist	Rich	Good	An acceptable major species
13	As above	*Coptis aspleniifolia* *Maianthemum dilatatum* *Lysichiton americanum*	Wet	Rich	Poor-medium	As above
14	Montane Very Wet Maritime CWH Variant (CWHvm2)	*Rhytidiopsis robusta* *Rhytidiadelphus loreus* *Kindbergia oregana* *Vaccinium ovalifolium*	Fresh	Poor-medium	Poor	An acceptable minor species on nutrient-medium sites
15	As above	*Rhytidiopsis robusta* *Vaccinium alaskaense* *Blechnum spicant* *Moneses uniflora*	Fresh	Poor-medium	Medium	As above
16	As above	*Dryopteris expansa* *Vaccinium ovalifolium* *Polystichum munitum* *Streptopus roseus*	Fresh	Rich	Medium-good	Sitka spruce and western hemlock can be alternatives to western redcedar and amabilis fir as major species

▶

◀ *Table 29*

	Subzone or variant[a]	Edatopic indicator group	Hygrotope	Trophotope	Site quality	Potential reforestation role of Sitka spruce
17	As above	*Dryopteris expansa* *Blechnum spicant* *Sphagnum girgensohnii* *Tiarella unifoliata*	Moist	Poor-medium	Medium	An acceptable minor species on nutrient-medium sites
18	As above	*Dryopteris expansa* *Tiarella unifoliata* *Streptopus roseus* *Rubus spectabilis*	Moist	Rich	Medium-good	An acceptable minor species or an alternative to amabilis fir
19	As above	*Coptis aspleniifolia* *Lysichiton americanum*	Wet	Rich	Poor	An acceptable minor species
20	Submontane Moist Maritime CWH Variant (CWHmm1)	*Athyrium filix-femina* *Lysichiton americanum*	Wet	Rich	Poor-medium	As above
21	Montane Moist Maritime CWH Variant (CWHmm2)	*Coptis aspleniifolia* *Lysichiton americanum*	Wet	Rich	Poor	Sitka spruce is an alternative to western redcedar as a minor species on this site unit
22	Central Moist Submaritime CWH Variant (CWHms2)	*Hylocomium splendens* *Vaccinium ovalifolium* *Polystichum munitum* *Gymnocarpium dryopteris*	Fresh	Rich	Medium-good	An acceptable minor species

▶

◀ *Table 29*

	Subzone or variant[a]	Edatopic indicator group	Hygrotope	Trophotope	Site quality	Potential reforestation role of Sitka spruce
23	As above	*Dryopteris expansa* *Polystichum munitum* *Gymnocarpium dryopteris* *Oplopanax horridus*	Moist	Rich	Good	As above
24	As above	*Athyrium filix-femina* *Lysichiton americanum*	Wet	Rich	Poor-medium	As above
25	Central Dry Submaritime CWH Variant (CWHds2) *Acer glabrum*	*Hylocomium splendens* *Rhytidiadelphus triquetrus* *Amelanchier alnifolia*	Dry	Rich	Poor-medium	Sitka spruce is an alternative to western redcedar as a minor species on this site unit
26	As above	*Dryopteris expansa* *Rubus parviflorus* *Gymnocarpium dryopteris* *Athyrium filix-femina*	Moist	Rich	Good	Sitka spruce is an alternative to Douglas-fir or western redcedar as a major species on this site unit
27	As above	*Athyrium filix-femina* *Lysichiton americanum*	Wet	Rich	Poor-medium	An acceptable minor species
28	Southern Very Wet Hypermaritime CWH Variant (CWHvh1)	*Vaccinium parvifolium* *Gaultheria shallon* *Rhytidiadelphus loreus* *Kindbergia oregana*	Fresh	Poor-medium	Low-poor	Sitka spruce is an alternative to western redcedar as a major species on nutrient-medium sites

▶

◄ *Table 29*

	Subzone or variant[a]	Edatopic indicator group	Hygrotope	Trophotope	Site quality	Potential reforestation role of Sitka spruce
29	As above	*Vaccinium parvifolium* *Gaultheria shallon* *Rhytidiadelphus loreus* *Blechnum spicant*	Fresh	Poor	Poor-medium	As above
30	As above	*Dryopteris expansa* *Kindbergia oregana* *Tiarella trifoliata* *Athyrium filix-femina*	Fresh	Rich	Medium-good	As above
31	As above	*Polystichum munitum* *Tiarella trifoliata* *Rubus spectabilis* *Lysichiton americanum*	Moist-wet	Rich	Good	As above
32	As above	*Coptis aspleniifolia* *Athyrium filix-femina* *Lysichiton americanum*	Wet	Rich	Poor-medium	As above

Source: Green et al. (1984). Names of biogeoclimatic subzones and variants used by Green et al. (1984) have been amended to agree with nomenclature in Green and Klinka (1994).

[a] See Appendix 4 for identity of biogeoclimatic zones, subzones, and variants.

seedlings, mostly from research in Britain (Deans et al. 1989; Coutts et al. 1990; McKay 1994a, 1994b; Townend and Dickinson 1995; and McKay and Howes 1996).

Site Diagnosis for Selection of Sitka Spruce as a Reforestation Species

The selection of appropriate species for forest renewal should be based on a match between ecologically viable species options and management objectives, using the selection criteria of maximum productivity, crop reliability, and silvicultural feasibility (Green et al. 1984). These criteria are based upon the observation and study of crop tree species in natural forests. To make this match, a forester must carry out a site diagnosis that, in British Columbia today, is usually based on a site-specific assessment of climatope, hygrotope, and trophotope within the province's system of biogeoclimatic zones, subzones, and variants (Meidinger and Pojar 1991; MacKinnon et al. 1992). In this approach to site diagnosis, *climatope* refers to areas of land that are under the influence of similar regional climates, as reflected by the distribution of similar climax or near-climax vegetation on zonal sites that are intermediate in slope positions, moisture regime, and nutrient regime. *Hygrotope* refers to the capacity of a soil to supply available water for plant growth. *Trophotope* is the capacity of a soil to supply nutrients for plant growth (Krajina et al. 1982; Green et al. 1984).

Using the Vancouver Forest Region as an example, the site diagnosis criteria used by Green et al. (1984) indicated that in this region there are 32 different site units in which Sitka spruce has reforestation potential, 16 in which Sitka spruce can be considered as a major potential reforestation species, and 16 where it could be of minor importance (Table 29). Within the five biogeoclimatic zones of the Vancouver Forest Region, the numbers of site units where Sitka spruce can be considered for reforestation, relative to the total number of site units described by Green et al. (1984), are summarized in Table 30. These judgments refer simply to site suitability for Sitka spruce and do not incorporate possible regional differences in white pine weevil hazard. The estimates are for Sitka spruce, excluding hybrids between Sitka spruce and interior spruces.

As shown in Table 29, virtually all ecosystems where Sitka spruce can be considered as a major or minor reforestation species occur in hygrotopes that are wet, moist, or fresh, and in trophotopes that are medium, rich, or very rich. Within the Vancouver Forest Region, most sites suitable for Sitka spruce as a reforestation species are within 5 of the 11 biogeoclimatic variants listed in Table 29: CWHvm1, CWHvm2, CWHms2, CWHds2, and CWHvh1.

Site Preparation

Site preparation is normally not required to achieve natural regeneration of

Table 30

Numbers of site units where Sitka spruce can be considered for reforestation within the Vancouver Forest Region

Biogeoclimatic zone	No. of site units where Sitka spruce can be considered for reforestation: Major	Minor	Total site units in zone	% of units possibly suitable for Sitka spruce
Mountain Hemlock (MH) Zone	2	0	16	12.5
Engelmann Spruce–Subalpine Fir (ESSF) Zone	0	0	9	0.0
Interior Douglas-fir (IDF) Zone	1	2	18	16.7
Coastal Douglas-fir (CDF) Zone	0	0	9	0.0
Coastal Western Hemlock (CWH) Zone	13	14	103	26.2
Entire Vancouver Forest Region	16	16	155	20.6

Source: Green et al. (1984)

Sitka spruce. Most of the literature to date suggests that western hemlock and Sitka spruce do not depend upon fire or other site preparation for their perpetuation (Ruth and Harris 1979). However, some investigators have noted that the regeneration and early growth response of Sitka spruce to slashburning in the CWH zone is generally beneficial (Feller 1982). The suggestion that slashburning is favourable for Sitka spruce regeneration concurs with observations by others (Embry 1963; Ruth 1964; Harris 1966; Krajina 1969; Phelps 1973). One reason to be cautious about site preparation for Sitka spruce regeneration is that soil disturbance encourages competition from alder, except on alluvial sites, where spruce planting can be more successful if preceded by site preparation.

Alluvial ecosystems are prime Sitka spruce sites, with high timber values, but they are also characterized by the most severe competition from other vegetation. Natural seedlings are usually insufficient to stock these sites, as seedlings can become established only on elevated microsites and on debris that is not subjected to flooding. In such ecosystems, seeding programs would normally be restricted to special situations, such as large burns where natural seed sources are insufficient. Generally, the inability to use rodent control chemicals to minimize seed losses, the high cost of seed, and the continuing development of better planting stock all favour planting instead of seeding for regeneration of Sitka spruce.

On a site dominated by devil's club, lady fern, and oak fern in the Skeena River valley, Coates et al. (1993) tested the effects of grass and legume seeding as a way to control vegetation that compete with planted Sitka spruce. Growth of Sitka spruce was best in unseeded control areas and in treatments where legumes had been seeded, but was poor where sod-forming grasses were tried. These grasses decreased the growth in both height and diameter of Sitka spruce regeneration. By decreasing red alder density, legume seeding will at minimum increase the length of time before spruce will need release from the overtopping alder.

Broadcast burning is practised for fire hazard abatement and to improve access to harvested sites for machinery and field crews involved in post-harvest silviculture. Broadcast burning also gives the manager some control over species choice, since burning favours Sitka spruce over western hemlock (Ruth and Harris 1979). In areas where serious weevil attacks occur, burning provides the option of converting to a different species, such as Douglas-fir. Broadcast burning also encourages action to achieve prompt regeneration, thereby eliminating the delay typically involved with natural regeneration.

Planting a burnt site also allows genetically superior stock to be used. Some broadcast burning is now being carried out on the Queen Charlotte Islands (Sanders and Wilford 1986). The greatest need for broadcast burning as a vegetation control method is on alluvial sites, where competing vegetation sometimes has to be desiccated by chemical treatment before burning. The fire-adaptive traits and fire-sensitivity ratings of Sitka spruce's key shrub and herb competitors are provided in the synopsis by Haeussler (1991). Wider-ranging guidelines for site preparation, applicable but not specific to Sitka spruce, are described in Chapters 11-13 of *Regenerating British Columbia's Forests* (Lavender et al. 1990), and in the current *Site Preparation Guidebook* (B.C. Ministry of Forests and B.C. Ministry of Environment, Lands and Parks 1995j).

Where Sitka spruce managers are involved with broadcast burning as part of site preparation, it is important to consider the potential for increased incidence of Rhizina root rot after burning (Callan 1993). Significant losses of spruce seedlings to *Rhizina* infection have been recorded in plantations in the Prince Rupert Forest Region. The incidence of *Rhizina* shows its greatest increase soon after a burn, and seedling mortality is most severe in the first year after a burn, declining to negligible levels by the third year (Baranyay 1972). Therefore in wetter regions, where *Rhizina* reproduces more abundantly than in dry interior regions, one possible control measure is to delay planting spruce for at least two years. Surveys for fruiting bodies from 10 to 16 months after a burn can indicate whether further delay is warranted. A disadvantage of such planting delays is increased weed competition. When planting commences, avoiding sites adjacent to stumps and large pieces of burnt wood may prevent the further spread of this fungus. No chemical or biological controls for *Rhizina* are known at present (Callan 1993). In Europe one control measure is to dig trenches (0.3 m deep and 0.3 m wide) around small burnt areas to reduce the radial spread of *Rhizina*. In western Britain the most prominent damage to groups of Sitka spruce occurs around spots where thinning debris has been burned.

Planting Requirements, Spacing, and Early Care of Plantations

As pointed out by Pollack et al. (1990), density control has long been recognized as an important tool for foresters who want to manipulate the physical characteristics of stands and increase crop tree size at harvest. For Sitka spruce, density control is the single most powerful tool that foresters can use on the Queen Charlotte Islands and north coast to:

- cover the age-class gap problem in wood supply (due to unbalanced, old-growth–dominated age-class structures of the whole forest) by accelerated operability of second-growth stands
- control wood quality at the juvenile stage of stand development
- provide, by using pre-commercial thinning, opportunities to do commercial thinning and thus potential recovery of mortality by harvesting
- provide pruning and fertilization opportunities
- maintain live crown ratios and individual tree vigour
- build resistance to wind, snow, and ice breakage
- remove competing trees and vegetation.

To date, nearly all research on stocking control in Sitka spruce has focused on thinning studies in which the stocking levels of existing plantations were reduced. These studies, all conducted in the British Isles (Brazier 1970; Jack 1971; Lynch 1980; Hamilton 1981; Kilpatrick et al. 1981; Lynch 1988; Rollinson 1988), demonstrate that early spacing control of Sitka spruce concentrates stand growth on the remaining crop trees. Most of these studies relied on stands that were originally planted at higher densities (2,500-3,000 stems per hectare) than are commonly used in North America, and thinnings were used to reduce stocking levels at 12-30 years of age.

In contrast, the project by Pollack et al. (1990) at Kitsumkalum Lake, 32 km northwest of Terrace, British Columbia, examined the effects of initial differences among plantations that remained unthinned for 27 years, and whose densities fell within the existing plantation targets of 1,000-1,100 stems per hectare now used in British Columbia. Such stands are typical of the less intensively managed forests common in North America. In this study, planting densities ranged from 478 to 2,990 stems per hectare. The study confirmed that wider spacing produced trees with larger stem diameters, larger crowns, and larger branches after 27 years. Total volume per hectare was greatest in the closest spacing when all trees were considered, although this relationship was reversed when only the largest 250 stems per hectare were considered. The results suggest that there is still flexibility in choosing an optimum stocking level, and a range from 800 to 1,400 stems per hectare is recommended. Criteria for assessing plantation performance in the Prince Rupert Forest Region had been published earlier by Pollack et al. (1985), including useful diagrams of total height versus age, and current annual height increment versus age, for up to 12 years following planting of Sitka spruce and hybrid spruce in the CWHws and ICHmc subzones.

The comprehensive review of responses to fertilization of planted seedlings carried out by Brockley (1988) focused mainly on spruce seedlings in the interior of British Columbia. He suggested that many of the findings may apply equally well to other coniferous species. The degree to which this is true for planted Sitka spruce seedlings on the coast is not known. However, the following findings of Brockley (1988) may be applicable to Sitka spruce:

- Planted seedlings show wide inconsistencies in their response to fertilization; seedling response potential is higher when fertilization is combined with site preparation and control of competing vegetation.
- The removal of surface organic matter and topsoil by site preparation may increase response to fertilization because lost nutrients are being replaced, soil surfaces are warmed, and competing vegetation is reduced.
- To reduce the negative effects of competing vegetation, seedling fertilization should be contemplated only on sites that have been recently prepared.
- The largest and most consistent responses to fertilization at planting have been obtained with nursery stock, as opposed to naturally regenerated seedlings.
- Initial seedling size has little effect on fertilization response potential, provided that planting stock is in good physiological condition.
- Application of soluble inorganic fertilizers at the time of planting generally reduces survival and does not improve the early growth performance of seedlings.
- Although commonly used slow-release fertilizers contain N, P, and K, there is little evidence that seedlings respond to nutrients other than N.
- Of the methods tested to date, broadcast application is generally the most effective method of applying fertilizer to seedlings at the time of planting.

On alluvial sites, brush competition is severe. Even after successful site preparation and planting on such sites, brush control is often required, either chemically or mechanically. The use of herbicides for vegetation control is more controversial on alluvial sites than on upland sites because of their potential ecological effects on adjacent bodies of water and their influences upon shrubs and herbs that are important as grizzly bear forage. The choice of herbicides and application methods must therefore be carefully evaluated. This involves trade-offs between vegetation management effort and desired regeneration densities of crop trees. These trade-offs are leading to constraints on herbicide use on alluvial sites, and there are new questions about optimal natural regeneration and planting densities on such sites.

After aerial application of glyphosate in 1984 in the Carnation Creek watershed on western Vancouver Island, conifer tolerance to the herbicide treatments and crop growth responses were evaluated in 1985, 1986, and 1987 (Reynolds et al. 1989). There was evidence of some die-back or death of the primary leaders in western hemlock and western redcedar, but not in Sitka spruce, amabilis fir, or Douglas-fir. In both Sitka spruce and western hemlock, root-collar diameter and diameter increment increased significantly. Competing vegetation was reduced for one growing season, the result mainly of salmonberry reduction after one post-spray growing season. Control of red alder was variable; trees ranged from completely healthy to dead. Salal and evergreen huckleberry were not controlled by glyphosate. Control of most shrubby species declined in 1986 and continued to decline in 1987. This is not surprising because maximum herbicide efficacy is attained when brush competition is young and not well established.

Reynolds et al. (1989) recommended that, for highly productive coastal sites, herbicides should be applied within one year of logging; if treatment is delayed, the root biomass of competing vegetation increases and becomes so well established that it is difficult to kill. Under such circumstances, it is typical for herbicidal control to appear high for one year and then to decline. This was the case at Carnation Creek, where logging ended in 1981, three years before herbicide treatment. In view of the advanced development of post-harvest brush, glyphosate provided better silvicultural results than expected during the first year after application.

Whether dealing with Sitka spruce or other conifers planted in British Columbia, current planting regulations and practices require the planter to scrape through the organic forest floor to mineral soil. The requirement that seedling roots be firmly planted in the mineral soil is starting to be questioned (Balisky et al. 1993). These researchers described cases where Sitka spruce seedlings planted in rotten wood in coastal Oregon grew significantly more in height than those planted in mineral soil. In Alaska, Sitka spruce seedlings do best when planted in undisturbed duff, and seedlings planted in exposed mineral soil experience considerable frost heaving. Naturally established seedlings root almost exclusively in the interface between mineral and organic layers, indicating that mixtures of humus and mineral soil may be a desirable germination medium for Sitka spruce.

An Overview of British Experience in Sitka Spruce Plantation Establishment

A review of spruce silviculture in western Scotland (Davies 1967) contained a summary of preferred practices for Sitka spruce plantation establishment. This experience, accumulated over half a century, is outlined here as a general guide for those involved with plantations of this species in its natural range. An important difference is that the procedures used in Scotland place considerable emphasis on improvement and maintenance of drainage, a problem that is generally of little concern in plantation sites for Sitka spruce in its natural range, except possibly for some lowland sites on the Queen Charlotte Islands. In addition to the practices recommended by Davies (1967), some recent analyses of British Sitka spruce plantations are described here. Published information on European Sitka spruce management is exceptionally abundant, and this subsection highlights only those features of plantation establishment that appear to be ecologically transportable back to the natural range of this species.

Sitka spruce is used on a wide range of sites in areas of western Scotland that have over 890 mm of precipitation per year. It is not used in sites with pseudo-fibrous peats, *Trichophorum/Calluna* knolls, sphagnum-dominated peats, or frost hollows. Lodgepole pine and Sitka spruce mixtures are used to a very limited extent. The worst sites are designated for planting of lodgepole pine, down to an area of 0.10 ha, and the remaining sites are planted with Sitka spruce, instead of entire areas being planted with 50/50 mixtures of pine and spruce. To break the monotony of conspicuous blocks of spruce, 10% of the area is sometimes planted with Japanese and hybrid larches. On infertile and very exposed sites, Sitka spruce is planted at 1.7 m × 1.7 m; on normal sites at 1.8 × 1.8 m, and on rich, sheltered sites at 2.1 × 2.1 m (Davies 1967).

In the establishment of Sitka spruce plantations in Britain, as much ground preparation as possible is done by machine. For example, out of 4,000 ha planted in 1966-67, about 3,035 ha were plowed. The ideal is to cultivate bare land uphill and downhill with a double-mouldboard plow, and then plow in drains as required by the soil type and terrain. These drains are deepened by hand to 75-90 cm if possible. Strong emphasis is placed on drain layout, and many of the receiving drains are levelled by instrument and pegged by supervisors before work starts. On steep slopes that are covered by bracken fern, direct planting without plowing is carried out. About 55 g of ground mineral phosphate per plant is applied at the time of planting on all sites except for grass/*Juncus,* bracken, and hardwood scrub areas. Slow-growing areas where lack of phosphate is obvious are broadcast-treated with 65 kg ground mineral phosphate per hectare. Spruce areas previously treated with phosphate that are growing poorly often suffer from potash deficiency. If this is so, potash is broadcast in the first half of the growing season. These techniques have resulted in a remarkable improvement in rates of Sitka spruce growth (Davies 1967).

Work in Scotland (Watson and Cameron 1995) evaluated the influence of nurse tree mixtures on stem form, branching habit, and compression wood content of Sitka spruce. Comparisons were made between Sitka spruce in the centre and on the edge of plantation blocks next to Japanese larch, lodgepole pine, and Scots pine. Japanese larch and Scots pine were associated with wider annual rings and greater branch area in the neighbouring spruce when compared with spruce trees next to lodgepole pine or with control spruce. Spruce trees adjacent to blocks of nurse trees had deeper live crowns compared with spruce in control areas.

It is predicted that Sitka spruce will remain the most important plantation species in western Scotland, and Sitka spruce is also considered to be the species most likely to repay intensive silviculture. There are several reasons for this. It withstands exposed sites better than any other species tried to date. It is very easy to handle as a transplant, and is cheap to produce. So far, Sitka spruce in Scotland has been almost pest-free. *Heterobasidion annosum* has caused some damage, but stump treatments with creosote or sodium nitrate have helped reduce the spread of this fungus. Frost has caused severe losses of Sitka spruce planted in hollows, but this can be prevented by avoiding hollows during planting. Sitka spruce is also popular because it regenerates freely in western Scotland. There are some excellent stands that have regenerated naturally. Given proper treatment, Sitka spruce grows well on a wide variety of infertile sites (Figure 35) where the only other prospect is lower-yielding lodgepole pine. Most important of all, Sitka spruce has proven to be a producer of good-quality saw timber, and pulp (Davies 1967) and wood-based panelboards (Thompson 1992).

In its natural environment, Sitka spruce is not very competitive with its companion tree species, has relatively high nutrient requirements, and occurs on a relatively limited range of site series. Yet this species has been remarkably successful as an introduced crop tree in windswept and waterlogged uplands, such as the Kielder Forest in Britain, even where there are difficult soil conditions involving surface-water gleys and peaty gleys (Mason and Quine 1995). Sitka spruce's tenacity in such relatively old human-made forests has allowed it to span a period of very marked changes in forest management objectives.

In the late 1920s, the original, rather narrow, objectives of the state's involvement in forestry were to provide a strategic reserve against the possibility of future wars and a source of employment in rural areas. These were gradually replaced by a broader vision of the benefits that forests could provide, as well as a greater awareness of the need for a balance between efficient lumber production and the impact of forests on landscape, wildlife conservation, and recreational opportunities. This debate continues, and has had a profound effect on the ways that areas such as Kielder Forest are being managed as they move from even-aged plantations towards mature forests (McIntosh 1995).

Kielder Forest is an example of a large human-made Sitka spruce forest; its 42,500 ha of plantations were planted with Sitka spruce (72%) and Norway spruce (12%) beginning in 1926. The intensive ecological studies carried out here make it an important site for assessing the effects of modern forest management. It is now recognized that carefully planned clearcutting within an even-aged human-made forest can provide several opportunities, such as: (1) diversifying forest age structure by varying the size of clearcut patches and the amount of separation in space and time between each patch; (2) relocating or increasing the amount of open space within forests; and (3) establishing tree species not included at the afforestation phase.

This approach is in stark contrast to the alternative of clearcutting the forest in the same sequence in which it was planted, treating Sitka spruce as though it were an agricultural crop (Petty et al. 1995). This largest and oldest of Britain's state-owned forests shows that it is possible to significantly increase structural diversity and multi-purpose uses of managed Sitka spruce plantations (McIntosh 1995; Mason and Quine 1995). British data suggest that pure Sitka spruce plantations close canopy when trees are about 10 m tall at a density of 1,000 stems per hectare, and about 5 m tall at a planting density of 2,000 stems per hectare. On this basis, British site index curves for Sitka spruce suggest that crown closure will take about 10 years on the best sites.

From studies in Britain, it is known that higher soil temperatures dramatically increase Sitka spruce root growth between 4° and 25°C (Tabbush 1986; Coutts and Philipson 1987). Mounds 20-50 cm high provided warmer conditions that favour the establishment of Sitka spruce. The average first-year height increment improved as the mound size increased and with carefully handled plants. Rough handling of plants resulted in significantly less first-year growth. Tabbush (1987a, 1987b) found that plant-handling systems that involve *brief* periods of root exposure are tolerated by Sitka spruce. Desiccation during handling is not a major problem for this species unless exposure is prolonged. A critical measure for Sitka spruce seedlings is a root moisture content above 180% of dry weight.

Stand-Level Treatments to Improve Timber Yield and Quality

The principles that govern the choice of stand-level silvicultural treatments for Sitka spruce stands and plantations are the same as those for other conifers. In this context, a matrix showing expected major positive effects, less than major positive effects, and little or no effects of various stand-level silvicultural actions is shown in Table 31 for stand attributes that are central

Table 31

A matrix of technical effects of stand-level silvicultural actions on stand attributes

			Density control measures					
Stand attribute	Tree improvement	Silviculture systems for natural regeneration	Successful 'basic' silviculture	Plantation spacing	Pre-commercial thinning	Commercial thinning	Pruning	Fertilization
Growth rate	e	e	E	e	–	–	–	E
Yield	E	e	E	e	E	E	–	E
Piece size	e	–	E	E	E	e	–	e
Log quality	e	e	e	E	e	e	E	e
Adjacency obligation for green-up	–	E	E	e	e	–	–	e

e = positive effect; E = major positive effect; – = little or no effect to forest management: growth rate, yield, piece size, log quality, and adjacency obligation for green-up.

In considering these possible effects of various stand-level silvicultural actions, the following assumptions must be made:

- On the Queen Charlotte Islands, harvesting will be in old-growth timber for several more decades because of the unbalanced age-class structure.
- Little can be done in the dense pole-sized second-growth stands originating from clearcuts 30-60 years ago; it is unlikely that commercial thinning will be economically feasible in such stands in the near future.
- Because of the white pine weevil hazard, large-scale plantations of Sitka spruce will not be established on the coast of British Columbia, except on the Queen Charlotte Islands.

The precedents for technical effects of incremental silviculture in Sitka spruce do not come from Canada or from the United States, where dense natural regeneration has been largely untended, but rather from Britain and Ireland. The intensive culture of large-scale plantations of Sitka spruce in the British Isles is somewhat like the culture of *Pinus radiata* in New Zealand, but without the commitment to grow knot-free logs by pruning.

Sitka spruce has been grown in Britain as a reliable, low-risk crop of short-rotation trees on moorlands and old pastures (Low 1987). The end product objective has usually been construction-grade lumber produced by planting spruce at about 2.5 m × 2.5 m spacing on most sites. These stands are often grown without thinning or pruning and have little competition-induced mortality, but many of them are on sites with high windthrow hazard. This large-scale afforestation program has been financed by British government tax allowances or direct grants that pay the full cost of site preparation and planting. Recently tax incentives to promote planting of Sitka spruce have been withdrawn, and the rate of new planting has declined.

Queen Charlotte Islands provenances of Sitka spruce seem best for northern England and Scotland, and Oregon-Washington provenances are favoured in southern England and Wales. Sitka spruce has been favoured because it grows well on a wide variety of sites and is relatively easy to establish, even in harsh, windy, poorly drained sites. It is also hardy and can withstand poor planting procedures. Sitka spruce production in Britain has also been favoured by reliability of yield, lack of serious pests and diseases, high demand, high prices, and government subsidies that attracted private investors and pension funds. The large wave of planting took place in the 1950s to 1990s. A 40-year rotation is widely used, largely because of high windthrow hazard as stands reach a top height of 20 m. A major harvest is now under way, to be followed by planting of the second crop of Sitka spruce on the same sites. Second rotations appear to have good vigour, although Dutch (1995) provided evidence of some growth reduction in the second rotation where whole-tree harvest removal had occurred. Although there is no weevil problem in Britain as there is in British Columbia, there have been suggestions that second-rotation stands may be more susceptible to insect and disease problems than first-generation stands.

Since the Queen Charlotte Islands are the only place in the world where Sitka spruce can be grown in an environment free of both white pine weevil damage and the severe windthrow hazard found in British plantations, it is currently the only place in the world where large-diameter, high-quality Sitka spruce can be grown on long rotations. If the problem of weevil damage is solved on an operational scale, this observation may someday also apply to other parts of coastal British Columbia.

Tree Improvement

In Britain, Sitka spruce has been the focus of an intensive program of tree improvement. The results of this program have been summarized by Lee (1992) and used to predict the potential increases in 10-year height growth when genetically improved spruce stock is used. These potential genetic gains in height growth are shown in Figure 37 as percentage increases over genetically unimproved seed sources (controls) from the Queen Charlotte Islands (Lee 1992). Lee's graph shows a time lag in the possible genetic gain because of the time required to produce seed and raise the improved planting stock. It takes 10-20 years to obtain seed from newly established seed orchards and 6 years to produce vegetatively propagated stock.

The projected genetic gains over a rotation are based on extrapolation of results in juvenile stands. The oldest progeny test data reported by Lee (1992) were volumes, heights, and diameters at age 27. A 15% genetic gain for final rotation volume is predicted for currently available genetically improved stock (the equivalent of one United Kingdom yield class, or about 2-3 $m^3/(ha \cdot yr)$. The genetic gain from improved form is estimated to be about 7% over a rotation.

Gains in wood quality cannot be quantified yet (Lee 1992). It has been found that in Sitka spruce there is 0.6 negative correlation between wood density and vigour compared with 0.3 for Norway spruce, so the most vigorous trees may have to be rejected because of low wood density. The desired economic traits are vigour, straightness, and higher wood density. Selection for stem straightness can begin as early as 6 years of age, but selection for wood density cannot take place before age 15, at which time a wood penetrometer can be used. To date, the aim has been to produce construction-grade timber, meaning straight trees with acceptable wood properties.

For Sitka spruce in Britain, Lee (1990) predicted that economic gains would continue to rise as more clones were screened for vigour, form, and wood density, and as the very best available clones were used in approved clonal seed orchards and in the production of stock from vegetative propagation. Ten-year height gains of 25% relative to unimproved Queen Charlotte Islands provenances are predicted by the end of the 1990s (Mason and Gill 1986); this potential gain is 10% higher than the potential gain of 15% mentioned by Lee (1992) because genetically superior stock will become increasingly available over the next few years.

Sitka spruce wood has a naturally low density, and there is a desire in breeding programs to seek genetic sources of higher wood density. Little is known about the effects of knot size and distribution on wood strength, but, as with other commercial tree species, knots can lower wood quality in Sitka spruce. Selections and reselections have been made for straight and fine-branched trees without a reduction in wood density. Populations that have shown yield gains of as much as 15% have not been tested for wood properties, but 1,100 trees from older tests were checked for wood density and 240 were selected. Based on 2,700 trees in the first-generation progeny tests, new breeding populations using 240 clones have been established in Britain.

To date, many British Sitka spruce plantations have been planted at about 2 m × 2 m spacing; thus the trees have a large content of juvenile core wood. Thompson (1992) pleaded for stiffer juvenile core wood to be made a criterion for selecting desirable genotypes. Longer fibre length and higher wood density in juvenile wood are important measures of superior wood quality in Sitka spruce. Silviculturally induced initial high regeneration density reduces the quantity of juvenile core wood. Juvenile wood cannot be avoided when Sitka spruce is propagated by cuttings, but vegetative propagation may have the advantage of reducing variability in wood properties. Genetic methods to improve wood properties have been successful in other widely used plantation species, such as radiata pine and eucalyptus, but there are fewer opportunities in Sitka spruce because of the relatively stronger influence of environmental factors on wood properties in this species.

Genetic improvement of Sitka spruce in British Columbia has not yet solved the white pine weevil problem through mass production of weevil-resistant genotypes for outplanting. Very little of the currently available non-resistant first-generation seed orchard seed is used at present except on the Queen Charlotte Islands. Selection and clonal screening of resistant genotypes have been in progress since 1989. About 250 putatively weevil-resistant trees have been selected in both provenance and test plantations and in natural stands of resistant provenances. However, there is still much to be learned about genetic differences in host-tree resistance where insect/host-tree relationships are involved. Until these differences are understood, and until it is known how modes of genetic transmission (seed reproduction versus vegetative propagation) influence the transfer of insect resistance from parents to offspring, there are obstacles to large-scale production of weevil-resistant planting stock. Current research is encouraging, however, because there are strong indications of a genetic basis for the resistance to weevil attack shown by some screened Sitka spruce genotypes (Ying 1991; Ying and Ebata 1994).

The question of preservation of genetic diversity in Sitka spruce breeding and domestication programs has been assessed by El-Kassaby (1992). He concluded that: (1) phenotypic selection successfully captured most of the variation present in the natural population; (2) seed orchard management practices have the potential to influence genetic diversity; (3) knowledge of seed biology is needed to support seed collection and storage methods; (4) growing seedlings from bulked seedlots requires further evaluation; and (5) tree improvement should be considered in the framework of a total delivery system to maintain acceptable levels of genetic diversity in Sitka spruce.

It may be very instructive to import genetically improved 'super Sitka' from Britain for testing on Queen Charlotte Islands sites from which the seed was originally derived. This would show whether the genetically selected strains would grow well on the Queen Charlotte Islands. Mason and Sharp (1992) have pointed out that 1.5 million cuttings of genetically improved Sitka spruce were planted in British forests in the 1988-89 planting season. The use of cuttings is recommended for British sites that have low windthrow hazard and that have a yield class of at least 12 $m^3/(ha \cdot yr)$. No loss of timber quality is expected if cuttings are used instead of plantings of seed origin. Such cuttings could be a way to test British 'super Sitka' genotypes in British Columbia.

Use of Silviculture Systems to Obtain Natural Regeneration

Within its natural range, Sitka spruce usually grows mixed with western hemlock and western redcedar, both of which are relatively shade tolerant, and both of which germinate and survive on organic seedbeds. By contrast, Sitka spruce natural regeneration is relatively more restricted in seedbed and light requirements, and thus tends to be more disturbance-dependent. For these reasons, Sitka spruce often regenerates naturally on landslides, on newly exposed alluvial sites, and on clearcuts. There are also reasons why clearcutting (with either subsequent planting or natural regeneration) is the recommended method of Sitka spruce regeneration in the alternatives outlined below by Weetman and Vyse (1990). Recent attempts at small-patch clearcutting involving Sitka spruce have involved the removal by helicopter of trees from clearings 25 m in diameter in an experimental mountain slope area near Rennell Sound, Queen Charlotte Islands (Figure 38), as well as partial cutting and helicopter yarding on environmentally sensitive floodplain sites near Naden Harbour, Queen Charlotte Islands (Moore 1991b).

Figure 38. An example of selective removal of Sitka spruce, western hemlock, and western redcedar by helicopter harvesting in 1992, involving small-patch clearcuts on erosion-prone terrain (see Figure 2), near Rennell Sound, Queen Charlotte Islands, British Columbia. (Photograph by F. Pendl)

Foresters involved with Sitka spruce in British Columbia can refer to the draft *Coastal Harvest Planning Guidelines* (B.C. Ministry of Forests 1990a) and the *Interim Forest Landscape Management Guidelines for the Vancouver Forest Region* (B.C. Ministry of Forests 1990c). Although not specific to Sitka spruce, these guidelines apply to spruce management in coastal ecosystems as much as they do to management of western hemlock, western redcedar, or Douglas-fir. In addition, a comprehensive set of guidebooks, cited in several other subsections of this text, was released in 1995 under the Forest Practices Code of British Columbia Act. Currently available guidelines provide detailed suggestions for manipulation of vegetation cover and ground surfaces, distribution of logging activities over time and space, visual simulations, and making landscape alteration acceptable to the public. Meeting these guidelines should be the primary focus of the forest manager involved with Sitka spruce or western hemlock–Sitka spruce ecosystems. To date, there are no particular criteria for making these operational guidelines specific to Sitka spruce. The current five-year strategic plan of the Ministry of Forests Research Branch is designed, in part, to develop the criteria needed for all managed forest species in the province (B.C. Ministry of Forests 1990d).

Sitka spruce was included among the species for which Weetman and Vyse (1990) reviewed the methods that are known to work in regenerating the forest type that was present before harvesting. Their assessment of various regeneration methods for re-establishing forest types in which Sitka spruce is a significant component can be summarized as follows:

To return harvested old-growth Sitka spruce to Sitka spruce:
Recommended: clearcut method with planting
Feasible: clearcut method with natural regeneration; seed tree method; shelterwood method
Not feasible: selection method; coppice

To return harvested young natural Sitka spruce to Sitka spruce:
Recommended: clearcut method with planting; clearcut method with natural regeneration
Feasible: seed tree method; shelterwood method
Not feasible: selection method; coppice

In summary, studies of clearcutting, shelterwood, selection, and seed tree silvicultural systems suggest that clearcutting is the best system for hemlock-spruce types if timber production is the primary use. Shelterwood and selection systems may be more appropriate if non-timber land uses are the highest priority (Harris and Farr 1974; Ruth and Harris 1979; Harris and Johnson 1983). The shelterwood system and other variations of clearcutting have been proposed in very unstable areas on the Queen Charlotte Islands to minimize blowdown and mass wasting (Sanders and Wilford 1986).

In Britain, an unexpected result has been the occurrence of up to 20,000 stems per hectare of natural regeneration of Sitka spruce seedlings following clearcutting of first-generation plantations on a variety of sites (Nelson 1990, 1991). The ideal seedbeds are mainly exposed organic and mineral soil surfaces that are moist and also relatively free of competition (Malcolm 1987b). This type of dense natural regeneration is rarely seen in Sitka spruce's natural range, except on some portions of landslides as documented on the Queen Charlotte Islands by Smith et al. (1986).

In British Columbia, where most Sitka spruce occur in mature and untended stands, the question of appropriate silvicultural systems is mainly speculative. Most of the site series in which Sitka spruce occurs naturally are very resilient for natural regeneration. There are, however, no precedents for 'geriatric' silvicultural manipulations in old-growth stands other than retrospective studies of historical disturbances. For old growth dominated by Sitka spruce, in mixture with western hemlock and western redcedar, the only feasible silvicultural system is clearcutting, perhaps in patches. It seems likely that attempts at true selection systems, using helicopter extraction, will lead to the dominance of western hemlock regeneration. Furthermore, there is also an economic incentive to remove only the best Sitka spruce, leaving behind lower-quality hemlock or hemlock infected with dwarf mistletoe.

Table 32

Recommended stocking standards for the biogeoclimatic zones, subzones, variants, and site series where Sitka spruce is considered a primary, secondary, or tertiary regeneration species in the Vancouver and Prince Rupert forest regions in British Columbia

Biogeo-climatic zone or variant[a]	Forest region	Site series[b]	Sitka spruce regeneration priority			Stocking standards[c] (well spaced per ha)		Regeneration delay[c] (yr)	Assessment		% tree over brush[d]
			Prim.	Sec.	Tert.	Target	Min.		Earliest (yr)	Latest (yr)	
CWHwh1	Vancouver	HwSs–Lanky moss	✓			900	500	6	11	14	150
CWHwh1	Vancouver	CwSs–Foam flower CwSs–Sword fern	✓			900	500	3	8	11	150
CWHvh1	Vancouver	HwSs–Lanky moss		✓		900	500	6	11	14	150
CWHvh1	Vancouver	CwSs–Sword fern CwSs–Foam flower CwSs–Devil's club		✓		900	500	3	8	11	150
CWHvh1	Vancouver	CwSs–Skunk cabbage			✓	800	400	3	8	11	150
CWHwh1	Vancouver	CwSs–Salal			✓	900	500	6	11	14	150
CWHvh1 CWHwh1	Vancouver	Ss–Trisetum Ss–Kindbergia Ss–Swordfern Ss–Lily-of-the-valley	✓			900	500	3	8	11	150

►

◀ *Table 32*

Biogeo-climatic zone or variant[a]	Forest region	Site series[b]	Sitka spruce regeneration priority			Stocking standards[c] (well spaced per ha)		Regeneration delay[c] (yr)	Assessment		% tree over brush[d]
			Prim.	Sec.	Tert.	Target	Min.		Earliest (yr)	Latest (yr)	
CWHvh1 CWHwh1	Vancouver	Ss–Salal Ss–Reedgrass Ss–Slough sedge Ss–Pacific crab apple	✓			400	200	3	8	11	150
CWHvh1 CWHwh1	Vancouver	CwHw–Salal			✓	900	500	6	11	14	150
CWHvh2	Vancouver,	HwSs–Lanky moss Prince Rupert	✓			900	500	6	11	14	150
CWHvh2	Vancouver, Prince Rupert	CwSs–Sword fern CwSs–Foam flower CwSs–Devil's club Ss–Lily-of-the-valley Ss–Trisetum Ss–Kindbergia Ss–Sword fern	✓			900	500	3	8	11	150
CWHvh2	Vancouver, Prince Rupert	Ss–Salal Ss–Reedgrass Ss–Slough sedge Ss–Pacific crabapple[e]	✓			400	200	3	8	11	150

▶

◄ *Table 32*

Biogeo-climatic zone or variant[a]	Forest region	Site series[b]	Sitka spruce regeneration priority			Stocking standards[c] (well spaced per ha)		Regeneration delay[c] (yr)	Assessment		% tree over brush[d]
			Prim.	Sec.	Tert.	Target	Min.		Earliest (yr)	Latest (yr)	
CWHvh2	Vancouver, Prince Rupert	CwHw–Salal			✓	900	500	6	11	14	150
CWHvh2	Vancouver, Prince Rupert	CwSs–Skunk cabbage			✓	800	400	3	8	11	150
CWHvm1 CWHvm2	Prince Rupert	HwBa–Blueberry HwBa–Deer fern		✓		900	500	6	11	14	150
CWHvm1 CWHvm2	Prince Rupert	CwHw–Sword fern BaCw–Foam flower BaCw–Salmonberry BaSs–Devil's club		✓		900	500	3	8	11	150
CWHvm1	Prince Rupert	Ss–Salmonberry Act–Red-osier dogwood		✓		900	500	3	8	11	150
CWHvm1 CWHvm2	Prince Rupert	CwSs–Skunk cabbage			✓	800	400	3	8	11	150
CWHwm	Prince Rupert	HwSs–Blueberry	✓			900	500	6	11	14	150
CWHwm	Prince Rupert	SsHw–Oak fern SsHw–Devil's club Ss–Salmonberry	✓			900	500	3	8	11	150

►

◄ *Table 32*

Biogeo-climatic zone or variant[a]	Forest region	Site series[b]	Sitka spruce regeneration priority			Stocking standards[c] (well spaced per ha)		Regeneration delay[c] (yr)	Assessment		% tree over brush[d]
			Prim.	Sec.	Tert.	Target	Min.		Earliest (yr)	Latest (yr)	
CWHwm	Prince Rupert	Ss–Skunk cabbage	✓			800	400	3	8	11	150
CWHwm	Prince Rupert	Act–Red-osier dogwood		✓		900	500	3	8	11	150
CWHwm	Prince Rupert	HwSs–Step moss			✓	900	500	6	11	14	150

Source: Silviculture Interpretations Working Group (1994)

[a] See Appendix 4 for the names of biogeoclimatic subzones or variants listed in this column in abbreviated form.

[b] See Appendix 4 for the identity of tree species whose abbreviations begin the name of each site series.

[c] Target and minimum numbers of well-spaced stems per hectare, regeneration delay (the minimum number of years regeneration must be established on a site before a free-growing assessment), and the earliest and latest times for regeneration assessment are as defined by the Silviculture Interpretations Working Group (1994).

[d] Tree/brush height ratio within a 1 m radius around the crop tree; for example, 150 refers to a crop tree seedling that is 150% as tall as nearby competing vegetation.

[e] Sitka spruce is a suggested primary regeneration species in the Ss–Pacific crabapple site series in the Prince Rupert Forest Region but not the Vancouver Forest Region.

Successful Basic Silviculture

In British Columbia the law requires that there be free-growing regeneration of crop trees within a specified time after harvesting. Table 32 lists the recommended stocking standards for the biogeoclimatic variants and site series in which Sitka spruce is a potential regeneration species, based on the judgment of the Silviculture Interpretations Working Group (1994). Target and minimum numbers of well-spaced stems per hectare, regeneration delay (the number of years regeneration units must be established on a site before a free-growing assessment), and the earliest and latest times for regeneration assessment are as defined by the Silviculture Interpretations Working Group. The last column, '% tree over brush,' refers to the tree/brush height ratio within a 1 m radius around the crop tree; for example, 150 refers to a crop tree seedling that is 150% as tall as nearby competing vegetation. Table 32 indicates that Sitka spruce is a primary species for forest managers to use in reforestation in the CWHwh1, CWHvh1, CWHvh2, and CWHwm biogeoclimatic subzones and variants of the Vancouver and Prince Rupert forest regions.

The relatively low target and minimum densities of less than 1,000 stems per hectare, which are expected to be achieved in 8-14 years (Table 32), imply delayed crown closure. Assuming no ingress of other species, these low minimum and target densities suggest that crown closure should not be expected until Sitka spruce achieves a top height of at least 15 m. Appreciable competition-induced mortality would not be expected until spruce reached a top height of at least 20 m.

The practical implications of these low recommended stand densities for Sitka spruce are large knot size, large diameters, delayed full-site occupancy, and loss of increment. At such low plantation densities, pruning is required for quality lumber. Depending on site index, much higher densities are required for quality sawlog production and greater stand growth rates. In practice, the natural ingress of hemlock regeneration, at densities as great as 5,000 stems per hectare, makes the low-density aspects of plantation Sitka spruce somewhat academic. If for some reason hemlock ingress were not to occur, then silviculturists would need to consider the consequences of the relatively low stand densities recommended for Sitka spruce.

In British Columbia, green-up is a relatively new requirement for each cutblock before an adjacent block can be cut. To assess this requirement, crown closure is a more realistic measure of green-up than an arbitrary height target. Ingress of naturally regenerating hemlock is common in many of the CWH site series where Sitka spruce is a component, so that green-up on these sites is accelerated. Current target stocking densities represent the minimum desired goal, which may be lower than that needed for stand or forest-level management objectives. When considerable areas of old-growth timber are reserved from harvest for long periods while awaiting adjacent cutblock green-up, allowable harvest levels may be greatly reduced. This situation not only creates pressure for very rapid green-up but also encourages use of partial-cutting systems to avoid the adjacency constraint entirely.

Fertilization

Sitka spruce is very responsive to fertilization, as indicated by over 50 years of experience with N and P fertilization of spruce plantations in Britain (McIntosh 1981; Malcolm 1987a; Miller and Miller 1987). In British Columbia, fertilization treatment has been limited to N and P application to chlorotic spruce plantations on salal-dominated clearcuts. Researchers in both Britain and British Columbia have noted a deterrent to Sitka spruce height growth, referred to as 'heather check' in Britain and 'salal-check' in British Columbia. In both cases, this phenomenon appears to be chlorosis associated with nutrient limitations and below-ground competition from ericaceous plants (*Calluna vulgaris* in Britain and *Gaultheria shallon* on Vancouver Island). It is suspected that ericaceous plants rely more than trees on ectomycorrhizae to obtain limited supplies of N and P.

The problem with *Calluna* was recognized in the 1930s in Britain and is now attributed to substances produced from the root system of *Calluna* that inhibit mycorrhizal development in spruce (Taylor and Tabbush 1990). The equivalent problem was diagnosed in British Columbia in the 1980s (Weetman et al. 1990). Research by Salal Cedar Hemlock Integrated Research Program (SCHIRP) scientists, based at Port McNeill, British Columbia, has shown similar mycorrhizal problems between salal and Sitka spruce, western hemlock, and western redcedar. In Britain, Taylor and Tabbush (1990) recognized four categories of sites to aid in the diagnosis of nutrient status and response to fertilization in Sitka spruce plantations (Figure 39). Because there have been few fertilizer trials in British Columbia, it is not possible to recognize the site series that require fertilization, except for the suggestions described below.

In the CWH zone on northern Vancouver Island, there are severe limitations to the growth of conifers that are naturally regenerated or planted, especially on areas that were dominated by western redcedar before being logged. Following logging of old-growth cedar-hemlock stands, salal becomes very prominent on northern Vancouver Island cutovers. This prominence appears to be correlated with severe chlorosis and reduced growth of conifer plantations. During the late 1960s, 1,900 ha of cedar-hemlock cutovers were planted with Sitka spruce by Western Forest Products Limited on northern Vancouver Island. Initially vigorous, the spruce plantations soon showed severe chlorosis and slow leader growth, coinciding with salal domination of the site after 5-8 years. In contrast, trees planted on exposed

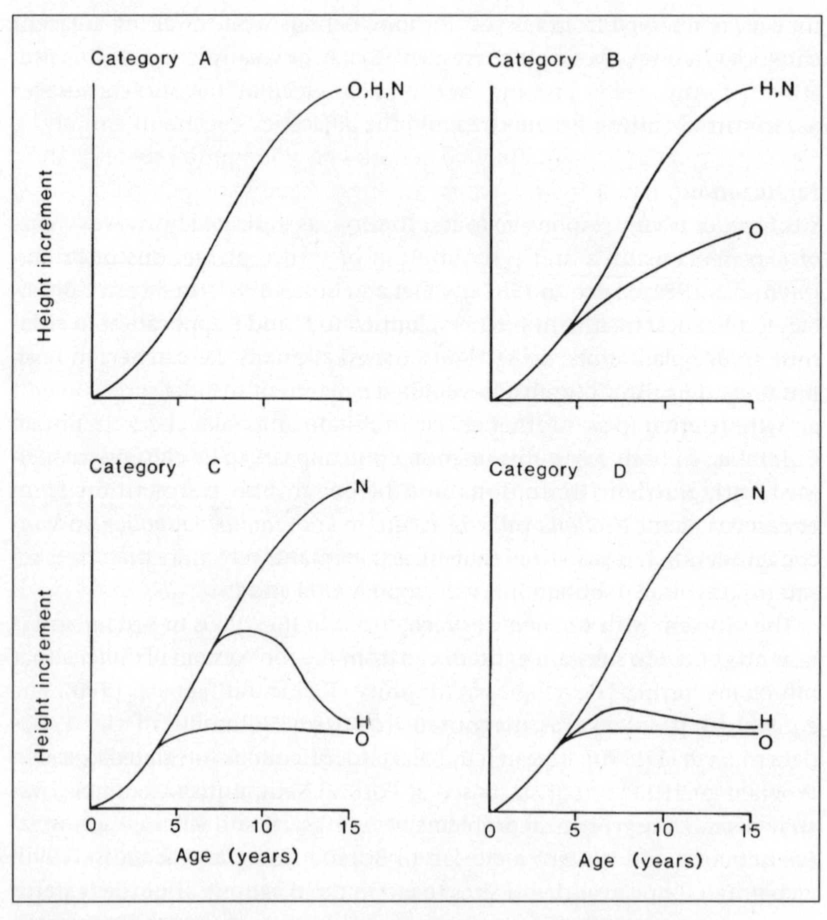

Figure 39. Five decades of research in Britain have revealed important relationships between early (0-15 years) Sitka spruce height growth on control sites (O), on sites where there has been regular application of nitrogen (N), and on sites with reduction of heather as a competing species (H).

Categories A to D in the graphs were defined by Taylor and Tabbush (1990) as paraphrased below:

Category A. Sites where there is sufficient nitrogen available for acceptable tree growth, despite the presence of heather. The inhibitory effect of heather seems to be reduced when soils are rich in available nitrogen, and Sitka spruce is unlikely to suffer any real check to growth, although there may be a slight yellowing of foliage in the 2 or 3 years prior to canopy closure. No herbicide or fertilizer is required. Normally these are sites where the heather is mixed with fine grasses, such as *Agrostis, Festuca,* and *Anthoxanthum* spp., in the transition zone between grassland and heathland; or weakly flushed moorland sites dominated by *Myrica* and vigorous *Molinia;* or sites heavily colonized by broom (*Cytisus scoparius*) or gorse (*Ulex europaeus*).

Category B. The sites in this category are those where heather is the principal cause of nitrogen deficiency and successful heather control would result in adequate availability of nitrogen for Sitka spruce. These are usually heathlands on more fertile lithologies (e.g., basic igneous, phyllites, pelitic schists) or western *Molinia/Eriophorum* moorlands where the heather is sub-dominant.

Category C. Heather is the dominant type of vegetation on these sites, but it is not the sole cause of nitrogen deficiency. The low mineralization rate is also a major factor and although heather control will result in a cost-effective growth response, it will not bring permanent relief from nitrogen deficiency and subsequent inputs of nitrogen fertilizer will be required to achieve full canopy closure. This category can include moorland sites where *Molinia* and *Trichophorum* are codominant with *Calluna* and certain heathland soils with low organic content.

Category D. The principal cause of nitrogen deficiency on these sites is the low mineralization rate. Heather control will not give a cost-effective growth response. In fact, on many of these sites heather is either not present or very sparse. Several inputs of nitrogen fertilizer will be required to maintain a reasonable growth rate in Sitka spruce and enable the crop to achieve full canopy closure. Normally this category will contain lowland and upland raised bogs but it also includes podzolic soils with low organic matter on quartzitic drifts.

mineral soil not dominated by salal showed no reduced growth rates. Also, trees growing on adjacent cutovers of western hemlock–amabilis fir stands of windthrow origin, and on the same parent material, and with little or no salal coverage, did not show reduced growth rates.

Based on the western redcedar–western hemlock (CH) and western hemlock–amabilis fir (HA) sites described on northern Vancouver Island (Lewis 1982; Germain 1985), there was evidence that the CH sites displaying nutrient supply problems influenced not only Sitka spruce but also western hemlock and western redcedar. These ecosystems are found on an estimated 100,000 ha throughout the wetter portions of the CWH zone on Vancouver Island, the coastal mainland, and the Queen Charlotte Islands. Salal is a normal understorey component in these old-growth forests. The salal cover expands rapidly from rhizomes left behind after clearcutting and burning. A decade of research summarized by Prescott and Weetman (1994) confirmed that on northern Vancouver Island, the reduced growth of Sitka spruce planted in dense salal could be rectified by the addition of a single application of N at 300 kg/ha and P at 50 kg/ha. Studies reported by Weetman et al. (1989) indicated that applications of lime would increase the rates of N mineralization in these sites. The addition of nutrients appears to be the only way to solve the nutrient supply problems on these salal-dominated sites (Prescott and McDonald 1994).

The rapid but temporary response of Sitka spruce on northern Vancouver Island to the application of N, whether from urea or ammonium nitrate, matches the responses noted by Taylor and Tabbush (1990). The following foliar nutrient concentrations for optimum and deficient levels in Sitka spruce, as suggested by Binns et al. (1976), are considered to be applicable to spruce on northern Vancouver Island:

nitrogen %	deficient < 1.2; optimum > 1.5
phosphorus %	deficient < 0.14; optimum > 0.18
potassium %	deficient < 0.5; optimum > 0.7
magnesium %	deficient < 0.03; optimum > 0.07

The research on northern Vancouver Island (Prescott and Weetman 1994) confirmed that adding 50 kg P per hectare was enough to keep P concentrations in the desired range of 0.14-0.20%. Values of 0.11-0.20% for P were found in the second and third year following N and P fertilization at rates of 100-300 kg N per hectare and 50-150 kg P per hectare. It appears that application of 50 kg P per hectare satisfies the P requirements for these young plantations.

In practice, because a long-lasting N effect is desired, 300 kg N per hectare is a suggested level of application for operational aerial fertilization. A single application of N and P at these levels is considered adequate to achieve crown closure and subsequent shading out of salal. If this is achieved, then an SI_{50} of 28 m and a mean annual increment (MAI) of 11 m^3/ha should be attainable. Because chlorotic reproduction is widespread on salal-dominated sites in this wet climatic zone, fertilization may be essential to restore site productivity. For example, fertilizer screening trials conducted by MacMillan Bloedel Ltd. (Kumi 1987) on the Queen Charlotte Islands in three chlorotic Sitka spruce plantations growing in dense salal confirmed that there were N and P deficiencies. Elsewhere in the CWH zone, chlorotic conifer reproduction has also been found growing on cutovers dominated by dense Alaskan blueberry (*Vaccinium alaskaense*). It is notable that salal in the understorey of Douglas-fir ecosystems in drier zones of coastal British Columbia does not appear to influence tree nutrition, although it competes for soil moisture (Price et al. 1986).

Researchers have focused on the side-by-side occurrence, on the same parent material, of old-growth cedar-hemlock forests characterized by a salal understorey, deep mor humus, and low N mineralization rates, versus adjacent western hemlock–amabilis fir of windthrow origin on sites with high mineralization rates. This has led to the hypothesis that windthrow, which results in the mixing of organic and mineral soil, is a requirement for revitalizing ecosystem productivity and avoiding immobilization of N in lithosols. There are current trials using backhoes to mix humus and mineral soil on planting spots on CH cutovers. The hope is that chlorosis and fertilizer requirements might be avoided, although control of salal may be required as well.

In summary, the chlorotic and slow-growing cedar and hemlock regeneration growing in dense salal respond to additions of fertilizer similarly to Sitka spruce planted in CH sites. There appears to be a requirement for both

N and P, as confirmed by foliar analyses and subsequent height growth response. An insufficiently researched question is the degree to which white pine weevils may be attracted to the more vigorous leaders of Sitka spruce after fertilizer application.

Pre-Commercial Thinning

Overstocked stands, ranging from 12,000 to 25,000 stems per hectare, are common in naturally regenerated sites of the Western Hemlock–Sitka Spruce forest type. Crown closure usually occurs within 15-20 years, and trees in these stands undergo intense competition. Individual tree growth is slowed under such dense stocking, but natural mortality is high and there is generally a good expression of dominance.

Western hemlock and Sitka spruce both respond well to stocking control and early commercial thinning, particularly up to the age of 40 years. Height growth response is similar for both species, but thinning produces greater diameter increments in Sitka spruce than in western hemlock. From the British experience, it is now known that Sitka spruce rotations can be as short as 40 years. With natural spruce regeneration in Britain, respacing is done when the trees are about 2 m tall because it is easier to cut small trees (T.C. Booth, pers. comm., May 1991).

Because pre-commercial thinning involves thinning from below, it favours Sitka spruce. It is therefore a useful method for increasing the spruce content of stands. Conversely, where weevil activity is high, spacing can be used to promote other species by removing the spruce. Spacing is usually done only to obtain the desired species composition and density. The density desired during pre-commercial thinning depends upon later objectives.

Spacing should not be carried out before canopy closure. Effective spacing requires an awareness of the potential for wind damage and the incidence of dwarf mistletoe and Annosus root rot on hemlock (Ruth and Harris 1979). Spacing generally improves the future windfirmness of a stand, since the root systems can develop better.

Thinning and pruning strategies for pre-commercial plantations should be based on an understanding of competition between individuals in a stand. There is considerable information available for Sitka spruce, thanks to the European experience with plantation management of this species. Some key features of competition in Sitka spruce plantations are outlined below.

The production and growth of branches change as the canopy of a Sitka spruce plantation passes through different stages of development. These changes were described by Ford (1985) and are summarized in Figure 40. Cochrane and Ford (1978) advanced the hypothesis that there was within-tree competition for resources during the early years of growth, which was related to the numbers of branches produced by a tree on its main stem. There was a marked decrease in the number of whorl branches produced each year after the branches of neighbouring trees met, which was related to an increase in leader growth rates. Before year 6, leader growth accelerated by about 5.2 cm/yr. Between years 7 and 10, the increase in leader increment was 8.2 cm/yr. While the leaders certainly exerted apical control over the whorl branches, the whorl branches also influenced leader growth (Ford 1985). From year 11 onward, mean annual height increment stabilized (Figure 40). Large trees produced the most branches and had the greatest leader growth. Cochrane and Ford (1978) suggested that the crown interlock stage in canopy development marked the onset of between-tree competition. However, competition in terms of trunk diameter increment was not evident until year 16. This delay may reflect the length of time required for changes in the top whorls of the tree to have an effect on wood production in the trunk (Ford 1982, 1985).

At year 16, new shoot production in Sitka spruce shifts from being evenly distributed throughout the crowns to being concentrated at the tops. The canopy changes quickly, from being akin to a collection of long but bush-like crowns to a two-tier mixture of a few dominant trees and many subdominant trees. Associated with this change is a recorded decrease in leaf

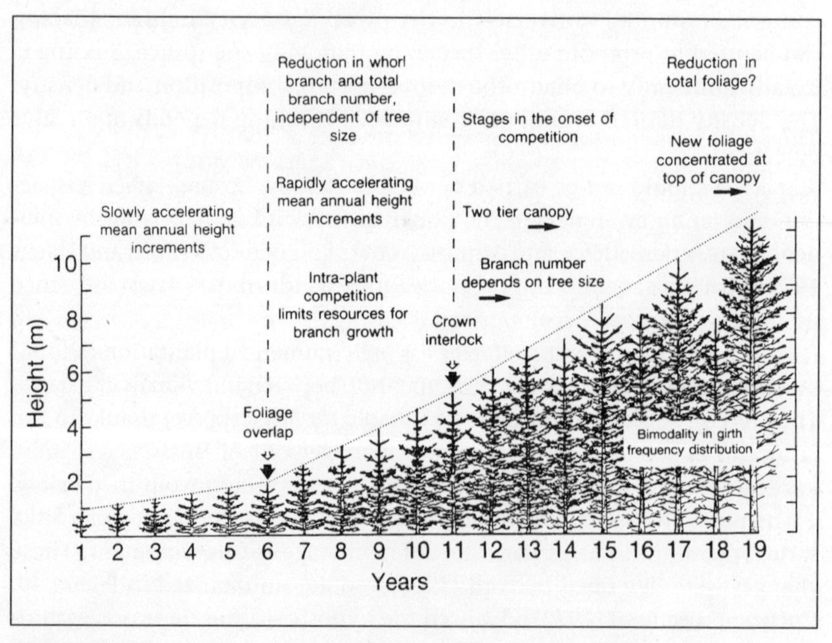

Figure 40. The production and growth of branches change as the canopy of a Sitka spruce plantation develops (adapted from Ford 1985).

area index of 11% between years 16 and 18 (Ford 1985). Similar decreases have been found in plantations of other coniferous species at the same stage of development.

Despite the range of responses available to trees that are dominated by others in a stand, high levels of competition unquestionably reduce individual tree growth. One way to quantify this is by using relative growth rate (RGR), defined as basal area tree growth over the past two years relative to tree basal area at the start of the two-year period (Perry 1985). In Sitka spruce, Ford (1975) showed that relative growth rates decline sharply with decreasing dominance of trees in the stand. When RGR was compared with tree height in Sitka spruce, Cannell et al. (1984) found that RGR was initially negatively correlated with tree height, but became positively correlated as competition intensified.

Competition has important influences on carbohydrate allocation and tree form aside from its effects on crowns. Mensurationists generally consider that growth in stem diameter is more affected by changes in stand density than growth in tree height (Perry 1985). In Sitka spruce there is evidence that the most dramatic effects on the height/diameter ratio occur at the highest initial stocking levels (Jack 1971).

Density Control

The practical questions about Sitka spruce density control relate the following decisions or events to stand management objectives: (1) initial plantation density; (2) pre-commercial thinning density; (3) the amount of fill-in ingress of actual regeneration in plantations; (4) the minimum operable piece size and volume per hectare needed to carry out a commercial thinning (CT) at a given location; (5) the need to carry out pruning to produce knot-free clear lumber; (6) the risks of loss of trees because of damaging agents; (7) the need for fast green-up.

The outcomes of different spacings for plantations and pre-commercial thinning can be readily explored by the use of the TASS-generated managed-stand yield tables contained in the Ministry of Forests computer program WinTIPSY (Mitchell et al. 1995) or the economic analysis module for WinTIPSY (Stone et al. 1996), or from stand density management diagrams (Figures 41 and 42). Pre-commercial thinning is considered a requirement for a feasible and early commercial thinning. Very wide spacings, besides reducing yield, produce very knotty Sitka spruce with large horizontal crowns.

The starting point for stand density control is the target and minimum numbers of well-spaced stems per hectare. These values are listed in Table 32 for site series in coastal British Columbia where Sitka spruce is considered by the Silviculture Interpretations Working Group (1994) to be a primary, secondary, or tertiary regeneration species. Target stocking standards range

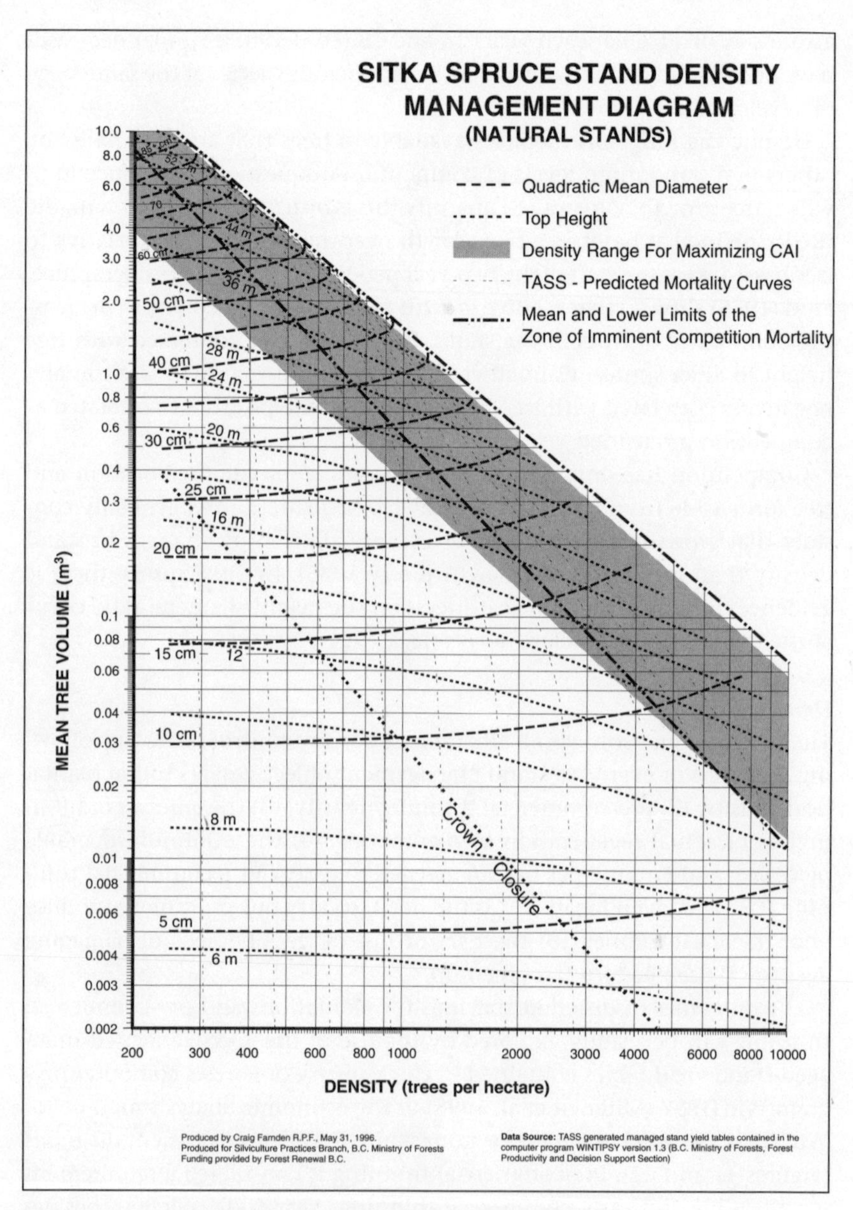

Figure 41. Sitka spruce stand density management diagram for natural stands. See following page for interpretation of the variables in this diagram. (Courtesy of P. Martin, Ministry of Forests, Victoria, from diagram produced by C. Farnden, 31 May 1996, from TASS-generated managed-stand yield tables contained in the computer program WinTIPSY, Forest Productivity and Decision Support Section, Ministry of Forests)

Legend for Figures 41 and 42. This guide to interpretation of stand density management diagrams is based on Farnden (1996).
Forest managers continue to seek new ways to predict the optimal density of natural or managed stands. Stand density management diagrams, as portrayed in Figures 41 and 42, are one way to predict the outcomes of silvicultural steps that influence stand density. Ironically, as foresters have increasingly recognized the usefulness of graphical expressions of tree size/tree density relationships, ways to portray these relationships have become increasingly complex. For example, the stand density management diagram for red alder (Hibbs and DeBell 1994; Puettmann 1994) is relatively easy to interpret compared with Figures 41 and 42 presented here. This legend outlines how various tree diameter, tree height, crown closure, and tree mortality rates change with increasing stand density in natural (Figure 41) and planted (Figure 42) stands of Sitka spruce in British Columbia.

The theory behind today's stand density management diagrams, derived from research by Drew and Flewelling (1979), is that at any given density of trees per hectare, there is a maximum average tree size that can be achieved. A key feature is the tree density at which there is an onset of competition-based mortality. In Figures 41 and 42, this important feature is shown by the two lines in the upper right portions of the graph labelled 'Mean and Lower Limits of the Zone of Imminent Competition Mortality' (ZICM). These lines define the ranges of densities at which self-thinning starts to accelerate dramatically.

Another set of lines defines the tree density range that is optimal for maximizing current annual increment (shown by the shaded band that slopes downward to the right on Figures 41 and 42). Users of these diagrams should note that the extreme upper line of Figures 41 and 42 is the mean limit of the ZICM. Thus this upper limit to the graphed information is not the absolute maximum size-density line; instead it is the maximum size/density relationship of the 'average' stand as modelled by the Tree and Stand Simulator (TASS). Therefore roughly half of all actual stands could have real trajectories that lie above the mean limit of the ZICM at the top of the diagram. As stand growth approaches the line for this mean limit, further increases in average piece sizes must be accompanied by decreases in density.

Figures 41 and 42 also show a crown closure line, which represents the points at which the crowns of trees in stands of different densities start to interact. Below this line, trees behave much like open-grown trees, with little or no inter-tree competition. Individual open-grown trees exhibit maximum vigour and diameter growth, but as trees grow above this line, they experience increasing inter-tree competition, decreasing rates of diameter growth, and increased natural pruning due to canopy shading. Maximum individual tree diameter growth occurs below the crown closure line shown in Figures 41 and 42, with decreasing individual tree vigour and diameter growth as one progresses from the crown closure line to the line representing the mean limit of the ZICM at the top of the diagram. Below the crown closure line, natural pruning is mainly due to self-shading; above this line, natural pruning is increasingly a result of canopy shading. Natural pruning is therefore maximized by growing a stand closer to the mean limit of the ZICM than to the crown closure line.

Figures 41 and 42 portray stand top heights that slope downward from left to right when they are plotted on a stand volume/tree density axis. Top height is important in these diagrams because under most conditions height growth is independent of stand density. These top-height isolines can also be used to incorporate time, and thus rates of growth, through the use of site index curves. Site index data are therefore critical for the application of stand density management diagrams.

In Figures 41 and 42, the quadratic mean diameter of Sitka spruce stands of increasing stand density is shown by a series of upward-sloping isolines. Quadratic mean diameter is the diameter of a tree of average basal area, calculated by squaring the diameters of measured trees, summing them, and taking the square root of the sum. These quadratic mean diameter isolines are silviculturally important, not only because tree stem diameter is highly sensitive to stand density but also because stem diameter is often a criterion for thinning and other silvicultural decisions.

The main differences between Figures 41 and 42 result from differences in spacing uniformity. The diagram for planted stands (Figure 42) is based on a TASS simulation that assumed a relatively uniform spacing of trees, in which each tree has roughly the same growing space and competitive conditions. In contrast, the natural stand diagram (Figure 41) assumes less uniform spacing with small openings and clumps. In this scenario, trees in the clumps experience competition-based mortality much earlier than the more open portions of the stand. A less vertical slope to the mortality curves below the lower limit of the ZICM is noticeable in the natural stand diagram (Figure 41) compared with the planted stand diagram (Figure 42). The result is a difference in the density/height/diameter relationships. Also, the ZICM is broader in the natural stand diagram, reflecting a more gradual transition from the lower portion of the diagram, which is free of mortality, to the area adjacent to the maximum density line. In addition, the zone of maximum current annual increment (shaded band on Figures 41 and 42) occurs higher in the natural stand diagram than in the planted stand diagram.

In a perfectly uniform stand of identical trees, the upper limit of the zone of maximum current annual increment might be expected to coincide with the lower limit of the ZICM, as any tree mortality would be a loss of merchantable volume and would leave unoccupied gaps in the canopy. In reality, the trees that are dying are much smaller than the remaining crop trees and leave little if any growing space unoccupied. Managed stands are more uniform in piece size than natural ones, which means that dying trees have more of an effect on current annual increment. This effect extends to natural stands that have been thinned, where the increase in stand uniformity results in both the lower limit of the ZICM being raised in the diagram, and a lower rate of mortality below the ZICM. For comparison, TASS-predicted mortality curves are superimposed on Figures 41 and 42.

It is emphasized that the stand density management diagrams shown in these two figures depend on the current data available to the TASS growth model as expressed in the managed-stand yield tables contained in WinTIPSY. No operational adjustment factors have been applied to Figures 41 and 42, which means that the diagrams reflect the growth relationships of fully stocked stands with no significant forest pest problems. Yields suggested by these diagrams are therefore *total potential yields* and may have to be scaled down to reflect *merchantable operational yields*. The relationships shown in Figures 41 and 42 also pertain to stands that grow to harvest age without mid-rotation silvicultural thinnings. It is also emphasized that these stand density management diagrams are suitable only for even-aged stands. They are currently available for single-species stands only, but they may also be useful in mixed stands where a single species is predominant.

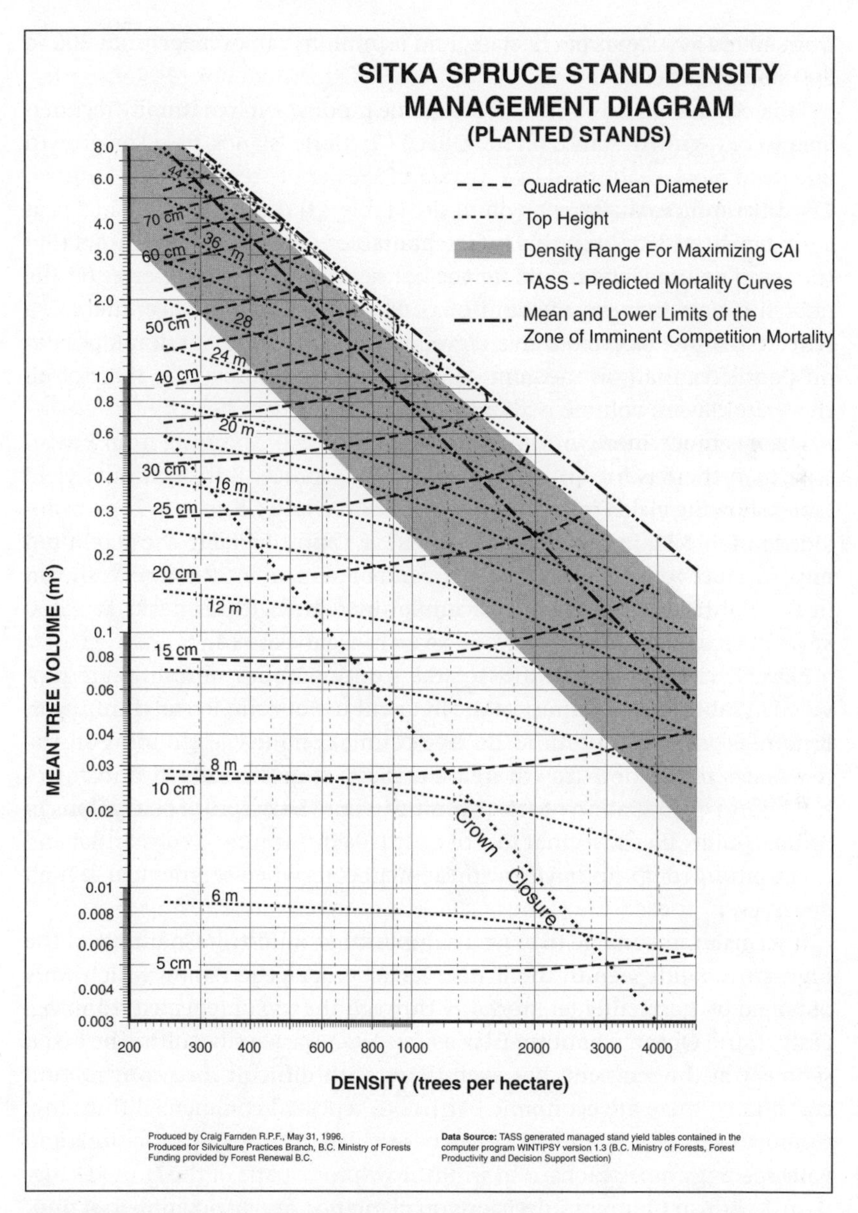

Figure 42. Sitka spruce stand density management diagram for planted stands. See explanatory legend following Figure 41 for interpretation of the variables in this diagram. (Courtesy of P. Martin, Ministry of Forests, Victoria, from diagram produced by C. Farnden, 31 May 1996, from TASS-generated managed-stand yield tables contained in the computer program WinTIPSY, Forest Productivity and Decision Support Section, Ministry of Forests)

from 400 to 900 stems per hectare, and minimum values range from 200 to 500 stems per hectare.

Table 33 and Figure 43 show net volume production in naturally regenerated second-growth stands on the Queen Charlotte Islands, based on growth and yield plots established by J. Barker of Western Forest Products Limited. The Sitka spruce data for a medium site (Table 33) show about 950 m^3/ha at a top height of 37 m, but this is merchantable volume. Table 33 shows that the stand volume increments for the last several decades are greater on the poor site class than on medium or good sites at the same age interval, a reflection of the fact that a given level of five-year increment develops later on poor sites than on medium or good sites. Note, however, that for all three site classes, volume is about the same for a given height.

There is much more information on the effects of thinning from British data than there is for British Columbia Sitka spruce. British normal yield tables showing yields from thinnings (Hamilton and Christie 1971) are provided in Table 34. In the table, yield classes 24 and 14 refer to the maximum mean annual volume increment of a stand of trees, in $m^3/(ha \cdot yr)$. As shown in the rightmost column, mean annual increment (MAI) peaks at 45-50 years in yield class 24 and at 55-60 years in yield class 14.

Normal yield tables for British Sitka spruce without thinnings are provided in Table 35. In contrast to the frequent use of commercial thinning in Britain, economic conditions do not permit commercial thinning in the few pole-sized second-growth hemlock–Sitka spruce stands in British Columbia. An examination of self-thinning in pure Sitka spruce plantations in Britain (Table 35) shows that loss of trees to self-thinning is substantial and that unthinned stands have low mean diameters when regenerated at high densities.

It seems unreasonable to expect a duplication in British Columbia of the high gross yields seen in the British Sitka spruce yield tables, which were obtained by harvesting all mortality through the use of repeated thinning. Only in the Queen Charlotte Islands can Sitka spruce silviculture be taken seriously at the moment, but even there, with difficult road construction conditions, there are economic barriers to repeated commercial thinning. Competition-induced mortality can be avoided only by short rotations and wide spacings, as is practised in windthrow-prone parts of the United Kingdom. In British Columbia, decisions on plantation and pre-commercial thinning densities are often driven by the objectives of achieving basic silvicultural stocking standards quickly and cheaply and achieving green-up quickly because of adjacency constraints.

In British Columbia, however, as in the Pacific Northwest hemlock-spruce stands documented by Kellogg et al. (1986), commercial thinning has a very low profit margin. In southeastern Alaska, Deal and Farr (1994) found that the amount of western hemlock and Sitka spruce regeneration increased with increasing thinning intensity. Most of this thinning-induced regeneration developed within three years of thinning. The rapid infilling of available growing space was especially noticeable in young stands (under 30 years) on upland sites. In older stands, heavy thinning also promoted

Table 33

Net merchantable volume production of Sitka spruce natural stands on the Queen Charlotte Islands, British Columbia

Good site class[a]			Medium site class			Poor site class		
Age	Stand height (m)	Stand volume[b] (m³/ha)	Age	Stand height (m)	Stand volume (m³/ha)	Age	Stand height (m)	Stand volume (m³/ha)
20	11.3	45	20	7.0	4	20	4.7	0
25	15.6	141	25	10.0	29	25	6.8	3
30	19.6	269	30	13.0	85	30	9.0	17
35	23.4	401	35	16.0	165	35	11.3	49
40	26.7	527	40	16.9	256	40	13.3	97
45	29.8	640	45	21.5	351	45	15.7	157
50	32.5	741	50	24.0	443	50	17.9	229
55	34.9	829	55	29.3	523	55	19.9	293
60	37.0	905	60	28.4	609	60	21.8	363
65	38.9	972	65	30.3	681	65	23.5	430
70	40.7	1,031	70	32.0	747	70	25.3	495
75	42.2	1,083	75	33.6	807	75	26.9	555
80	43.5	1,128	80	35.1	861	80	28.4	612
85	44.8	1,169	85	36.5	909	85	29.8	665
90	45.9	1,204	90	37.7	954	90	31.2	715
95	46.9	1,236	95	38.8	993	95	32.4	760
100	47.9	1,265	100	39.5	1,030	100	33.5	803
105	48.5	1,291	105	40.8	1,063	105	34.6	842
110	49.4	1,314	110	41.7	1,093	110	35.6	878
115	50.0	1,335	115	42.5	1,120	115	36.5	912
120	50.7	1,354	120	43.3	1,146	120	37.4	943
125	51.2	1,372	125	44.0	1,169	125	38.2	972
130	51.5	1,388	130	44.6	1,190	130	39.0	999
135	52.3	1,402	135	45.3	1,210	135	39.7	1,024
140	52.7	1,416	140	45.8	1,228	140	40.4	1,048
145	53.1	1,426	145	46.3	1,245	145	41.0	1,070
150	53.5	1,440	150	46.8	1,261	150	41.6	1,090
155	53.9	1,450	155	47.3	1,275	155	42.2	1,109

Source: Unpublished data courtesy of Western Forest Products Limited, Vancouver

[a] Site class is based on top height at breast-height age 50 for three general site classes: good, 35 m; medium, 27 m; poor, 21 m.

[b] Volume is for 12 cm or greater dbh, to a top inside bark diameter of 10 cm and a stump height of 30 cm, with no allowance for decay, waste, or breakage.

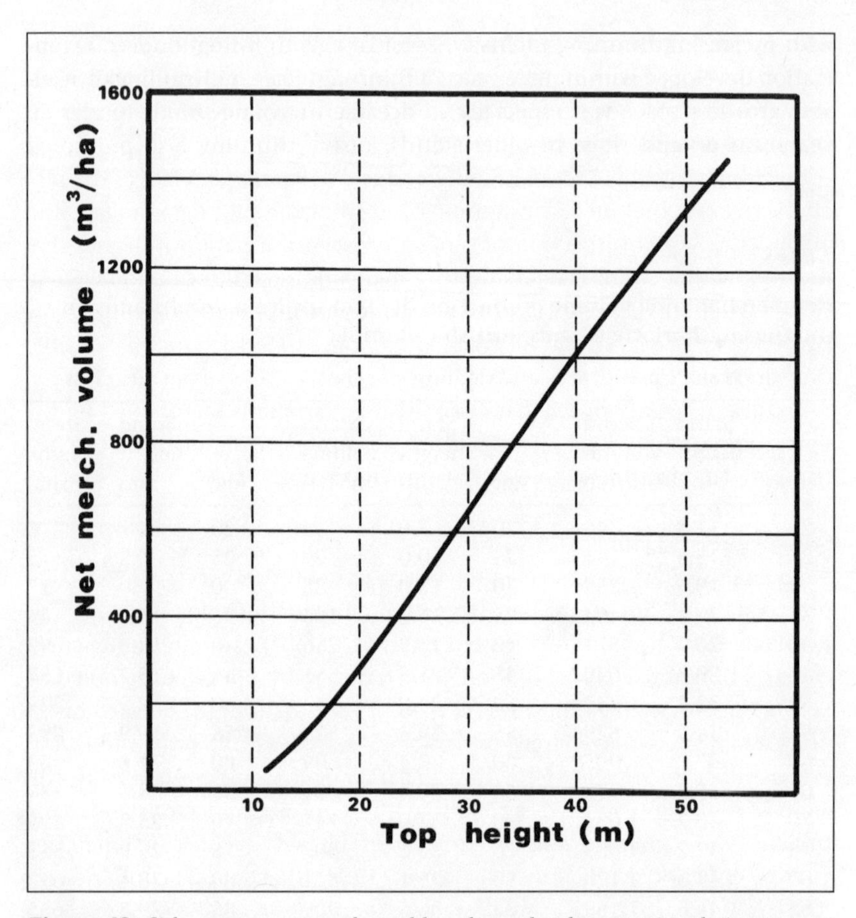

Figure 43. Sitka spruce net merchantable volume development in relation to top height, based on natural stands on the Queen Charlotte Islands, British Columbia, from data assembled by Western Forest Products Limited, Vancouver. Top height is defined as the average height of the 100 trees of largest diameter at breast height (dbh) on 1 hectare. Net volume is based on all stems with a dbh of 12 cm or more, a top inside bark diameter of 10 cm, and a stump height of 30 cm, with no allowance for decay, waste, or breakage.

understorey conifer regeneration sufficiently dense that it was difficult for other understorey plants to become established.

It is interesting to note the amount of gross volume production seen under British conditions with repeated light thinning. For example, the top yield class of 24 $m^3/(ha \cdot yr)$ produces 1,605 m^3/ha at age 75 (Table 34). In contrast, the same yield class with no thinning and spacing produces only 1,424 m^3/ha at age 76 (Table 35). A comparison of British yield tables for lodgepole pine and TASS (Tree and Stand Simulator) output has indicated

Table 34

Normal yield tables for Sitka spruce in Britain, showing yields from thinnings

	Main crop after thinning							Yield from thinnings						Cumulative production		Increment		
					Volume in m^3 to top diameter of:						Volume in m^3 to top diameters of:					CAI[a]	MAI[a]	
Age (yr)	No. of trees	Top ht (m)	Mean dbh (cm)	Basal area (m^2/ha)	7 cm	18 cm	24 cm	No. of trees	Mean dbh (cm)	Mean vol/tree (m^3)	7 cm	18 cm	24 cm	Basal area (m^2/ha)	Vol. to 7 cm (m^3)	Basal area (m^2/ha)	Vol. to 7 cm (m^3/[ha · yr])	Vol. to 7 cm (m^3/[ha · yr])
Yield Class 24[b]																		
10	3,342	5.9	7.7	15.4	32	0	0	0	0.0	0.000	0	0	0	15.4	32	5.20	16.8	3.2
15	2,510	9.9	11.5	26.3	96	0	0	832	12.6	0.047	39	1	0	36.6	135	3.76	24.8	9.0
20	1,365	14.1	15.5	25.8	157	18	0	1,145	13.6	0.073	84	4	0	52.9	279	3.01	31.4	14.0
25	875	18.2	20.5	28.8	242	108	18	490	16.8	0.171	84	15	0	66.7	448	2.55	34.5	17.9
30	625	22.0	25.4	31.7	333	239	99	250	21.2	0.337	84	42	8	78.5	624	2.18	34.4	20.8
35	477	25.3	30.2	34.1	417	354	230	148	25.7	0.570	84	61	26	88.5	791	1.85	32.1	22.6
40	383	28.2	34.6	36.0	487	444	350	94	30.0	0.879	82	70	45	97.0	944	1.57	28.8	23.6
45	324	30.6	38.6	37.9	549	517	446	59	34.1	1.238	72	66	51	104.2	1,079	1.35	25.4	24.0
50	285	32.7	42.2	39.9	606	580	525	39	37.8	1.635	63	59	50	110.5	1,198	1.17	22.4	24.0
55	258	34.4	45.4	41.8	656	635	589	27	41.2	2.057	54	52	46	115.9	1,302	1.01	19.7	23.7
60	238	36.0	48.2	43.5	701	682	641	20	44.3	2.469	47	46	42	120.6	1,395	0.86	17.2	23.2

▶

◀ *Table 34*

	Main crop after thinning							Yield from thinnings						Cumulative production		Increment CAI[a]	MAI[a]	
					Volume in m^3 to top diameter of:						Volume in m^3 to top diameters of:							
Age (yr)	No. of trees	Top ht (m)	Mean dbh (cm)	Basal area (m^2/ha)	7 cm	18 cm	24 cm	No. of trees	Mean dbh (cm)	Mean vol/ tree (m^3)	7 cm	18 cm	24 cm	Basal area (m^2/ha)	Vol. to 7 cm (m^3)	Basal area (m^2/ha)	Vol. to 7 cm (m^3/ [ha · yr])	Vol. to 7 cm (m^3/ [ha · yr])
65	223	37.3	50.6	44.9	739	722	683	15	47.0	2.877	42	41	38	124.5	1,475	0.74	14.9	22.7
70	211	38.5	52.7	46.0	770	754	717	12	49.4	3.231	38	37	35	127.9	1,544	0.64	13.0	22.1
75	201	39.5	54.6	47.2	797	782	747	10	51.5	3.696	34	33	32	131.0	1,605	0.56	11.4	21.4
80	193	40.3	56.3	48.1	818	804	770	8	53.4	4.160	31	31	29	133.5	1,657	0.49	9.6	20.7
Yield Class 14[b]																		
15	3,331	6.2	8.3	18.0	39	0	0	0	0.0	0.000	0	0	0	18.0	39	2.62	8.6	2.6
20	2,681	8.8	10.6	23.7	75	0	0	650	12.5	0.044	28	1	0	31.7	103	2.52	14.8	5.1
25	1,809	11.5	13.0	23.9	109	3	0	872	12.8	0.056	49	1	0	43.2	187	2.17	18.3	7.5
30	1,299	14.3	16.1	26.3	159	22	0	510	14.0	0.096	49	3	0	53.4	285	1.93	20.6	9.5
35	982	16.9	19.4	29.0	217	79	10	317	16.0	0.155	49	7	0	62.5	392	1.68	21.3	11.2
40	780	19.3	22.6	31.2	274	159	42	202	18.6	0.242	49	15	1	70.2	499	1.43	20.7	12.5
45	639	21.5	25.6	32.8	325	235	99	141	21.3	0.347	49	25	5	76.8	599	1.22	19.1	13.3
50	545	23.4	28.3	34.2	371	300	169	94	23.8	0.475	45	29	9	82.4	689	1.04	17.1	13.8
55	483	24.9	30.7	35.6	413	354	237	62	26.0	0.607	38	28	13	87.2	769	0.88	14.9	14.0
60	438	26.2	32.7	36.9	448	399	296	45	28.0	0.744	33	27	15	91.2	838	0.73	12.8	14.0

▶

◀ *Table 34*

	Main crop after thinning							Yield from thinnings						Cumulative production		Increment		
																CAI[a]		MAI[a]
					Volume in m^3 to top diameter of:						Volume in m^3 to top diameters of:							
Age (yr)	No. of trees	Top ht (m)	Mean dbh (cm)	Basal area (m^2/ha)	7 cm	18 cm	24 cm	No. of trees	Mean dbh (cm)	Mean vol/ tree (m^3)	7 cm	18 cm	24 cm	Basal area (m^2/ha)	Vol. to 7 cm (m^3)	Basal area (m^2/ha)	Vol. to 7 cm (m^3/ [ha · yr])	Vol. to 7 cm (m^3/ [ha · yr])
65	404	27.3	34.5	37.8	478	436	342	34	29.8	0.854	29	25	15	94.4	897	0.61	11.1	13.8
70	377	28.3	36.1	38.6	503	466	382	27	31.3	0.992	26	23	16	97.2	948	0.54	9.9	13.5
75	356	29.1	37.5	39.4	527	493	417	21	32.8	1.125	23	21	15	99.8	995	0.47	8.7	13.3
80	339	29.8	38.7	40.0	545	514	444	17	34.0	1.241	22	20	15	102.0	1,035	0.39	7.0	12.9

Source: Hamilton and Christie (1971). The Hamilton and Christie (1971) yield tables have been superseded by Forestry Commission Booklet 48 (Edwards and Christie 1981; see also Rollinson 1985). As of December 1996, Booklet 48 was being revalidated by mensurationists in the Forestry Commission (J. Parker, pers. comm., Dec. 1996).

[a] CAI = current annual increment; MAI = mean annual increment

[b] Yield classes 24 and 14 refer to the maximum mean annual volume increment of a stand of trees, in m^3/ (ha · yr). As shown in the right-hand column of this table, mean annual increment (MAI) peaks at 45-50 years in yield class 24 and at 55-60 years in yield class 14.

Table 35

Normal yield tables for Sitka spruce in Britain with no thinning

Age (yr)	Top ht (m)	Trees per ha	Mean dbh (cm)	Basal area (m²/ha)	Mean vol. per tree (m³)	Vol. (m³/ha)	Percent mortality	MAI (m³/ [ha · yr])	Age (yr)
Yield Class 24									
16	9.6	1,069	17	23	0.09	91	0	5.7	16
21	14.0	1,018	23	41	0.23	232	3	11.0	21
26	18.1	885	27	52	0.43	384	5	14.8	26
31	21.8	771	31	60	0.70	538	6	17.4	31
36	25.0	690	35	66	0.99	685	7	19.0	36
41	27.8	632	38	72	1.30	821	7	20.0	41
46	30.1	591	41	77	1.59	942	7	20.5	46
51	32.2	562	43	81	1.87	1,050	7	20.6	51
56	33.9	540	45	84	2.12	1,145	7	20.4	56
61	35.3	525	46	88	2.34	1,229	7	20.1	61
66	36.6	513	47	91	2.54	1,303	7	19.7	66
71	37.8	504	49	93	2.72	1,368	6	19.3	71
76	38.8	497	50	96	2.87	1,424	6	18.7	76
Yield Class 14									
18	6.6	1,085	11	11	0.03	29	0	1.6	18
23	9.5	1,069	16	22	0.08	84	0	3.6	23
28	12.3	1,040	20	34	0.16	170	1	6.1	28
33	15.0	985	24	44	0.27	265	3	8.0	33
38	17.5	912	27	51	0.40	362	4	9.5	38
43	19.7	842	29	57	0.54	455	5	10.6	43
48	21.7	779	31	61	0.69	539	6	11.2	48
53	23.4	729	33	64	0.84	612	6	11.5	53
58	24.8	690	35	66	0.98	674	7	11.6	58
63	26.0	661	36	69	1.10	727	8	11.5	63
68	27.0	638	38	70	1.21	774	8	11.4	68
73	27.9	618	39	72	1.32	817	8	11.2	73
78	28.7	601	39	73	1.42	854	8	10.9	78

Source: Hamilton and Christie (1971). The Hamilton and Christie (1971) yield tables have been superseded by Forestry Commission Booklet 48 (Edwards and Christie 1981; see also Rollinson 1985). As of December 1996, Booklet 48 was being revalidated by mensurationists in the Forestry Commission (J. Parker, pers. comm., Dec. 1996).

that the British tables are good predictors of lodgepole pine yields in British Columbia (Kovats 1993); it is reasonable to assume that Sitka spruce yields in British Columbia are also well predicted by the British yield tables.

In northern England, the main limiting factor to Sitka spruce growth at present is wind disturbance compounded by shallow rooting on gleyed soils. To avoid risk of windthrow, stands are left unthinned and are clearcut at 35-40 years of age. However, the deterministic nature of the windthrow hazard classification used to predict the onset of wind damage means that the possibility of retaining stands for longer rotations may have been underestimated. Recent evidence suggests that, provided stands are planted using cultivation techniques that promote a stable root architecture, and with thinning at an early age to promote stem diameter growth, it should be possible to maintain some stands for at least 75-80 years to enhance structural diversity (Mason and Quine 1995).

Recommended times of first thinning for Sitka spruce are between 18 and 30 years depending on growth rate, when the trees have a top height of 8.5-12.0 m (Hamilton and Christie 1971). An alternative strategy is to carry out early pre-commercial thinning before trees reach heights of about 12 m. If trees grow in height at between 0.5 and 1.0 m/yr and it takes 3-5 years for canopy closure to recur after thinning, then thinning needs to be carried out when the trees are between 5 and 8 m in height. For wind stability, stands of spruce trees that are thinned at a relatively early age, when the root systems are capable of adaptive response to increased wind loading, should, in theory, be capable of closing canopy before reaching critical height. This strategy assumes that adaptive root growth compensates for increased wind loading on the crowns so that the risk of windthrow is not increased, an assumption that requires adequate verification in the field (Mason and Quine 1995).

Pruning

Sitka spruce sheds dead branches poorly, and, unless pruned, it produces only clear wood on very long rotations. In Britain, pruning is considered essential for the production (without repeated thinning) of clear grades of sawtimber on trees with 40-55 cm diameter and 30-45 m top heights. Experiments with pruning Sitka spruce were started in 1938 at Inverliever in Argyll, but there are few recorded pruning criteria specific to Sitka spruce. These early studies have been reviewed by Henmanns (1963), who made general recommendations for pruning of British conifers under the following headings:

- reasons for pruning
- utilization aspects of pruning

- economic considerations
- choice of species
- selection of stands (with the greatest priority on young stands on sites with low risks and with high site productivity)
- selection of the largest and most vigorous stems
- size of tree
- number of trees to prune (usually from 300 to 350 trees per hectare)
- amount of pruning of green branches, as opposed to simply pruning dead branches (usually no more than two whorls of live branches should be pruned)
- season of year for pruning (typically March to May)
- length of stem to prune (typically 3-6 m)
- criteria for thinning stands that have already been pruned (there may still be a need to concentrate growth on pruned trees by use of crown thinnings, but also to retain some lower canopy trees to reduce the incidence of lower-stem epicormic shoots on Sitka spruce)
- needs for recordkeeping
- recommended tools, techniques, and schedules.

In Henmanns' summary of pruning experience in Britain, Norway spruce and Sitka spruce were grouped together, neither species experiencing any problems of disease in the knots or wood; stem diameter growth after pruning wounds were healed resulted in production of normal clear wood.

The forest pruning bibliography by O'Hare (1989) recorded 1,129 references to pruning, of which only a few were for Sitka spruce. The healing process following pruning of Sitka spruce grown in Norway was described in detail in Vadla (1990a). After mainly live branches were pruned, an average of 11.4 years was required to complete healing. As expected, length of branch stubs was the most important single variable in the early stages of healing; ring width was the most important variable in the later stages. In a sample of 306 knots, all were free of decay and stain. About 37% of the knots had fragments of bark that had grown into the healing zone. Pruning time varied substantially between stands, mainly because stands with the largest average branch diameters take longer to prune than stands with smaller branch diameters (Vadla 1990b).

The computer model SYLVER for Douglas-fir, which simulates the growth of pruned trees and predicts lumber quality, has not been calibrated for Sitka spruce, and neither has the model DFPRUNE for Douglas-fir in the Pacific Northwest (Fight et al. 1992). In general, however, the literature does not suggest any particular problems with pruning Sitka spruce provided that excessive epicormic branching can be avoided. On this basis there appears to be a case for pruning Sitka spruce at present premium values for clear grades of lumber.

Pruning may have another beneficial effect. In Alaska, Hard (1992) observed that pruning live branches of long-crowned Lutz spruce baited with frontalin, an attractant pheromone, reduced attacks by the spruce beetle (*Dendroctonus rufipennis*) in pruned sections of stems. The benefits of pruning were thought to result primarily from increased temperature and light intensity on pruned boles. Pruning may be a useful tool for reducing the number of suitable attack sites in trees that are at high risk for spruce beetle attack. This could be especially important as a way to protect spruce with high commercial or aesthetic value.

As of 1993, British Columbia Ministry of Forests pruning guidelines were not specific to particular species of coniferous crop trees, but they did recommend an initial pruning of up to 2.9 m of branch removal (referred to as 'lift' of the lower edge of the tree crown), and then a second pruning that would raise the total lift of the crown base to 5.5 m. Based on TASS stand growth projections, Massie (1992) analyzed financial returns from pruning British Columbia coastal conifers and concluded that pruning could be feasible if the first pruning (to 2.9 m above ground level) could be done for $2.00 per stem or less, and if the second pruning (to 5.5 m) could be done for $4.00 per stem or less. Massie's estimates were based on a 4% per year rate of return for 1986-90 price differences between knotty and clear butt logs 2.5-5.0 m in length, with no specified core diameter in which knots would be expected.

Pruning of Sitka spruce is not yet widely performed in British Columbia, but in 1993-94 the South Moresby Forest Replacement Account financed the pruning of 95 ha of Sitka spruce on the Queen Charlotte Islands in cooperation with the Sandspit Division of TimberWest Forest Ltd. (SMFRA 1994).

Crop Planning

Decisions regarding density control by initial spacing in plantations (Table 32) and residual densities in pre-commercial thinning and commercial thinning are key components of crop planning. Decisions are usually based on stand yield projections using classical yield tables or stand simulations. Crop planning is a general concept that can be defined and developed in several alternative ways. Here we define a crop plan as a management regime for an individual stand or group of stands. Crop planning, therefore, consists of activities undertaken to determine the management regime.

A crop plan is developed to achieve a particular set of management objectives. In broad terms, these objectives are for the provision of specific resource values over time. Interpreted in terms of timber characteristics, the objective of a crop plan is to produce a specified series of stand structures over time by following a particular pattern of stand development. At each stage of stand development, the stand structure that the crop plan has helped produce will provide some of the resource values specified in the management objectives.

To illustrate this definition of a crop plan, consider a simple example with two management objectives and a recently harvested stand. Assume that the objectives for the regenerated stand are to extend the period of maximum browse production and to reduce the period required to produce merchantable timber. One way to achieve these management objectives would be to regenerate the stand to a wide spacing (perhaps 600-800 stems per hectare). Under this crop plan, a set of stand structures would be produced over time on the site from which the desired resource values flow. By controlling establishment density, we can encourage the stand to progress along a specified pathway of stand development. More detailed examples could involve objectives that are more accurately specified quantitatively.

A crop plan may include: (1) manipulation of the site (such as fertilization and site preparation); (2) manipulation of the stand (such as harvesting, planting, thinning, and other silvicultural activities); and (3) protection of the stand from damage and disturbance due to fire, wind, insects, disease, animals, and other agents.

As indicated above, the concept of a crop plan and crop planning applies to both individual stands and groups of stands. At a strategic level, crop plans are developed for broad stand types. At the operational level, crop planning pertains to individual stands, and the crop plan describes the management regime proposed for a specific stand. In most crop-planning processes, the operational and strategic levels are linked.

To understand how specific growth and yield information relates to crop planning for Sitka spruce, it is useful to briefly consider the process of developing a crop plan. Two recent Ministry of Forests publications (B.C. Ministry of Forests 1990b, 1991) provide examples of the required steps. Crop planning at the strategic level generally involves the use of both growth and yield prediction models and forest-level models. The characteristics of the existing forest are projected into the future. The selected models are used to simulate alternative regimes of harvesting, silviculture, protection, and natural losses. These simulation results are interpreted relative to the management objectives in order to identify regimes that best meet the objectives.

Crop planning at the operational level generally involves designing a specific plan for a single stand within the framework of the general crop plans developed by the forest-level, strategic analysis. Since the crop plan has meaning at both strategic and operational levels, growth and yield information is required for both levels. Each of the types of growth and yield information considered in the sections 'Estimating Site Quality' (see page 208) to 'Growth and Yield Data' (see page 228) later in this chapter can be required for either strategic or operational crop planning.

More than growth and yield information is required for comprehensive crop planning. For example, economic and other resource values must be considered. However, the subsections that follow are limited to a consideration of traditional timber values. The recent report *Wood Production Strategies and the Implications for Silviculture in B.C.* (H.A. Simons Strategic Services 1992) is a good example of the kind of global economic information that could be considered during comprehensive crop planning. The recently developed economic analysis module for WinTIPSY (Stone et al. 1996) is another useful aid in crop planning.

Biogeoclimatic Ecosystem Classification and Crop Planning

British Columbia's system of biogeoclimatic ecosystem classification (BEC) is the most important source of crop-planning information in the province today (MacKinnon et al. 1992). Key features of British Columbia's BEC zones, subzones, variants, and site series where Sitka spruce is most likely to occur are outlined in the subsections 'Latitudinal and Altitudinal Distribution of Sitka Spruce' (see page 40) to 'Summary of Biogeoclimatic Zones, Subzones, Variants, and Site Series Where Sitka Spruce Occurs in British Columbia' (see page 51) in Chapter 2, particularly in Tables 14, 15, and 16. Such a classification system helps to highlight differences between various ecosystems or sites; such differences influence most subsequent silvicultural activities, including harvesting, site preparation, regeneration, and stand tending.

The BEC provides a taxonomic classification of forest ecosystems, but the provision of growth and yield information is not one of its explicit objectives. Growth and yield information from other sources must be related to the classification to enable additional silvicultural interpretations. Fortunately a set of ecosystems grouped together by the BEC will generally have many similar growth and yield characteristics. For Sitka spruce, important examples of the latter are presented in Tables 14, 15, and 16, where estimated site index values for spruce at breast-height age 50 are given, on a site series basis, for the Vancouver and Prince Rupert forest regions, based on data from Green and Klinka (1994) and Banner et al. (1993), respectively.

Growth and Yield Forecasts

Information on the growth rate and current yield typically involves the following traditional timber parameters: stand mean diameter, stand top height, merchantable volume per hectare, number of stems per hectare, and site index. These parameters describe the distribution of trees on the site and the capability of the site to produce trees. Growth and yield information describes how timber parameters change in response to stand, site, damaging agents, and treatment factors. Such information is reviewed in subsequent sections in terms of site quality estimation, predicting regeneration, density management, prediction of volume, growth and yield prediction models, and an overview of Sitka spruce crop planning.

Estimating Site Quality

Attributes of trees as crop plants (Cannell and Jackson 1985) have been referred to in earlier sections of this chapter. One key attribute of crop trees is mean annual height increment. This variable was shown in a generalized way for Sitka spruce in Figure 40, where data from Ford (1985) were used to portray canopy development in a spruce plantation. For crop planning, it is necessary to have estimates of the influence of site quality upon measures such as years to reach breast height and top-height growth curves. The following subsections focus on approaches to obtaining site index estimates for Sitka spruce. Fortunately in British Columbia this has been a subject of recent significant data refinement. Good predictions of Sitka spruce site index in the province can now be made with up-to-date information on site index conversion equations for mixed Sitka spruce–western hemlock stands (Nigh 1995), relationships between site index and soil moisture–soil nutrient regimes for western hemlock and Sitka spruce (Kayahara and Pearson 1996), calibration of Sitka spruce site indices for a variable growth intercept model (Nigh 1996b), and new site index curves and tables for Sitka spruce (Nussbaum 1996). The types of growth and yield information considered here support the full range of decisions for crop planning, including decisions pertaining to species selection, establishment method, brushing and weeding, spacing, and thinning.

Site Index and Height Growth in Sitka Spruce

Some of the earliest site index curves prepared for Sitka spruce in its natural range are those by Taylor (1934), Meyer (1937), Stephens et al. (1969), and Hegyi et al. (1981). In the last decade, Farr and Harris (1983), Farr (1984), Mitchell and Polsson (1988), Thrower and Nussbaum (1991), Nigh (1995), and Nussbaum (1996) have also provided improved site index curves for this species. The construction of site index curves is an ongoing process, and the latest published site index curves and tables for Sitka spruce are those prepared by Nussbaum (1996). The curves by Nussbaum are a refinement of curves by Thrower and Nussbaum (1991), which had been developed largely from an unpublished report by Barker and Goudie (1987). Barker and Goudie's height growth curves were developed from stem analyses of trees in 48 plots located throughout the Queen Charlotte Islands. Sample trees ranged up to 150 years at breast height, and in site index from 17 to 38 m at age 50. Barker and Goudie considered the curves to be suitable for second-growth coastal Sitka spruce. Estimates of site index from known height and age are obtained from iteration of the height curve equation.

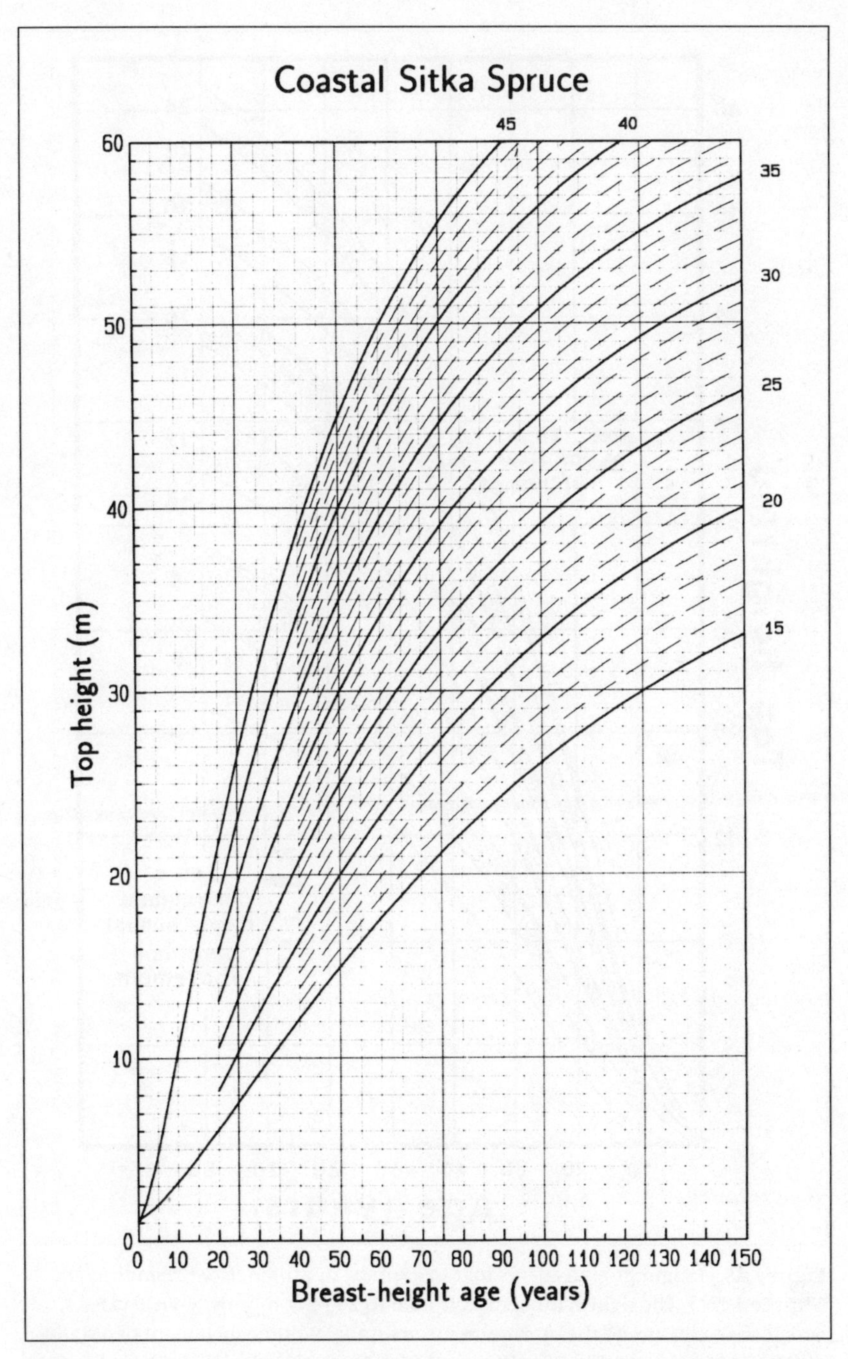

Figure 44. Site index curves for Sitka spruce in British Columbia (from Nussbaum 1996).

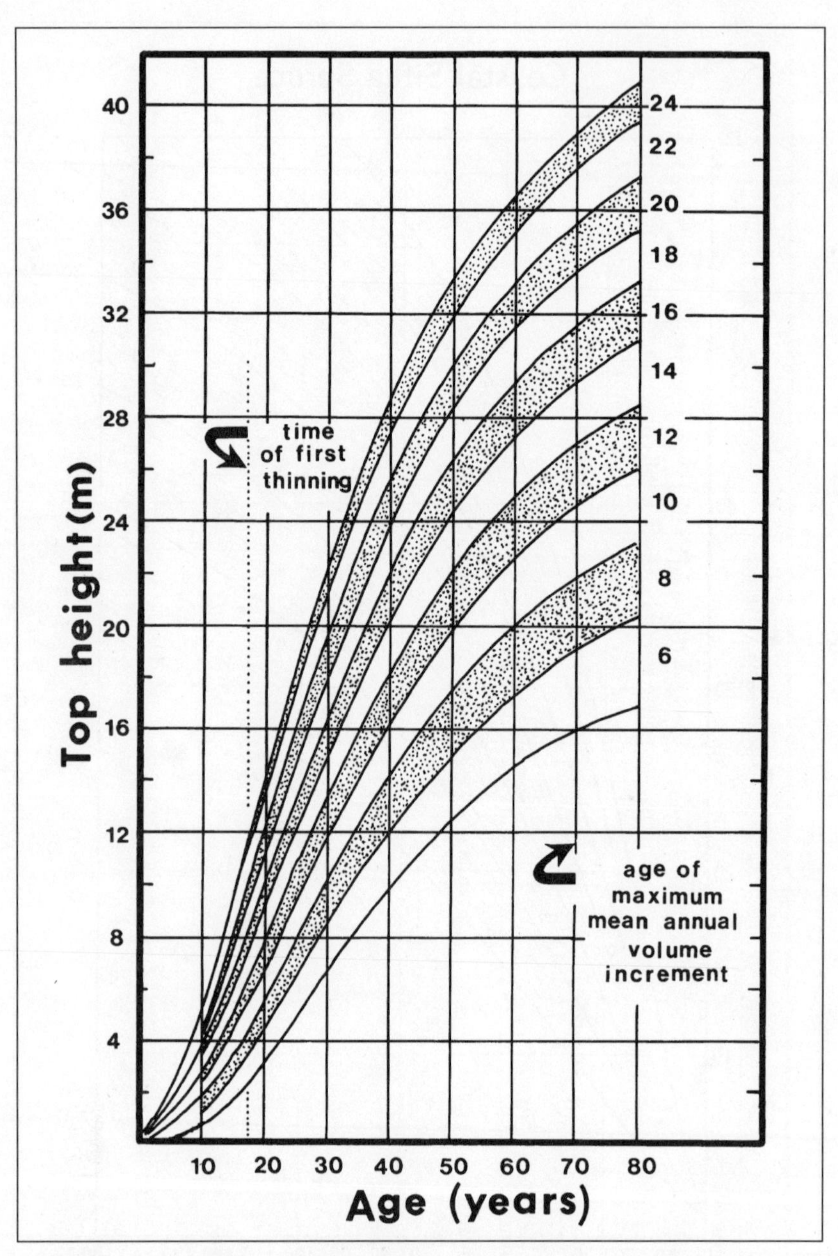

Figure 45. Height growth curves for Sitka spruce in Britain (from Hamilton and Christie 1971). The right-hand figures from 6 to 24 refer to general yield classes, which are a measure of the maximum mean annual volume increment of a stand of trees in $m^3/(ha \cdot yr)$. The Hamilton and Christie (1971) yield tables have been superseded by Forestry Commission Booklet 48 (Edwards and Christie 1981; see also Rollinson 1985). As of December 1996, Booklet 48 was being revalidated by mensurationists in the Forestry Commission (J. Parker, pers. comm., Dec. 1996).

Nussbaum's refined site index curves for Sitka spruce in British Columbia, derived from a height-age model by Nigh (1996a), are reproduced in Figure 44; the corresponding 1996 site index table for Sitka spruce is provided in Table 36. The curves and the table are for second-growth stands and should not be applied to old-growth forests. For comparison, the height-over-age curves for British plantations with the equivalent yield classes are given in Figure 45.

The site index values for Sitka spruce in Table 36 are based on the following equations:

$$H = 1.3 + (SI - 1.3)(b_1/b_2)$$
$$b_1 = 1 + \exp\ [8.947 - 1.357\ln 49.5 - 1.013\ln(SI - 1.3)]$$
$$b_2 = 1 + \exp\ [8.947 - 1.357\ln(A - 0.5) - 1.013\ln(SI - 1.3)]$$
$$ytb = 11.7 - 0.185SI$$

where:

H = top height (m)
SI = site index (height at breast-height age 50)
A = breast-height age (years)
exp = exponential function to the base e
ln = natural logarithm
ytb = number of years to reach breast height (1.3 m)

The values in Table 36 are the basis of the height-age (site index) curves shown in Figure 44. These data were developed from 40 stem analysis plots established in ecologically uniform areas of Sitka spruce stands on the Queen Charlotte Islands. All plots were in the CWHwh1 biogeoclimatic variant. Plot ages ranged from 50 to 121 years at breast height, and site index ranged from 13.6 to 40.3 m (Nigh 1996a).

The user of Figure 44 or Table 36 should note that the site curves recommended by the Coastal Forest Productivity Council are based on top height, breast-height age, and an index age of 50 years. Top height, derived from the largest 100 trees per hectare, is an objective measure of site height, unlike those based on subjectively selected dominants and codominants. It is employed in Europe and is being accepted in Canada and the United States.

Breast-height age is preferred because it is convenient to obtain; it also tends to ignore the early development of site trees when non-site factors (such as animal damage and brush competition) can affect height growth. It has been adopted by most agencies in the western United States. The choice of breast-height age does not preclude the use of the total age for other management purposes.

Table 36

Site index table for Sitka spruce in British Columbia

bh age (yrs)	Top height (m)																											
	6	8	10	12	14	16	18	20	22	24	26	28	30	32	34	36	38	40	42	44	46	48	50	52	54	56	58	60
	Site index (m)																											
10	29	37	44	50	–	–	–	–	–	–	–	–	–	–	–	–	–	–	–	–	–	–	–	–	–	–	–	–
15	20	25	31	36	40	45	49	–	–	–	–	–	–	–	–	–	–	–	–	–	–	–	–	–	–	–	–	–
20	15	19	24	28	32	35	39	42	46	49	–	–	–	–	–	–	–	–	–	–	–	–	–	–	–	–	–	–
25	12	16	19	23	26	29	32	35	38	41	44	47	50	–	–	–	–	–	–	–	–	–	–	–	–	–	–	–
30	10	13	16	19	22	25	28	31	33	36	39	41	44	46	49	–	–	–	–	–	–	–	–	–	–	–	–	–
35	–	11	14	17	19	22	24	27	29	32	34	37	39	41	44	46	48	50	–	–	–	–	–	–	–	–	–	–
40	–	10	12	15	17	20	22	24	26	29	31	33	35	38	40	42	44	46	48	–	–	–	–	–	–	–	–	–
45	–	–	11	13	15	18	20	22	24	26	28	30	32	35	37	39	41	43	45	47	49	–	–	–	–	–	–	–
50	–	–	10	12	14	16	18	20	22	24	26	28	30	32	34	36	38	40	42	44	46	48	50	–	–	–	–	–
55	–	–	–	11	13	15	17	18	20	22	24	26	28	30	32	34	36	38	40	42	43	45	47	49	–	–	–	–
60	–	–	–	10	12	14	15	17	19	21	23	24	26	28	30	32	34	36	37	39	41	43	45	47	49	–	–	–
65	–	–	–	–	11	13	14	16	18	19	21	23	25	26	28	30	32	34	36	38	39	41	43	45	47	49	–	–
70	–	–	–	–	10	12	13	15	17	18	20	22	23	25	27	29	30	32	34	36	38	40	41	43	45	47	49	–
75	–	–	–	–	10	11	13	14	16	17	19	21	22	24	26	27	29	31	33	34	36	38	40	42	44	45	47	49
80	–	–	–	–	–	10	12	13	15	16	18	19	21	23	24	26	28	30	31	33	35	37	39	40	42	44	46	48
85	–	–	–	–	–	10	11	13	14	16	17	19	20	22	23	25	27	28	30	32	34	36	37	39	41	43	45	46
90	–	–	–	–	–	–	11	12	13	15	16	18	19	21	23	24	26	27	29	31	33	34	36	38	40	42	43	45
95	–	–	–	–	–	–	10	11	13	14	16	17	19	20	22	23	25	27	28	30	32	33	35	37	39	41	42	44
100	–	–	–	–	–	–	10	11	12	14	15	16	18	19	21	22	24	26	27	29	31	33	34	36	38	40	41	43
105	–	–	–	–	–	–	–	10	12	13	14	16	17	19	20	22	23	25	27	28	30	32	33	35	37	39	41	42

►

◄ *Table 36*

bh age (yrs)	Top height (m)																											
	6	8	10	12	14	16	18	20	22	24	26	28	30	32	34	36	38	40	42	44	46	48	50	52	54	56	58	60
	Site index (m)																											
110	–	–	–	–	–	–	–	10	11	12	14	15	17	18	20	21	23	24	26	28	29	31	33	34	36	38	40	42
115	–	–	–	–	–	–	–	10	11	12	13	15	16	17	19	20	22	24	25	27	29	30	32	34	35	37	39	41
120	–	–	–	–	–	–	–	–	10	12	13	14	16	17	18	20	21	23	25	26	28	30	31	33	35	36	38	40
125	–	–	–	–	–	–	–	–	10	11	12	14	15	16	18	19	21	22	24	26	27	29	31	32	34	36	38	39
130	–	–	–	–	–	–	–	–	10	11	12	13	15	16	17	19	20	22	23	25	27	28	30	32	33	35	37	39
135	–	–	–	–	–	–	–	–	–	10	12	13	14	16	17	18	20	21	23	25	26	28	30	31	33	35	36	38
140	–	–	–	–	–	–	–	–	–	10	11	12	14	15	16	18	19	21	22	24	26	27	29	31	32	34	36	38
145	–	–	–	–	–	–	–	–	–	10	11	12	13	15	16	17	19	20	22	24	25	27	29	30	32	34	35	37
150	–	–	–	–	–	–	–	–	–	10	11	12	13	14	16	17	19	20	22	23	25	26	28	30	31	33	35	37
155	–	–	–	–	–	–	–	–	–	–	10	11	13	14	15	17	18	20	21	23	24	26	28	29	31	33	35	36
160	–	–	–	–	–	–	–	–	–	–	10	11	12	14	15	16	18	19	21	22	24	26	27	29	31	32	34	36
165	–	–	–	–	–	–	–	–	–	–	10	11	12	13	15	16	17	19	20	22	24	25	27	29	30	32	34	36
170	–	–	–	–	–	–	–	–	–	–	10	11	12	13	14	16	17	19	20	22	23	25	26	28	30	32	33	35
175	–	–	–	–	–	–	–	–	–	–	–	10	12	13	14	15	17	18	20	21	23	24	26	28	30	31	33	35
180	–	–	–	–	–	–	–	–	–	–	–	10	11	12	14	15	16	18	19	21	23	24	26	27	29	31	33	34
185	–	–	–	–	–	–	–	–	–	–	–	10	11	12	13	15	16	18	19	21	22	24	26	27	29	31	32	34
190	–	–	–	–	–	–	–	–	–	–	–	10	11	12	13	15	16	17	19	20	22	24	25	27	29	30	32	34
195	–	–	–	–	–	–	–	–	–	–	–	10	11	12	13	14	16	17	19	20	22	23	25	27	28	30	32	34
200	–	–	–	–	–	–	–	–	–	–	–	–	10	12	13	14	15	17	18	20	21	23	25	26	28	30	32	33
Site index					≤ 11				12-18			19-22			23-28				29-33			34-38				39 ≥		
Years to bh					10				9			8			7				6			5				4		

Source: Nussbaum (1996)

The index age at which site productivity is referenced was lowered to 50 years from the old-growth standard of 100 years to conform to current management practices in second-growth forests, and to make stem analysis data more compatible with modelling techniques. The bottom line in Table 36 provides an estimate of the number of years for Sitka spruce to reach breast height on various sites. These estimates range from 10 years on poor sites to 4 years on the most productive sites.

Farr's (1984) Sitka spruce curves were developed from data collected on 71 plots established in unmanaged even-aged stands of Sitka spruce and western hemlock in southeastern Alaska. Site index (breast-height age 50 basis) ranged from 12 to 35 m. Maximum breast-height age was 180 years. Farr provided both height growth curves and site index curves. When

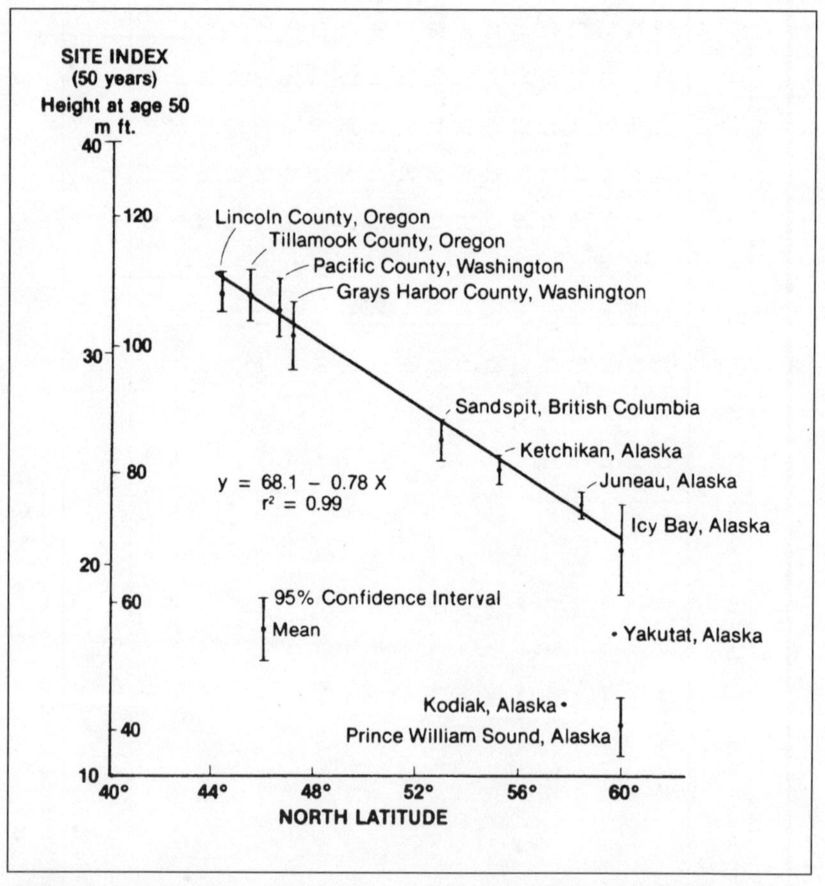

Figure 46. Relationship between Sitka spruce site index and latitude along the Pacific coast from Lincoln County, Oregon, to Icy Bay, Alaska (from Farr and Harris 1983).

compared with Sitka spruce from further south, the Alaska analyses indicate that within the hemlock-spruce type from the vicinity of Coos Bay, Oregon, northwest to the Alaska Peninsula (Farr and Harris 1979, 1983), mean site index for spruce at 50 years decreases about 0.8 m per degree of northward latitude (Figure 46). When this latitudinal decrease in Sitka spruce site index is expressed in terms of degree-days greater than 5°C (Figure 47), a similar relationship is revealed (Farr and Harris 1983).

Relating site index to degree-days was done, in part, because there is speculation that growth in Sitka spruce in a coastal region that generally lacks a pronounced summer drought is more strongly influenced by temperature than by soil moisture. Sitka spruce shares with other conifers of the Pacific Northwest several functionally important structural characteristics that

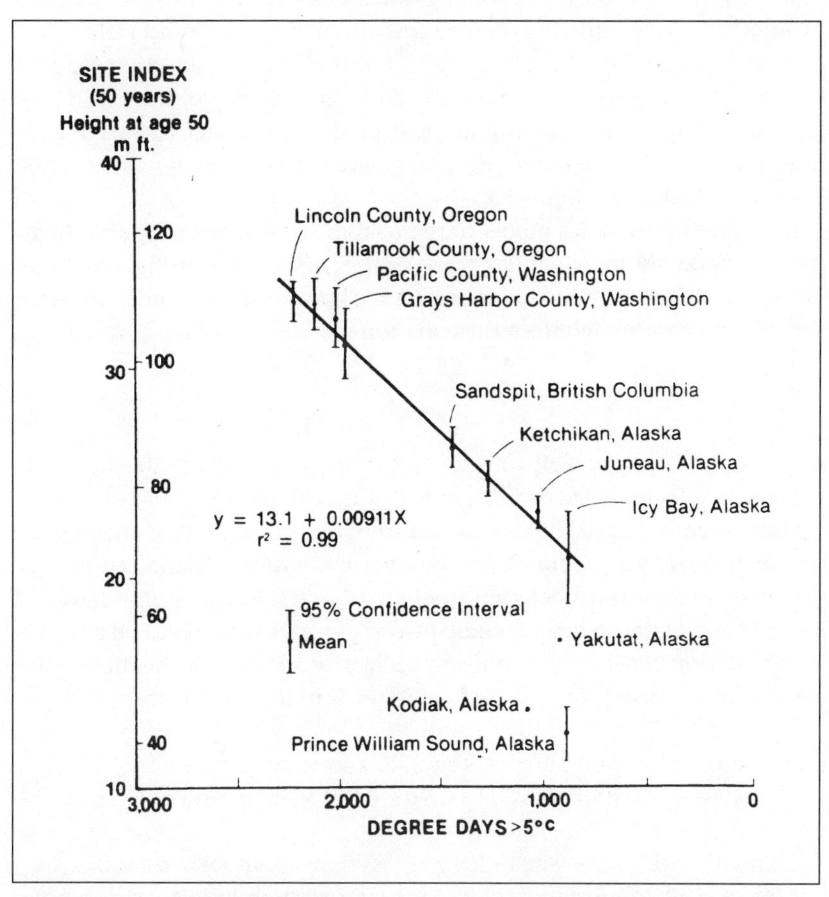

Figure 47. Relationship between Sitka spruce site index and degree-days greater than 5°C for areas along the Pacific coast from Lincoln County, Oregon, to Icy Bay, Alaska (from Farr and Harris 1983).

Waring and Franklin (1979) considered to be advantages under the moisture, temperature, and nutrient regimes of this region, including massiveness of mature forests, evergreenness, large leaf areas, and needle-shaped leaves. Mild winter temperatures permit winter photosynthesis, and cool summer nights result in relatively low respiratory losses from the large standing crop of foliage, branches, and wood. In addition, considerable carbon uptake is possible at temperatures below freezing, even by coastal species such as Sitka spruce (Waring and Franklin 1979).

Farr (1984) cited several sets of earlier site index curves developed with the guide curve method. The most relevant to this study are the site index curves used by the Ministry of Forests in the 1980s (Hegyi et al. 1981). Frequently, two species growing on the same site can achieve different heights at a given site index reference age. Relative site index information describes the site index achieved by one species relative to that achieved by another species growing on the same site. This growth and yield information is useful for crop planning in terms of species selection, species conversion, and predicting stand dynamics. For example, the Stand Projection System (SPS) growth and yield prediction model (see page 227) uses one measure of relative site to regulate interspecies dynamics while simulating stand development.

Farr (1984) provided a model of the relationship between western hemlock and Sitka spruce site index when both species occur on the same site. When the site index for western hemlock is known, the site index for Sitka spruce can be estimated from the equation

$$\text{Estimated } SI_{Ss} = -6.62 + 1.152 \times SI_{Hw}$$

where SI_{Ss} is Sitka spruce site index (feet) at breast-height age 50, and SI_{Hw} is western hemlock site index (feet) at breast-height age 50.

More recently, Nigh (1995) used data from 65 sample plots of mixed Sitka spruce and western hemlock on the Queen Charlotte Islands to predict site index relationships between these two species. Based on stand ages of 28-110 years at breast height, comparisons of top heights revealed a strong linear relationship between the site indices of spruce and hemlock. The conversion equations for site index (metres at index age 50) are:

$$\text{Estimated Sitka spruce } SI = -4.94 + 1.25 \times \text{western hemlock } SI$$
$$\text{Estimated western hemlock } SI = 3.96 + 0.801 \times \text{Sitka spruce } SI$$

On poorer sites, when site index was less than about 20 m for either species, Nigh (1995) found that the site index of western hemlock was higher than the site index of spruce. However, as the productivity of sites increases, spruce grows at an increasingly faster rate than hemlock. A possible management implication is that it may be preferable to regenerate poor sites with hemlock and better sites with Sitka spruce.

Alternative Ways to Estimate Site Index

Site index is often poorly estimated in both very young and very old stands. Improved estimates of site quality for these stands are required for crop planning at both the forest level and the stand level of analysis. Many analysts contend that site index estimates produced by standard inventory procedures underestimate site index in old stands. This bias can have a substantial impact on forest-level crop planning for Sitka spruce. If site index is underestimated, the allowable annual cut may be underestimated, and crop-planning decisions may be less than optimal.

Growth intercept models allow site index to be accurately estimated in young stands, based on the average annual height growth immediately above breast height (Nigh 1996b). Typically, the height growth is identified from the annual branch whorls and is averaged over a five-year period. This average annual height growth is called the *growth intercept.* An alternative approach, developed through what is called a *variable growth intercept model,* is to estimate site index from the average annual height growth of all of the tree height above breast height, not just the first five annual whorls above breast height.

The latter approach has been applied to Queen Charlotte Islands spruce data by Nigh (1996b). Although not reproduced here, Nigh's variable growth intercept model allows Sitka spruce site index to be estimated from a knowledge of only top height (in metres) and number of annual growth rings at breast height. The site index table can be used for spruce with breast height ages of 1-30 years. Nigh cautioned that the model's results could be biased if measurements are taken in spruce stands where there has been leader damage from white pine weevil. However, site index estimates can still be obtained by this method if the growth intercept is measured below the weevil damage. A field guide with details on growth intercept sampling, including sampling to avoid bias in spruce stems with leader damage from weevils, has been prepared by B.C. Ministry of Forests (1995).

In the last decade, British Columbia's coastal clearcut areas that had no residual trees whose top height and breast-height age could be used to calculate site index has led to an interest in using ecological variables of biogeoclimatic site series, especially soil moisture and soil nutrient differences between site series, to predict site index. The earliest work to establish a biogeoclimatic basis for site index estimation in British Columbia (Green et al. 1989; Klinka et al. 1989; Klinka and Carter 1990; Courtin 1992) did not deal specifically with Sitka spruce. Recently, however, Pearson (1992) and Kayahara and Pearson (1996), with support from the South Moresby Forest Replacement Account, focused on the opportunities for predicting

Sitka spruce and western hemlock site index from soil moisture and soil nutrient regimes. At about the same time, Wang et al. (1994a, 1994b) and Wang (1995) directed their attention to the same research objective for white spruce and lodgepole pine.

Kayahara and Pearson's (1996) research was based on data from near Gold River, Tahsis, Zeballos, and Port McNeill on Vancouver Island, and from the Seymour River valley on the mainland. It indicated that Sitka spruce mean site index values increased from the slightly dry soil moisture regime to a maximum on moist and very moist sites, and decreased on wet sites. Along the soil nutrient gradient, mean site index values increased from poor to very rich, with maximum site index values (40 m at breast-height age 50) occurring on very rich sites. These researchers concluded that site index relationships derived for the various combinations of soil nutrient and soil moisture regimes can be used to estimate site index for western hemlock in the CWHvm1 variant and for Sitka spruce in the CWHwh1 variant.

To improve the estimates of site index used in MacMillan Bloedel Ltd. (MB) timber supply analyses, Smith (1993) related biophysical attributes available in the MB inventory to known site index. The data set used to develop the model used only stands with precise estimates of site index. The model can be used to estimate site index in stands where the traditional approach to estimating site index is not precise. Input to the model can be extracted from information carried for all stands in the MB inventory. Smith (1993) used an approach to split the data set into homogeneous strata, in which stands within a single stratum have a similar site index. Five biogeoclimatic variables provided optimal stratification: species, elevation, BEC variant, longitude, and latitude. Aspect, slope, and slope position did not add to the predictive capability of the model.

Besides providing improved estimates of site index for timber supply analyses, Smith's model indicated several things about Sitka spruce site index in coastal British Columbia. Stands in which Sitka spruce is a leading species tend to have a mean site index similar to stands where Douglas-fir is the leading species. On the other hand, stands with red alder or western hemlock as leading species differ moderately, and stands with balsam fir or western redcedar as the leading species differ substantially, from those with Douglas-fir or Sitka spruce as the leading species. Tables 14, 15, and 16 indicate the degree to which Sitka spruce site indices vary by BEC variant in the Vancouver and Prince Rupert forest regions, based on site index estimates from Green and Klinka (1994) and Banner et al. (1993), respectively.

In addition to the approaches described above, other methods for classifying site quality have been examined. Van Hees (1988) used tree volume growth percent as an index of timber productivity in old-growth forest ecosystems in southeastern Alaska. His index generated the same ranking of

Sitka spruce ecosystems by timber productivity as that produced by site index. Depending on the ecosystem, 60-80% of the variation in tree volume growth percent was explained by two variables that were extracted from the inventory database: mean DBH and live-tree density. Ford et al. (1988) and Farr and Ford (1988) summarized recent efforts to relate soil properties to Sitka spruce site quality in southeastern Alaska.

Years to Breast Height

Predictions of years to breast height can assist crop planning for species selection and brush control. Years-to-breast-height equations are used in most growth and yield prediction models to initialize simulations of stand development.

Barker and Goudie (1987) provided the following years-to-breast-height equation: $ytb = 11.7 - 0.185SI$, where ytb = years to breast height and SI = site index (height in metres at breast-height age 50). This equation was not revised in the latest Sitka spruce site index tables prepared by Nigh (1996a) and Nussbaum (1996).

Predicting Regeneration

In the following subsections, the period of stand development under consideration is the interval between a stand-regenerating disturbance and crown closure. Growth and yield information for this life stage is required for both forest-level and stand-level crop planning. This information is particularly important since most silvicultural effort is expended in the stages of stand development up to crown closure.

Desirable Regeneration Densities Based on Plantation Trial Results

Weetman et al. (1989) concluded that nutrient deficiencies can substantially reduce growth in Sitka spruce plantations established on particular sites. Sites with a deep forest floor that had been logged, burned, planted with Sitka spruce, and subjected to extensive invasion by salal were associated with nutrient deficiencies. These deficiencies were alleviated, and growth rates were accelerated, by fertilization with N and P, although fertilized trees were frequently subject to weevil attack, possibly because of their succulent, rapidly growing leaders.

Omule (1988) analyzed the data collected over 26 years in a species and planting density trial on western Vancouver Island. His results were summarized under 'Early Growth and Comparisons with Other Species' (page 84) in Chapter 2 and in Table 19. In relation to Sitka spruce, Omule's main conclusions were as follows:

- Initial spacing (2.7 m × 2.7 m, and 3.7 m × 3.7 m, versus 4.6 m × 4.6 m) did not affect Sitka spruce survival; overall survival averaged 87%.

- Initial spacing effects were delayed or confounded by attacks of white pine weevil.
- Weevil damage was not related to initial spacing but increased with tree size and vigour.
- Published height growth curves do not properly represent top-height development in Sitka spruce stands affected by weevils.
- Because of weevil damage, the top height of Sitka spruce was substantially less than the top height of Douglas-fir, western hemlock, and western redcedar on all but the site unit that was moist to very moist as well as medium to very rich in nutrients.
- Competition from salal for nutrients may have impeded Sitka spruce growth.

Omule (1987) also reported results from 28 years of remeasurement of a species trial on western Vancouver Island in which Sitka spruce had not been affected by weevil damage. For these Sitka spruce, he concluded that Barker and Goudie's (1987) height growth curves provided a good model of the height growth pattern of Sitka spruce in the study location. This recommendation is assumed to apply also to the growth curves produced by Nussbaum (1996) as a refinement of the earlier Barker and Goudie (1987) or Thrower and Nussbaum (1991) curves.

From remeasurement of a 27-year-old plantation density trial, involving Sitka spruce and hybrids of Sitka spruce in the Prince Rupert Forest Region, Pollack et al. (1990) concluded that the following guidelines should be applicable to spruce crop planning in that forest region:

- An initial density of 800-1,400 stems per hectare is suggested for spruce sawtimber production.
- Undesirable increase in branch size can be dampened if initial density is not allowed to fall below 500 stems per hectare.
- Volume of non-crop trees can be reduced if initial density is not allowed to exceed 2,000 stems per hectare.
- Spacing should occur before age 20 to maintain optimal crown area.

Currently recommended stocking standards are summarized in Table 32 for the biogeoclimatic subzones, variants, and site series in which the Silviculture Interpretations Working Group (1994) considered Sitka spruce to be a primary, secondary, or tertiary regeneration species in the Vancouver and Prince Rupert forest regions.

MacMillan Bloedel Ltd. Regeneration Model

As of 1994, MacMillan Bloedel Ltd. staff were developing a regeneration model for use in forest-level simulation. For the preliminary version, the opinions of field foresters were translated into a model that partitioned the harvested area in the Sitka spruce cover type into six alternative regeneration classes (Table 37). Each density regeneration class was linked to a specific yield table produced by XENO (see 'XENO' on page 227). Subsequent versions of the model will be refined with historical silvicultural data. This level of modelling detail is uncommon in traditional timber supply analyses, where harvested timber volume flow is emphasized, but is essential for effective forest-level crop planning. Early assignment of cutblocks to yield curves puts the cutblock area back into inventory.

Density Management

Growth and yield information is required to support crop-planning decisions that pertain to control of stand density. Stand density is managed through silvicultural activities and decisions regarding stand regeneration, spacing, and thinning. Stand density management diagrams were referred to under 'Density Control' (page 191), where Figure 41 portrays a variety of mensurational variables in relation to number of trees per hectare for natural stands of Sitka spruce. Figure 42 portrays the same density-dependent variables for planted stands of Sitka spruce.

The theory behind these diagrams is that there is a remarkably constant relationship between the mean size and the density of survivors in a dense population. This relationship simplifies complex interactions between density, yield, and mortality, and thereby permits the use of stocking tables.

Table 37

The portions of Sitka spruce in each regeneration density class in the MacMillan Bloedel Ltd. Regeneration Model, Queen Charlotte Islands, British Columbia

	Regeneration class					
Model attribute	A	B	C	D	E	F
Portion of Ss cover type (%) allocated to regeneration class	25	20	10	30	10	5
Natural regeneration all species, stems per hectare	300	500	500	1,200	3,000	6,000
Brush problem	yes	no	yes	no	no	no
Currently planting	yes	yes	yes	yes	no	no
Currently spacing	no	no	no	no	no	yes on high sites

Source: P. Kofoed, MacMillan Bloedel Ltd., Nanaimo, B.C. (pers. comm., 1994)

The practical importance of stand density control diagrams, such as those described by Drew and Flewelling (1979), is that size/density relationships are predictable, so that yield is a product of mean tree volume and density, independent of site. Therefore, forecasting of ultimate yield is possible when the rate of progress of the stand is known.

A more detailed description of the Sitka spruce stand density management diagrams is provided in the explanatory legend for Figures 41 and 42. The diagrams in these figures were produced by C. Farnden (pers. comm., May 1996) from TASS-generated managed-stand yield tables contained in the computer program WinTIPSY, prepared by the Forest Productivity and Decision Support Section, Ministry of Forests, Victoria. Interested users can obtain a guide for producing managed-stand yield tables with WinTIPSY Version 1.3, prepared by the Ministry of Forests (Mitchell et al. 1995).

Other Recent Analyses of Stand Density/ Growth Relations in Sitka Spruce

Measures of stand density are an important type of growth and yield information. *Crown competition factor* (CCF) is one common measure of stand density. CCF is computed by summing (over all trees on a plot) predicted maximum crown width and dividing this by plot area. Farr et al. (1989) fit the following model of maximum crown width to a data set derived from 226 Sitka spruce trees ranging from Kodiak Island, near the northwestern limit of this spruce, through southeastern Alaska, to the Queen Charlotte Islands: $MCW = 1.23 + 0.268DBH^{0.8700}$, where MCW = maximum crown width (in metres) and DBH = diameter at breast height (in centimetres).

The broad latitudinal range of the data set, from Kodiak Island to the Queen Charlotte Islands, did not reveal regional differences when tested in the model. Consequently, this equation is probably appropriate for Sitka spruce in all of its natural range in British Columbia.

Farr and Ford (1988) reported that in the mid-1970s, a long-term cooperative study was started to measure the effects of stand density on the growth and development of even-aged stands of hemlock and spruce in southeastern Alaska. By 1988, 278 one-acre (0.40 ha) plots had been established over a range of stand ages, site quality, and stand density. These data have been used to produce a variant of PROGNOSIS called SEAPROG.

Kellogg and Olsen (1988) examined four alternative density management regimes in a spaced western hemlock–Sitka spruce stand in the Oregon Coast Range. Stand characteristics after thinning were projected to final harvest with the Stand Projection System (SPS) described by Arney (1984). Based on an analysis of the present net worth of timber following growth and yield simulation, Kellogg and Olsen concluded: 'These study results suggest that if density of managed western hemlock–Sitka spruce stands has been controlled with precommercial thinning, these stands will develop adequately to rotation age without the need for commercial thinning. The reduced logging costs and log premiums at final harvest will not be sufficient to compensate for the high cost of intermediate thinning entries.'

Stand Density in Relation to Growth and Yield Prediction Models

Density-related growth and yield information can be used to develop growth and yield prediction models. However, density-related information can also be produced by growth and yield simulators. Growth and yield prediction models will be described under 'Growth and Yield Prediction Models for Sitka Spruce' (page 224). For example, TIPSY (page 224) can be used to examine the effect of initial density and spacing on Sitka spruce stand development. Variable density yield prediction (VDYP) (page 226) can be used to examine the effect of alternative levels of crown closure upon yields of natural Sitka spruce stands.

Prediction of Tree Volume

Tree volume equations are used to predict the volume of a tree stem from some related variables, generally tree height and diameter. This class of growth and yield information is required for accurate estimates of stand yield. Tree volume equations are embedded within many of the growth and yield prediction models that are used to assess silvicultural alternatives.

Phelps (1973) provided the following summary of the historical development of tree volume equations for Sitka spruce. The first tree volume tables for Sitka spruce on the west coast were published by Meyer (1937). Between 1937 and 1970, various authors published tree volume tables for Sitka spruce in Alaska, British Columbia, and Washington (see, for example, Puget Sound Research Centre Advisory Committee 1953; Forest Club of the University of British Columbia 1959; Skinner 1959; Bones 1968).

Today, predictions of tree volume are based on tree profile equations (taper equations). Phelps (1973) reported that Sitka spruce butt taper tables were published in 1966 (B.C. Forest Service 1966) and taper curves for the entire bole were published in 1968 (B.C. Forest Service 1968). Following the development of Demaerschalk and Kozak's (1977) whole-bole system, the Ministry of Forests began computing tree volume with these new taper curves. With the most recent advances in tree profile prediction (Kozak 1988), the Ministry of Forests is in the process of switching to Kozak's new variable-exponent taper equation.

Kozak (1988) fit his variable-exponent taper equation to spruce trees sampled in coastal British Columbia. It can be assumed that the sample was predominantly Sitka spruce, although hybrids of Sitka spruce and interior spruces may have been included. The sample was split into two groups, one consisting of 318 immature coastal spruce and the other of 354 mature coastal spruce.

Growth and Yield Prediction Models for Sitka Spruce

Most of British Columbia's natural forest involving Sitka spruce is relatively old, with only limited areas of second-growth forest. Also, there are not many growth and yield plots for either the older natural stands or the second-growth stands of Sitka spruce. Thus comprehensive crop planning involves the use of computerized models to predict stand growth and yield. These growth and yield prediction models incorporate many of the previously discussed sources of growth and yield information, such as site quality estimates, regeneration submodels, density management information, and tree volume estimates. By calibrating these models, available growth and yield data can be generalized and the applicable range of such data can be somewhat extended. The data used to calibrate these growth and yield prediction models are described in the section 'Growth and Yield Data' (page 228). It is beyond the scope of this book to describe each growth and yield prediction model in detail, but they are summarized here in the context of Sitka spruce data.

PROGNOSIS

PROGNOSIS is a widely used, distance-independent individual tree model developed and supported by the United States Forest Service (Stage 1973). Farr and Ford (1988) reported that data collected in Alaska had been used to produce a variant of PROGNOSIS called SEAPROG. The latter is calibrated for Sitka spruce.

PROGNOSIS was originally developed to predict yields from the complex stands in the intermountain and Rocky Mountain regions of the western United States. The Inventory Branch of the Ministry of Forests has calibrated PROGNOSIS for some stand types in British Columbia. We have obtained no further information on the status of SEAPROG or on the prospects for a calibration for Sitka spruce in British Columbia. Such a calibration would substantially improve the growth and yield information available to support Sitka spruce crop planning.

Tree and Stand Simulator (TASS) and WinTIPSY

TASS is a distance-dependent individual tree model developed and supported by the Research Branch of the Ministry of Forests (Mitchell and Cameron 1985). TASS results have been published as yield tables. More recently, they are distributed as a computer database accessed by a second computer program called WinTIPSY (Mitchell et al. 1995). An economic analysis module for WinTIPSY is now available (Stone et al. 1996). As of early 1994, TASS had been calibrated for Sitka spruce using the data set for 144 permanent sample plots that was available to the Research Branch at that time (see Table 38).

Table 38

Ministry of Forests Research Branch permanent sample plots in which Sitka spruce makes up at least 80% of the trees in a sample stand

Source[a]	Stand origin[b]	Treatment[c]	No. of plots	No. of meas.	Top ht. (m)	Total vol. (m^3/ha)	Stems per ha	Mean dbh (cm)	Basal area (m^2/ha)	Total age (yr)
EP 368	P	S	7	4-5	8-25	18-621	790-1581	7-34	5-76	12-30
EP 570	P	ED	12	5	1-13	0-169	478-2991	0-22	0-38	7-27
EP 571	P	ED	38	3-6	3-18	0-559	430-1328	0-30	0-82	8-29
EP 712	P	ED	18	?	?	?	?	?	?	?
Old EP	N	?	28	2-5	25-50	149-1,991	195-1,195	14-68	17-129	63-109
WFP	N	none	41	1-2	7-51	57-2416	334-40,500	2-59	19-164	18-156

Source: Based on a summary by Goudie (1991)

[a] The numbers EP 368 to EP 712 refer to Ministry of Forests experimental plots by which remeasurement projects are identified; Old EP refers to a discontinued experimental plot; WFP refers to a permanent sample plot established by Western Forests Products Limited.

[b] For stand origin, P = planted and N = natural regeneration.

[c] For stand treatment, S = spaced; ED = espacement trial; ? = data not available.

WinTIPSY yield predictions are available for the following ranges of site and stand conditions:

- Establishment method and density: planted, 331-4,444 stems per hectare; natural, 331-10,000 stems per hectare
- Site index: 0-50 m at breast-height age 50
- Spacing at 6 m top height: residual density of 331-4,100 stems per hectare.

TASS is linked to a series of models that process the yields it predicts. These models can simulate alternatives regarding bucking stem logs, sawing logs, and grading lumber, from which it is possible to obtain estimates of financial return. Taken together, this series of simulators is called SYLVER. The type of information provided by SYLVER is essential for comprehensive crop planning. At the time of writing, TASS and WinTIPSY did not link to the attributes of inventory polygons in the Ministry of Forests' inventory system, and SYLVER was not yet calibrated for Sitka spruce.

Appendix 1 shows a sample page of a printout from a 1994 version of TIPSY (Version 2.1.2 Gamma), showing managed-stand yields and mean annual increments for pure Sitka spruce stands to age 150. This sample page is for an assumed site index of 24 m at breast-height age 50, and an assumed stand density of 500 trees per hectare.

Persons interested in using WinTIPSY for Sitka spruce should consult Mitchell et al. (1995) for a user's guide to WinTIPSY Version 1.3 running under Microsoft Windows®, and Stone et al. (1996) for the economic analysis module for WinTIPSY. This model can be accessed by contacting the Ministry of Forests, Research Branch, Forest Productivity and Decision Support Section, Victoria, British Columbia. Upon request, the predicted yield parameters available include volume to several limits of merchantability, basal area, mean diameter, stem count, mean annual increment, and various prime tree characteristics. The user can also modify predicted yields by specifying a regeneration delay and various operational adjustment factors.

Variable Density Yield Prediction (VDYP)

Variable Density Yield Prediction (VDYP) is a whole-stand model developed and supported by the Inventory Branch of the Ministry of Forests (Smith 1990); it is linked to forest inventory polygons. In its common operating mode, VDYP provides estimates of yield from data on stand species composition, geographic location, age, height, and crown closure. VDYP is based on data collected in natural, untended stands and is expressed as net merchantable volume. The yield estimates represent historical unmanaged-stand performance. Anyone attempting to compare the volume yields or mean annual increments shown in the WinTIPSY sample page in Appendix 1 with those in the VDYP sample page in Appendix 2 will note that, for a given stand age, yields predicted by VDYP are substantially lower. This is because they are based on averages, whereas WinTIPSY focuses on maximum potential yields.

VDYP is used to estimate the volume and mean diameter of timber stands in Ministry of Forests inventory files. It is also used to generate yield tables for timber supply analyses. When using VDYP to generate yield tables, the user accepts the modelling assumptions that species composition and crown closure are not variable over time and that stand top-height growth in any particular inventory area will follow the top-height curve shown in the model. No management interventions can be specified in simulation runs of the model. Reports may be produced by species component and for the whole stand. VDYP was calibrated from a large database of inventory temporary sample plots.

Appendix 2 shows a sample page from VDYP Version Prod 4.5, listing heights, diameters, volumes (close utilization less decay volume, for diameters over 17.5 cm), and mean annual increments for pure Sitka spruce stands aged 10 to 250 years. This sample page is for an assumed site index of 24 m at breast-height age 50 and an assumed crown closure of 25%. Persons interested in using VDYP for Sitka spruce should contact the Ministry of Forests, Research Branch, Forest Productivity and Decision Support Section, Victoria, British Columbia.

Stand Projection System (SPS)

The Stand Projection System (SPS) is a distance-independent individual tree model developed and described by Arney (1984). While the SPS architecture is well suited to crop planning, the publicly available version is not fully calibrated for Sitka spruce. As a result, only provisional yield estimates are available for this species. However, the literature does suggest that this model gives good results when used to predict Sitka spruce growth and yield (Kellogg and Olsen 1988).

With SPS, site characteristics are described by geographic region, site index, and a site-specific modifier that can alter the shape of site index curves. The stand is described by entering a tree list, stand table, or stand average values. Simulations of management regimes can include establishment method, intermediate harvests (spacing and thinning), and fertilization. Reports may be produced for a wide range of yield parameters.

XENO

XENO is a distance-dependent individual tree model developed and supported by MacMillan Bloedel Ltd. (Northway 1989). Although it has not been calibrated for Sitka spruce, based on a comparison between Sitka spruce yields and XENO predictions, MacMillan Bloedel foresters use hemlock yields as a surrogate predictor of Sitka spruce yields. XENO's model architecture is well suited to crop planning, which, in coastal British Columbia, would be greatly enhanced by the calibration of XENO for all major coastal species. As it is not specifically calibrated for Sitka spruce, XENO is not described here in any further detail.

Growth and Yield Data

Sitka spruce growth and yield information includes both completed analyses and unanalyzed data. Completed analyses were outlined previously. This section reviews currently unutilized data sources that are available to support Sitka spruce crop planning, including data from temporary and permanent sample plots, as well as stem analysis data. To fulfil outstanding needs, growth and yield data may be collected as part of either a designed experiment or a sample survey.

Temporary Sample Plots

Information from temporary sample plots (TSPs) includes data collected from silviculture plots, cruise plots, inventory plots, and other samples. Data from TSPs are an important source of growth and yield information pertaining to regeneration. There are several important data sources, mainly from the history record files in the Silviculture Branch of the Ministry of Forests. Another data source is the Silviculture Branch's Silviculture Experiment demonstration plots. These are informal demonstration areas of species trials, stock type trials, espacement trials, and other investigations. Regeneration performance data maintained by the larger forest products companies are an additional source of TSP data.

Temporary sample plots are also an important source of growth and yield information for mature stands. The Inventory Branch of the Ministry of Forests maintains a large data set of inventory ground plots that are located mainly in mature stands. VDYP data have been assessed in relation to this TSP data set, supplemented with Ministry of Forests permanent sample plot data. Cruise plots are another data source for mature stands. For this book, Sitka spruce information was not collected from these unanalyzed data sources.

Permanent Sample Plots

Temporary sample plots provide data on yield at the stand age when trees on the plot were measured, but remeasurements on permanent sample plots (PSPs) provide information on growth rates (rates of increase of yield over time). PSPs are more costly to install and maintain than TSPs; partly because of cost differences, TSP data are often more abundant than PSP data. Permanent sample plot data are maintained by both the Research Branch and the Inventory Branch of the Ministry of Forests. Research Branch PSP data on Sitka spruce top heights and yields (m^3/ha), which do not include data from MacMillan Bloedel Ltd. PSPs, are listed in Table 38; additional growth and yield data for Sitka spruce are likely available from PSP records maintained by the Inventory Branch.

MacMillan Bloedel Ltd. maintains a large PSP database. A few of the plot characteristics that are easily summarized from the plot data are listed in Table 39. A visual inspection of plot summary information indicates that most plots that contained Sitka spruce mensurational information were in stands aged 30-100 years, and that a wide range of stand densities were involved. Most MacMillan Bloedel PSPs relating to Sitka spruce are located on the Queen Charlotte Islands.

A current registry of PSP data in British Columbia is maintained by the Forest Productivity Councils of British Columbia, which can be reached through the Ministry of Forests, Inventory Branch, Victoria.

Stem Analysis Data

Stem analysis data are collected to construct site index curves and tree volume equations, and to estimate tree volume reductions due to decay. The Research Branch maintains a stem analysis database for site index curve construction and TASS calibration. This data set includes trees sampled by Farr (1984) and Barker and Goudie (1987). Farr's data were collected on 71 plots established in unmanaged even-aged stands of Sitka spruce and western hemlock in southeastern Alaska. Barker and Goudie's data were collected on 48 plots located throughout the Queen Charlotte Islands. The Inventory Branch maintains a stem analysis data set for construction of tree volume equations. When used by Kozak (1988), this data set contained information from 672 Sitka spruce stems.

Table 39

MacMillan Bloedel Ltd. Sitka spruce permanent sample plot data

Type of stand	Number of plots[a]	Site index (m) at age 50 years	Number of measurements
Natural	70	23-53	1-8
Planted	15	28-36	2-6
Spaced[b]	14	23-37	4-7
Other[c]	3	29-38	3-7

Source: B. Wilson, MacMillan Bloedel Ltd., Nanaimo, B.C. (pers. comm., 1993)

[a] The number of plots listed here includes only those permanent sample plots where Sitka spruce is a leading species.

[b] Spaced stands were not specified as to planted or natural regeneration origin.

[c] The definition of stands defined as 'other' was not specified.

Overview of Sitka Spruce Growth and Yield Prediction and Crop Planning

This section provides an overview of the growth and yield information that is available to support Sitka spruce crop planning in British Columbia. A general conclusion is that the required information base needed for Sitka spruce crop planning is incomplete. Information needs are highlighted below.

New growth and yield tools for estimating site quality are required for some stand conditions. The Barker and Goudie (1987) and Nussbaum (1996) site index curves are an excellent tool for site quality estimation when they are correctly applied to appropriate stands. However, these curves are not suitable for very young stands or very old stands. For crop planning at the forest level, the important issues of site quality estimation relate to possible underestimation of mean site index in forest-planning exercises, including underestimation of the timber supply analyses that support the determination of annual allowable cuts. If there is bias, the conclusions drawn from forest-level crop planning will be incorrect. To address this issue, two investigations are required. First, it is necessary to determine whether bias exists and, if so, to quantify its magnitude and extent. Second, alternative procedures, such as those developed by Smith (1993), need to be adopted to estimate site quality where traditional methods are inadequate.

Although most silvicultural effort is expended to bring harvested areas to a free-growing condition, only a relatively small proportion of available growth and yield information is targeted at supporting silvicultural decision-making at this stage. For predicting regeneration characteristics, the growth and yield information base is underdeveloped. Some improvements can be obtained from relatively simple analyses of currently available data. For example, crop-planning exercises that employ growth and yield prediction models require appropriate input values to initialize the growth model simulations. Better estimates of appropriate average values could be extracted from the substantial silvicultural history data sets maintained by the Ministry of Forests.

Although currently available growth and yield information for Sitka spruce tree volume prediction meets all crop-planning requirements, in relation to regeneration there is a need to assemble a growth and yield information base to test the BEC field guide recommendations that are so widely used. One way to organize such tests is to develop regeneration models of the kind linked to PROGNOSIS (Ferguson et al. 1986). Also, management strategies must be formulated to resolve the problems of Sitka spruce damage from white pine weevil and spruce growth limitations when harvested areas are replanted (see 'White Pine Weevil Damage' on page 120 and 'Fertilization' on page 185). Particularly at the stand regeneration stage, improved quantitative growth and yield information is required for crop planning.

Much of the important growth and yield information pertaining to density management is already available to support crop planning for Sitka spruce. Density management information related to establishment density and post-spacing density is provided by WinTIPSY, even though it is supported by a relatively small PSP data set (see Figures 41 and 42). One refinement would be to quantify the response of Sitka spruce to partial-cutting regimes.

Only some of the crop-planning information needs can be satisfied by the growth and yield models currently available. VDYP is well suited for its intended use (the estimation of stand volume in the inventory), but it is inadequate for all but the most rudimentary crop-planning exercises. Several good individual tree growth models are available, but only TASS is both fully calibrated for Sitka spruce and widely available in British Columbia (through WinTIPSY). An economic analysis module is also available for WinTIPSY (Stone et al. 1996). Crop planners would benefit from having access to other appropriate models such as XENO, PROGNOSIS, and SPS, which have proven useful for crop planning for other species in other locations. It may be possible to obtain SEAPROG, a variant of PROGNOSIS calibrated for Sitka spruce in Alaska, for use in British Columbia. The available Sitka spruce data set is large enough to calibrate SPS for Sitka spruce in British Columbia.

Improved growth and yield prediction capabilities are required since stand-level crop-planning needs commonly exceed WinTIPSY's current capabilities. There are four important limitations to WinTIPSY's usefulness for crop planning:

- Growth projections are frequently required for established stands. Following a stand examination, data collected in the existing stand are input to the model and alternative management regimes are simulated.
- Stand table output is required for many analyses. For example, crop planning often involves an economic analysis where the yields are assigned an economic value by summing values assigned to the predicted stand table.
- Sitka spruce is commonly found in mixed-species stands. Crop planning for Sitka spruce would benefit from the development of silvicultural strategies for such stands.
- Besides establishment density and post-spacing density, crop planning for Sitka spruce should consider fertilization and partial-cutting regimes. The silvical characteristics of Sitka spruce suggest that it is suitable for intensive management, including commercial thinning.

Because TASS models tree growth in considerable detail, it is likely that TASS can provide growth and yield information for several crop-planning issues that are not well understood. Issues peculiar to Sitka spruce, such as weevil attack and epicormic branching, can likely be examined within the existing model architecture. In this respect, TASS provides an excellent growth and yield information base for Sitka spruce crop planning. SYLVER should be calibrated for Sitka spruce as soon as possible.

In summary, improved growth and yield prediction capabilities are required to support crop planning for Sitka spruce in British Columbia. The growth and yield data sets for Sitka spruce are much smaller than those available to support Douglas-fir and western hemlock crop planning. To reduce this imbalance, it may be possible to substantially augment the existing British Columbia PSP data set with Sitka spruce data from Alaska. Second, MacMillan Bloedel Ltd. and the Ministry of Forests may be able to supplement their respective data sets through additional data sharing. Third, data collection programs to support model calibration for partial-cutting regimes involving Sitka spruce should be initiated.

Management to Minimize Losses to Insects and Disease

The biology and control of the white pine weevil were summarized under 'White Pine Weevil Damage' (page 120) in Chapter 2. Key features of Sitka spruce's relationship with this weevil are reiterated below, with the recommendation that this economically important insect pest be managed through integrated pest management (IPM).

There is evidence that the current high level of weevil damage to Sitka spruce has been encouraged by the practice of planting nearly pure spruce for reforestation of clearcuts. The white pine weevil is always present in natural stands of Sitka spruce. Under natural conditions, however, pure stands of Sitka spruce usually occur only close to the ocean, where heat accumulation is insufficient to encourage weevil populations. In natural stands inland, Sitka spruce typically occurs in mixed stands with other conifers and hardwoods, a circumstance not conducive to the build-up of weevil populations (Alfaro et al. 1994). Foresters now believe that the widespread adoption of clearcutting followed by plantations of nearly pure Sitka spruce has created conditions extremely favourable for weevil outbreaks. It is particularly significant that productivity losses to weevils are magnified by the fact that weevils prefer the fastest-growing trees in a stand (Alfaro and Ying 1990).

The current approach to reducing weevil damage involves an integrated pest management system that combines silviculture-driven and resistance-driven tactics. Such a system relies on accurate hazard ratings of plantation sites, continuous monitoring of attack levels, and forecasting of productivity losses using a decision support system (Alfaro et al. 1994; Alfaro 1996). The key feature of integrated pest management is that its objective is to use a combination of tactics to *reduce damage* rather than eliminate the pest (MacLean 1996).

The two most promising silvicultural tactics involve changes to stand microclimate that make Sitka spruce stands less suitable for weevil attack or weevil survival. One tactic makes use of the observation that there are lower rates of weevil damage in Sitka spruce under a red alder overstorey compared with Sitka spruce in the open (McLean 1989). Another tactic is based on evidence of lower weevil attack rates for Sitka spruce planted at relatively close spacing (Alfaro and Omule 1990). Closer spacing reduces but does not eliminate the ability of the white pine weevil to cause damage. Alfaro and Omule (1990) recommended that Sitka spruce plantations be started at close spacing (2.74 m × 2.74 m) and then be pre-commercially thinned at 25 years. During this thinning, the largest trees of good form should be retained.

After a white pine weevil attack, a spruce tree may take one to several years to recover. During this time, branches from the uppermost whorl below the damaged terminal compete for dominance, and the tree remains for one or more years with multiple leaders. Depending on the number of internodes destroyed and the growth characteristics of the tree, a permanent stem defect can form at the point of injury (Alfaro 1989; Alfaro and Omule 1990). In severely infested stands, losses due to reduced growth and defects can be as high as 40% of stand volume (Alfaro 1992, 1994, 1996). Outbreaks on monoculture Sitka spruce plantations begin when plantations are about five years old. First, a few trees are attacked, then the weevil population rapidly increases to levels where 30-50% of trees are attacked per year. This rapid increase in infestation is partly a result of the multiple leaders produced in the first stage of attack, which increase the number of oviposition sites and the supply of food for weevil larvae. A period of stability is reached 10-20 years after initial attack. During this period of stability, the stability level is higher on nutrient-rich, warm sites, where vigorous growth is accompanied by rapid development of new leaders, than on nutrient-poor, cool sites (Alfaro 1996).

It is notable that white pine weevil is not abundant in natural, undisturbed stands. The characteristics of such stands that prevent outbreaks include:

- shaded, cool habitats that are unfavourable for weevil feeding, oviposition, and development (for example, McMullen [1976a] calculated that a minimum heat accumulation of 888 degree-days above 7.2°C was necessary for weevil development in Sitka spruce)
- low food supply and suboptimal oviposition sites, mainly because spruce regeneration under shade grows slowly, producing thinner leaders that contain less food for weevil larvae

- trophic webs that are more complex in natural stands than in pure Sitka spruce plantations; natural stands are characterized by a wider variety of vertebrate and invertebrate predators that consume weevils at various stages of their life cycle (Alfaro 1996).

Today's key to controlling the white pine weevil problem in Sitka spruce is an IPM system (Alfaro 1995, 1996; MacLean 1996) that relies on restoring ecosystem balance by reducing the frequency of conditions leading to weevil outbreaks. One practical tactic is to silviculturally discourage heat accumulation in the stand by using tree species that are not weevil hosts, including encouraging an overstorey of broadleaf species, such as red alder. In low-hazard sites, reduction of spruce stem defects by sanitation thinning and by planting spruce at close spacing is recommended (Alfaro 1996). In high-hazard sites, these silvicultural approaches probably need to be coupled with the use of Sitka spruce stock with above-average resistance to attack by the white pine weevil. Alfaro (1996) stressed that Sitka spruce genotypes resistant to weevil damage had been noted in British Columbia as long ago as the early 1930s. However, there is still no large-scale production of Sitka spruce planting stock resistant to weevil attack.

The weevil/spruce relationship is the product of a very long period of coevolution during which several defences and counter-defences have developed. Today's active research by Alfaro (1996), Ying (1991), and others suggests that the best approach to weevil control is to use intensive silviculture involving increased spruce plantation density, mixed-species plantings, creation of as much shade as possible, pruning of infested leaders to reduce weevil populations, sanitation thinning to remove trees with defects from weevil attack, and concentration of spruce growth on trees free of weevil damage. There is optimism that in the future Sitka spruce will be available with increased levels of weevil resistance, as a result of vegetative propagation of resistant trees (Ying 1991) and production of such trees from seed-orchard seed.

The natural distribution of Sitka spruce in British Columbia is much greater than the range over which the species is managed to counteract damage by white pine weevil (MacSiurtain 1981). A 1984 review of Sitka spruce growth in relation to weevil attack, only recently published (Smith and McLean 1993), indicates that the way we view Sitka spruce silviculture in relation to weevil damage has not changed much in the past decade. Except on the Queen Charlotte Islands and along the west coast fog belt on Vancouver Island, foresters are still reluctant to plant Sitka spruce. Weevils commonly attack the tallest trees in young stands, causing enough leader damage to significantly reduce site quality estimates if they are based on Sitka spruce height growth. Furthermore, crooks in stems due to loss of the current leader result in reduced internal wood quality and in irregular bole shape and taper. Sommers (1983) reported that repeated weevil attacks resulted in shorter trees with larger diameters at breast height. Weevil-damaged trees taper more rapidly as new growth is shifted from height extension to bole and branch development.

These circumstances are unfortunate because economic losses are serious when insect damage is focused on the leaders of the tallest and best trees in a stand, and especially when a side effect is the redirection of growth into development of large branches. Economic losses from weevil damage are magnified by the fact that Sitka spruce can outgrow other tree species on certain sites. These problems are also unfortunate because, without weevil damage, Sitka spruce can produce high-quality wood and is moderately resistant to deer browsing compared with other conifers on the Queen Charlotte Islands.

In relation to the white pine weevil, an important development in the past decade has been the discovery of genetic evidence of geographic patterns of variation in weevil resistance, growth rate, and frost-hardiness (Lester 1993). To date, the most promising areas for finding Sitka spruce provenances resistant to the weevil are in the drier parts of spruce's natural distribution, particularly in the Strait of Georgia lowland. Ying (1995) singled out the Haney area, east of Vancouver, and the Big Qualicum area of eastern Vancouver Island as the most promising of such sites. The search for weevil-resistant spruce is being carried out in conjunction with a search for geographic differences in spruce growth rate. Lester (1993) reported that outstanding increases in growth rate are evident in coastal provenances from Oregon and Washington, but at present these provenances are useful only on the Queen Charlotte Islands. Material from drier parts of the species' range in Oregon and Washington show a combination of increased weevil resistance and high growth rate. As more is learned about the inheritance of weevil resistance, such resistance may be combined with increased growth rate in plantations.

Hulme (1995) demonstrated seasonal variation in Sitka spruce's resistance to white pine weevil by advancing or retarding the period when weevil oviposition occurs; when oviposition is advanced, normally resistant trees become more susceptible, and when oviposition is retarded, susceptible trees become more resistant. This resistance may be seasonally transient rather than permanent. Observed variations in the weevil resistance of translocated Sitka spruce trees at different planting sites may reflect changes in tree phenology under different environmental conditions. The evidence that resistance to weevil attack may be largely site-specific led Hulme to stress the importance of matching the tree to the planting site.

As summarized by Morrison (1989), spores of Annosus root rot (*Heterobasidion annosum*) are present in the air throughout the year, although their numbers are lowest during summer and winter and are reduced by precipitation (Reynolds and Wallis 1966; Redfern 1982). Except for western redcedar, freshly cut stumps of all conifers on the coast are susceptible to infection by spores of *Heterobasidion*. Following infection, the stump and its roots may be colonized by Annosus root rot, which spreads to adjacent residual trees at root contacts (Morrison and Johnson 1978). In most stands, particularly those that are naturally regenerated, pre-commercial thinning creates large numbers of susceptible stumps. If even a small percentage of these stumps are colonized by *Heterobasidion,* the amount of inoculum on the site increases markedly (Morrison and Johnson 1978).

From sample Sitka spruce stands on the Queen Charlotte Islands, Morrison (1989) noted a sharp increase in the percentage of stump surface

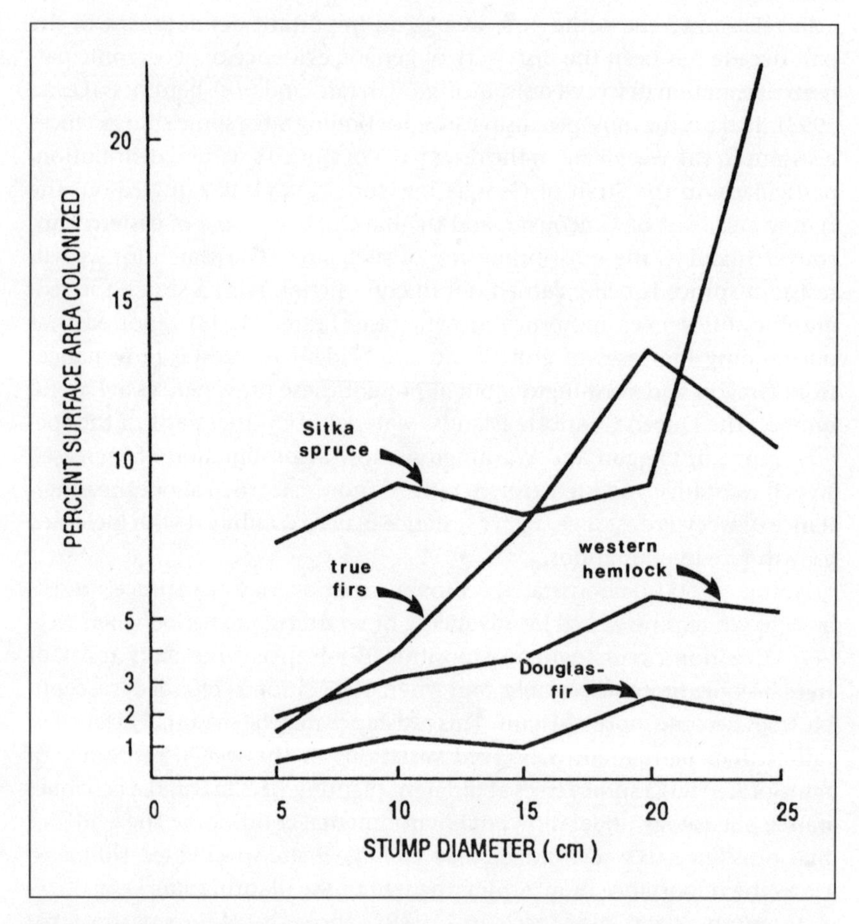

Figure 48. Tree species differences in the relationship between stump diameter and percentage of stump surface area colonized by Annosus root rot, based on data from the Queen Charlotte Islands (from Morrison 1989).

area colonized by Annosus root rot as stump diameter increased from 5 to 25 cm (Figure 48). It is evident that target size (stump diameter) is an important determinant of Annosus root rot infection, particularly when inoculum levels are low. For the forest manager, the important concern is being able to assess circumstances where stump treatment is warranted in order to reduce infection. From studies in southeastern Alaska, Shaw (1989) suggested that there was little chance that Annosus root rot would damage young, managed stands of Sitka spruce and western hemlock within a 90- to 120-year rotation.

There are well-developed management guidelines for reducing the risk of Armillaria root disease (Morrison 1981). These guidelines do not refer specifically to Sitka spruce but the point is made that all spruces, together with western redcedar, lodgepole pine, western white pine, and all of British Columbia's main broadleaf species, are only moderately susceptible to *Armillaria.* The true firs, Douglas-fir, and western hemlock are the most susceptible. In the regeneration phase, therefore, a forest manager can use host susceptibility as a criterion for selecting species for planting in *Armillaria*-infested sites. Morrison (1981) suggested that when disease centres are few and widely scattered, the centres and a 20 m wide surround should be planted with less susceptible species; for many sites on the coast, this could be Sitka spruce. Areas away from the infection centres could be planted with species less resistant to *Armillaria* infection. When *Armillaria* infection centres are large or are distributed throughout the stand, the entire area will require special treatment. Because no coastal coniferous species – not even Sitka spruce – is completely resistant to *Armillaria,* some mortality can be expected in such stands (Morrison 1981).

As described under 'Diseases' (page 127) in Chapter 2, Sitka spruce managers probably have little need to be concerned about most of the root diseases in British Columbia, although all spruces, together with lodgepole pine, are considered to be the most susceptible to tomentosus root rot. This disease occurs primarily as a root and butt rot of interior spruce north of about Williams Lake in British Columbia. Laminated root rot and black stain root disease are primarily diseases of Douglas-fir. *Rhizina undulata,* while potentially infecting all coniferous species, tends to be limited to newly planted sites that have been burned.

Protection of Old-Growth Sitka Spruce in British Columbia

MacKinnon and Eng (1995) defined British Columbia coastal 'old forests' as stands with a mean canopy age over 120 years. Coastal British Columbia was defined as the area covered by the CWH, MH, and CDF biogeoclimatic zones south of latitude 54°N (see Plate 7). Based on interpreted satellite imagery obtained between 1989 and 1991, MacKinnon and Eng make the following estimates:

- 53% (5.4 million ha) of the CWH zone is covered with old forest; spruce makes up about 4% of this zone's area of old forest.
- 34% (1.3 million ha) of the MH zone is covered with old forest; spruce makes up about 3% of this zone's area of old forest.
- 0.5% (1,100 ha) of the CDF zone is covered with old forest; spruce is not recorded as a component of this zone's area of old forest.

There is substantial biological uncertainty about the growth and yield benefits of actions associated with second-growth silviculture and with old-growth protection and liquidation. These uncertainties have been summarized by Weetman (1991) for British Columbia's forest resource as a whole. For Sitka spruce, it is notable that, compared with British Columbia's other main tree species, there is a low risk of loss associated with holding stands of old-growth trees because the main natural cause of loss for old-growth Sitka spruce is windthrow. British Columbia data available for managed yield prediction is relatively poor. The available data are derived from proprietary corporate databases, stand density diagrams, and data derived from British Sitka spruce plantations. The most encouraging point is that for Sitka spruce in British Columbia, the uncertainty of managed yield prediction is low (Table 40).

A recent inventory of old-growth forests in coastal British Columbia revealed that the two most common protected old-growth types are western redcedar–western hemlock (25%) and western hemlock–amabilis fir (23%). Douglas-fir–western hemlock stands make up 10% of protected old growth, and Sitka spruce–western hemlock stands are in fourth place at 9.5% (Roemer et al. 1988). As reviewed by Roemer and co-workers, the reasons for old-growth conservation in British Columbia are no different from those for old-growth areas in the United States, where Sitka spruce is the species of interest less commonly than in British Columbia. Old-growth forest is of interest because of heritage values, recreational values, landscape and scenery enhancement, wildlife habitat requirements, opportunities for scientific research, gene pools, educational values, and watershed functions (Franklin et al. 1981; Meslow et al. 1981; Meehan et al. 1982; Harris 1984; Norse et al. 1986; Norse 1990; B.C. Ministry of Forests 1992b; Franklin 1993). There is increasing concern that the harvesting of coastal old-growth forests is reducing wildlife diversity, and there have been calls for increased research to compare wildlife diversity in old-growth forests and managed forests (Rosskam and Hyde 1995). In this context, riparian corridors, which are typical Sitka spruce habitats, play a particularly important role.

Attention has been focused on the old-growth Sitka spruce stands in the Carmanah Creek valley on Vancouver Island. To place the Carmanah Sitka spruce ecosystems in context, MacMillan Bloedel Ltd. recently surveyed other

Table 40

Sitka spruce compared with other British Columbia coastal tree species in terms of biological risk and uncertainty in holding old-growth timber and predicting yields of managed stands

	Old growth		Managed yield prediction		
Species	Risk of loss	Main natural agency of loss	Availability of B.C. data	Technique used[a]	Uncertainty of prediction
Western hemlock	low	mistletoe, rots	fair	4, 7, 10, 11	low
Douglas-fir	low	fire	good	1, 2, 4, 5, 7, 11	low
Red alder	high	rots	poor	8	low
Sitka spruce	low	windthrow	poor	4, 7, 11	low
Western redcedar	very low	none	none	4, 7, 11	high
Mountain hemlock	low	rots	none	none	high
Amabilis fir	moderate	rots, windthrow	poor	4	high

Source: Weetman (1991)

[a] Data sources for managed yield prediction: (1) DFSIM Simulator; (2) TASS Simulator; (4) Proprietary corporate yield data; (5) Consultant stand models; (7) stand density diagrams; (8) U.S. Forest Service yield data; (10) U.S. corporate yield tables; (11) U.K. plantation yields. See Weetman (1991) for identity of these information sources. Information sources 3, 6, 9, and 10 are not listed here because they pertain to tree species inland from the natural range of Sitka spruce.

Table 41

Some areas on Vancouver Island with large Sitka spruce on alluvial sites

Location	Area[a] (ha)	Description
Provincial Parks		
Carmanah Walbran Provincial Park	128	Although this was not a provincial park at the time of Beese's 1989 review, Carmanah Creek Sitka spruce were described as follows: concentrated in a 4 km section of a floodplain 6 km long and 0.4-0.8 km wide; mixed with western hemlock and amabilis fir; 9 spruce trees > 80 m tall with many trees over 70 m; large trees are mostly 1.5-2.5 m dbh; scattered individuals are over 3 m dbh; world record Sitka spruce is located in the lower valley of Carmanah Creek.
Moyeha River valley within Strathcona Provincial Park	–	Stands of 75% pure spruce in lower 8 km of drainage; heights over 70 m are common. At least two trees were measured at heights of 82 m; 2.5 km upstream there are smaller-diameter spruce in a stand that is about 40% spruce (H.L. Roemer, pers. comm., Sept. 1996).
Tahsish-Kwois Provincial Park	40	At the time of Beese's 1989 review, Tahsish-Kwois, now a provincial park, was a proposed ecological reserve. Beese's description of the proposed ecological reserve was as follows: total 140 ha reserve in which Sitka spruce occurs on a small portion of the floodplain; best spruce stands are concentrated in the lower 5 km of the floodplain; spruce are mostly 60-70 m tall, with 11 trees 70-74 m tall.

►

◄ *Table 41*

Location	Area[a] (ha)	Description
Ecological Reserves		
Clanninick Creek	15	Total 37 ha reserve, logged on perimeter except south edge; 15-20% of area windthrown; mostly western hemlock in SE corner. Most Sitka spruce are undamaged. About 35 spruce trees are in the range 2.5-3.0 m dbh and about 75 m tall.
Klaskish River Ecological Reserve	40	Total 376 ha reserve where spruce up to 75 m tall have been measured.
Other Locations		
Megin River (outside Ecological Reserve)	–	Stands of 60% spruce in lower 5 km of drainage; heights over 70 m are common.

Source: Beese (1989)

[a] Alluvial area supporting Sitka spruce.

Sitka spruce stands on western Vancouver Island. Although many areas with spruce 60-80 m tall were located, few areas can compare with the Carmanah Creek valley in terms of the size of the largest spruce and the total area on which spruce occurs. Based on this survey, the most comparable spruce stands are in the Megin, Tahsish-Kwois, and Moyeha river valleys. These surveyed stands were summarized by Beese (1989), as reproduced in Table 41.

Areas in British Columbia's ecological reserves system that are examples of ecosystems with alluvial-site Sitka spruce include the Clanninick Creek Ecological Reserve near Kyuquot and the Megin River Ecological Reserve north of Flores Island (B.C. Parks 1994). Both reserves are quite small. The Clanninick Creek Ecological Reserve has experienced significant windthrow damage, although most of the spruce were not affected. Spruce in the Megin River Ecological Reserve are concentrated on an island in the river, together with large western redcedar. There are also some alluvial sites supporting Sitka spruce in the Mace Creek valley and at the head of Mercer Lake within the 9,834 ha Vladimir J. Krajina (Port Chanal) Ecological Reserve on the west coast of Graham Island, Queen Charlotte Islands. The Mace Creek site is atypical, having developed on landslide materials subsequently moved by the creek (Beese 1989, based on personal communication with H. Roemer).

Tahsish-Kwois Provincial Park protects significant floodplain stands that contain Sitka spruce, and the Klaskish River Ecological Reserve near Quatsino Sound includes about 40 ha of spruce in a stand from which the best trees have been harvested. Sitka spruce also occurs within Cape Scott Provincial Park, in which 6,400 ha of its total area of 15,054 ha is old growth (Roemer et al. 1988). Within the approximately 25,000 ha of old growth in Pacific Rim National Park, there are several small areas with Sitka spruce 60-80 m tall; however, none occur on a large floodplain such as in the Carmanah Creek valley (Beese 1989).

Two aspects of the Carmanah Sitka spruce groves make them unique:

- In both size and areal extent, the alluvial spruce stands in the Carmanah Creek valley are the best remaining on Vancouver Island. None of the Sitka spruce stands in today's ecological reserves are comparable. Only the Megin, Moyeha, and Tahsish-Kwois river valleys have Sitka spruce stands that approach the stature or extent of spruce in the Carmanah Creek valley.
- Because Carmanah Creek is near the southern limit of the range of Sitka spruce in British Columbia, spruce's growth potential there probably exceeds that of any other remaining alluvial spruce old growth in the province. Except for the Carmanah Creek valley, within British Columbia there are no sizable alluvial spruce stands in existing or proposed parks or reserves south of Strathcona Provincial Park (Beese 1989), although

excellent stands of spruce occur on alluvial sites further south, in Washington and Oregon.

The lower valley of Carmanah Creek was declared a provincial park in 1990, and Walbran Creek valley and the upper valley of Carmanah Creek were added in 1995. As a result, Carmanah Walbran Provincial Park now protects 16,450 ha of diverse forest ecosystems, including the large Sitka spruce of the Carmanah Creek valley. The valley's largest Sitka spruce, the 95 m tall Carmanah Giant, is the tallest tree known in Canada, and is thought to be the tallest Sitka spruce in the world even though it is less than 400 years old. In the same valley there are much older western redcedar, estimated to be over a thousand years old. This park's spruce stands attain a biomass per hectare nearly twice that of undisturbed tropical forests. No other provincial park in British Columbia protects as large an area of temperate rain forest as Carmanah Walbran Provincial Park (B.C. Parks 1992). Aside from its magnificent spruce, the presence of Marbled Murrelet and Northern Pygmy-Owl in this valley has highlighted the park's importance as wildlife habitat (B.C. Parks 1996).

Other areas further north along the British Columbia coast were observed by Beese (1989) to have alluvial sites with very large spruce. Examples are the Koeye and Nootum rivers near Namu, and several river valleys in the Queen Charlotte Islands. Besides the substantial difference in latitude from Vancouver Island, Sitka spruce sites on the Queen Charlotte Islands differ from mainland and Vancouver Island spruce sites in two major ways: (1) the absence of grand fir or amabilis fir mixed with Sitka spruce on the Queen Charlotte Islands; and (2) the occurrence, on the Queen Charlotte Islands, of grassy, parklike understoreys with plant cover and composition adversely affected by introduced deer.

Roemer et al. (1988) indicated that the 21,100 ha of old growth in the 72,641 ha Naikoon Provincial Park, Queen Charlotte Islands, contains good examples of these distinctive Sitka spruce and Sitka spruce–western hemlock stands. They reported that the most northerly protected area in British Columbia containing Sitka spruce is Gingietl Creek Ecological Reserve; 600 ha of its 2,873 ha area are covered with old-growth forest containing Sitka spruce. Sitka spruce also occurs widely within Gwaii Haanas/South Moresby National Park Reserve on the Queen Charlotte Islands. It is reassuring that old Sitka spruce are present in a relatively large number of forests that are not scheduled for harvest; thus Gwaii Haanas/South Moresby National Park Reserve will not be the only location in the province where large Sitka spruce are found.

There is another aspect of Sitka spruce protected areas that relates to conservation of coniferous genetic resources. Yanchuk and Lester (1996) stressed the importance of gene conservation for safeguarding the future evolutionary potential of tree species to respond to climate change and to provide new biotic products or uses, and for commercial genetic improvement programs. The tactical options, which can involve native Sitka spruce, include: (1) maintaining existing protected areas, (2) creating new reserves for in situ management, and (3) ex situ collections of various genetic types (Edwards and El-Kassaby 1993). Part 1 of the strategy presented by Yanchuk and Lester (1996) includes a survey of the frequency of each of British Columbia's 23 native coniferous species in current protected areas in the province. These researchers considered that reserves of 250 ha or greater were needed to assure a reasonably large number of individuals of the target coniferous species to meet gene conservation concerns. The threshold of 250 ha results in an underestimate of the number of reserves where some level of gene protection already exists. Yanchuk and Lester used the following criteria to define 'adequate protection' of coniferous genetic resources:

- Conservation of alleles is highly dependent on the number of trees in the population. Assuming a random distribution of alleles, alleles with a frequency as low as 0.5% will be conserved if 1,200 trees are present.
- For most species, Yanchuk and Lester (1996) considered a 'population' in a reserve to be adequately protected if it is represented by more than 5,000 m^3 of wood. Despite species differences in average volume per tree and volume per hectare, in most cases the number of trees would easily reach 5,000, which is large enough to address the recent concerns of Lande (1995) regarding population sizes needed for long-term population viability.
- Population sizes estimated by this approach would not have considered trees below merchantable size; therefore, the estimates are conservative. Small trees have the same value as large trees in in situ gene conservation, assuming that their probability of survival is similar to that of the larger trees.
- Lester and Yanchuk (1996) and Yanchuk and Lester (1996) also assumed that because the census numbers for individual species will be relatively high in most reserves that are 250 ha or larger, genetic drift that could result from small numbers of individuals of a given species should be minimal.

With the gene conservation priority rating system proposed by Yanchuk and Lester (1996), Sitka spruce was ranked with a higher need for gene protection than some of its companion coastal conifers (Douglas-fir, western redcedar, and western hemlock), but with a lower priority than others (mountain hemlock, grand fir, amabilis fir, yellow-cedar, Pacific yew, and western white pine).

Management of Sitka Spruce Ecosystems for Fish, Wildlife, Recreation, and Water-Related Uses

Ecosystem management techniques to protect and manage values for fish, wildlife, recreation, and water use are usually not specific to any one coniferous or broadleaf tree species occurring in habitats where these non-timber values are high. This section therefore adds very few management suggestions beyond those found under 'Soil and Soil-Water Relationships' (page 95) and 'Some Fish, Mammalian, and Bird Habitat Features of Sitka Spruce Ecosystems' (page 136) in Chapter 2. Practical suggestions for protecting non-timber values in sites where Sitka spruce may be involved are provided by a comprehensive series of guidebooks released under the Forest Practices Code of British Columbia Act (B.C. Ministry of Forests and B.C. Ministry of Environment, Lands and Parks 1995a, 1995b, 1995d, 1995e, 1995f, 1995g, 1995l), as well as the *British Columbia Coastal Fisheries/Forestry Guidelines* (B.C. Ministry of Forests et al. 1993). The following subsections merely highlight some of the key ecosystem processes that are described in more detail in Chapter 2.

Fish and Wildlife Considerations in Sitka Spruce Ecosystems

The scientific details of fish-related aspects of ecosystem management involving Sitka spruce are summarized under 'Fish Relationships' (page 137) in Chapter 2. Other water-related aspects of Sitka spruce ecosystems are outlined in the next subsection. Here we focus on additional points of practical value for protecting fish habitat in these ecosystems.

Detailed stream studies on Etolin Island, near Wrangell, southeastern Alaska, showed the importance of considering both summer and winter habitats when assessing the effects of Sitka spruce–western hemlock management on salmonid production. Studies of the impacts of logging on salmonid habitat may be misleading if data are gathered only in summer (Murphy et al. 1982). On the western slopes of the Cascade Range in Oregon, clearcut logging has been shown to increase stream productivity and salmonid standing crop during summer (Murphy and Hall 1981). However, these increases can be nullified if salmonid winter habitat is damaged. In coastal forests that contain Sitka spruce, deterioration of salmonid winter habitat can be expected if stream banks are damaged, if natural large organic debris is removed or disturbed, if channel stability is reduced, or if sedimentation is increased as a result of logging and road construction (Bustard and Narver 1975). Small tributary streams and pools in secondary channels that may be dry during summer, and therefore easily overlooked in management prescriptions, need special protection to ensure continued high production of salmon in low-altitude west coast streams (Murphy et al. 1982).

There is circumstantial evidence that evapotranspiration from rapidly growing forests may markedly reduce minimum summer stream flows. This process is of concern in the Sitka spruce–western hemlock forests of southeastern Alaska, where in many streams minimum summer stream flows limit the spawning success of pink and chum salmon, and may limit the habitat of coho salmon that rear in streams (Myren and Ellis 1982).

A recent review of Carnation Creek studies on Vancouver Island by Hetherington (1996) indicated that logging had only a small negative effect (–5%) on returning numbers of adult coho, but a larger adverse impact (–26%) on chum salmon. Unlike the coho, which rear in the stream, the chum leave the creek after emerging in spring and appear to have been mainly affected by the deterioration of the quality of spawning gravel. The results suggest that the impacts of logging on salmon are strongly conditioned by environmental and climatic factors. The importance of side channels in the Carnation Creek floodplain as overwintering habitat for young coho salmon was a major finding of these studies (Bustard and Narver 1975).

The subsection 'Mammal Relationships' (page 139) in Chapter 2 outlined the key biological relationships of mammals that are common in Sitka spruce habitats. The following paragraphs describe some additional management concerns for grizzly bear, Sitka black-tailed deer, and porcupine.

Hamilton et al. (1991) suggested that if regional stocking targets and free-growing requirements are met on specific site associations in the CWH biogeoclimatic zone, there will be a shortage of grizzly bear forage on these sites for most of the rotation; little forage is available for grizzlies under a closed, second-growth canopy. Therefore, the recent *Interim Guidelines for Integrating Coastal Grizzly Bear Habitat and Silviculture in the Vancouver Forest Region* (Hamilton and Berry 1992) defined the ecosystems for medium-rich to very rich nutrient regimes only, in which priority should be given to encouraging canopy conditions conducive to the survival and productivity of grizzly bear forage throughout the rotation. These ecosystems, most of which have Sitka spruce as either the primary or secondary tree species, are listed in Table 42 in relation to biogeoclimatic subzones, soil moisture regime, and position on floodplain (high, medium, and low benches).

The 1992 guidelines apply only to the ecosystems listed in Table 42. On these ecosystems, where grizzly bear forage production is to be maintained or enhanced, the following species are acceptable, in descending order of preference as forage for grizzlies:

devil's club	*Oplopanax horridus*
red elderberry	*Sambucus racemosa*
Ribes spp.	*Ribes* spp.

Table 42

Medium-rich to very rich ecosystems in which forage production for grizzly bears is now a management guideline throughout a rotation involving production of Sitka spruce as a crop tree

Soil moisture regime	CWHvh (outer, inner)	CWHwm	CWHvm (submontane, montane)	CWHdm	CWHds (southern, central)	CWHms (southern, central)	CWHws (submontane, montane)
	Site series, by biogeoclimatic subzone, on medium-rich to very rich soil nutrient regimes						
moist	CwSs–Foamflower	SsHw–Devil's club	BaCw–Salmonberry[a]	Cw–Lady fern[c]	Cw–Devil's club	BaCw–Devil's club	
very moist	CwSs–Devil's club		BaSs–Devil's club[b]				
wet	CwSs–Skunk cabbage	Ss–Skunk cabbage	CwSs–Skunk cabbage				
FPL[d] high bench	Ss–Lily-of-the-valley	Ss–Salmonberry					
FPL medium bench	Ss–Trisetum	Act–Red-osier dogwood					
FPL low bench	Dr–Lily-of-the-valley	Act–Willow					

Sources: Banner et al. (1990), Hamilton et al. (1991), Hamilton and Berry (1992). See Appendix 4 for full names of abbreviated biogeoclimatic units and for tree species whose abbreviations begin the names of each site series.

[a] In the southern portion of the CWHvm.

[b] In the northern portion of the CWHvm.

[c] Cw–Lady fern is no longer recognized as a site series name in Green and Klinka (1994); this former site series name is now best represented by Cw–Foamflower in the CWHdm zone.

[d] FPL = floodplain

skunk cabbage	*Lysichiton americanum*
cow parsnip	*Heracleum sphondylium*
Angelica spp.	*Angelica* spp.
highbush-cranberry	*Viburnum edule*
black twinberry	*Lonicera involucrata*
salmonberry	*Rubus spectabilis*
Vaccinium spp.	*Vaccinium* spp.
red-osier dogwood	*Cornus sericea*
lady fern	*Athyrium filix-femina*
Pacific water-parsley	*Oenanthe sarmentosa*
Pacific hemlock-parsley	*Conioselinum pacificum*
thimbleberry	*Rubus parviflorus*

Regarding logging in the lower Kimsquit River valley, Hamilton (1987) noted that grizzlies fed on berries in recently logged clearcuts. In the sites observed, shrubs began producing berries 2-3 years after logging. Work in similar sites in southeastern Alaska indicated a rapid reduction in understorey shrubs about 25 years after logging, a condition that may persist for at least another 125 years if the site is not managed to reduce the tree canopy (Alaback 1984). Partly for this reason, British Columbia wildlife managers have an interest in lowering the stocking standards for trees on alluvial sites so that more open overstorey canopies will allow the higher light intensities and therefore the more vigorous growth of understorey shrubs and herbs important to grizzlies (Hamilton et al. 1991). This interest applies particularly in sites that are designated as Grizzly Bear Management Areas under the British Columbia Grizzly Bear Conservation Strategy (B.C. Ministry of Environment, Lands and Parks 1995).

Grizzly bear management requires careful planning of forestry activities, especially in relation to shrub species that compete with crop trees. Care is required in forestry-related glyphosate use. Shrubs that compete with Sitka spruce and other crop trees have different forage values for grizzlies. Where possible, forage of the highest value should be left untreated (Table 42). Hamilton et al. (1991) recommended that managers be selective about which sites and shrub species to treat with herbicide. The following shrubs are listed in descending order of value as grizzly bear forage: red elderberry, red raspberry, salmonberry, *Vaccinium* spp., *Ribes* spp., black twinberry, highbush-cranberry, red-osier dogwood, and thimbleberry. Skunk cabbage, lady fern, cow parsnip, *Angelica* spp., Pacific water-parsley, and Pacific hemlock-parsley should also be avoided during herbicide treatments because of their value as grizzly bear forage.

There are records that squirrels, beavers, porcupines, and rabbits can all cause problems by feeding on Sitka spruce (Baumgartner et al. 1987; Loucks et al. 1990). However, these species have relatively little influence on Sitka spruce ecosystems compared with the impact of Sitka black-tailed deer. Because deer have no natural predators on the Queen Charlotte Islands, their population there is very high, and there is widespread evidence of browsing on Sitka spruce. In general, porcupines inflict most of their damage on juvenile stands, whereas black-tailed deer do most damage to seedlings. Where the range of Roosevelt elk overlaps with that of Sitka spruce, their damage to spruce, as with deer, occurs at the seedling or small sapling stage.

The abundant Sitka black-tailed deer on the Queen Charlotte Islands affect forest vegetation in two important ways. Of most direct concern to forest managers is the intense deer browsing of coniferous seedlings, especially western redcedar but also Sitka spruce when it first flushes. Seedlings and saplings are at risk of heavy browsing until they are at least 1.2 m tall. In the case of Sitka spruce, one protective measure is to treat seedlings with a repellent, such as the Swedish product called Plant Skyd (Stirling 1996). Although such repellents are effective for only a few months, even temporary protection during spruce's vulnerable period, when there is a flush of soft new leader and branchlet growth, can be effective. Later in the growing season, when spruce needles harden and become sharp, they are less attractive as deer browse, and repellents are less needed. Sitka spruce needs less of the labour-intensive regeneration practices required to protect western redcedar seedlings. Practices such as fertilizing seedlings at planting time to encourage rapid growth to 1.2 m (above the typical browse line), and encasing seedlings in corrugated plastic tubes, are unnecessary for Sitka spruce.

A second way that heavy deer browsing on the Queen Charlotte Islands affects vegetation management is through deer use of understorey species. The very favourable growing conditions on these islands would pose challenging brush-control problems for managers if heavy deer browsing did not effectively limit brush competition in many cases (Stirling 1996).

Recent studies make clear that a Sitka spruce manager interested in maximizing wildlife values needs to be concerned about more than the size of clearcuts and the species composition of ecosystems that provide browse before and after logging. For example, stand-age differences in nutritional quality of forage for black-tailed deer in coastal forest ecosystems were described by Hanley et al. (1987) for Sitka spruce–dominated ecosystems in southeastern Alaska. They documented the chemical composition of blueberry (*Vaccinium alaskaense*) and bunchberry (*Cornus canadensis*) from May through October in three young stands resulting from clearcuts and in two older stands with understorey vegetation exposed to little direct sunlight. There were pronounced differences in chemical composition of blueberry and bunchberry between the young and older stands. Plants in the young stands had greater astringency, concentrations of phenolics, and total nonstructural carbohydrates, but lower concentrations of nitrogen, than plants in the older stands. In vitro dry-matter digestibility did not differ among stands, however. Chemical analyses of plants from clearcut and adjacent forest at another study area indicated that concentrations of digestible nitrogen were 2.0-2.3 times greater in leaves from the forest than in those from the clearcut. Palatability trials with three captive black-tailed deer showed a consistent preference for leaves from the forest. The results indicate that the carbon/nutrient balance of these plants, controlled by the relative availability of light and nutrients in their environment, has major implications for nitrogen availability to deer but may be less important for the availability of digestible energy.

In Washington and Oregon, strychnine salt blocks have been used for porcupine control, but not very successfully. Alternative methods recommended by Sullivan (1990) include the following steps:

- Identify areas with high hazard.
- In thinning programs, leave species that are not so readily attacked by porcupines, such as amabilis fir and western redcedar, and encourage stand density of under 1,000 stems per hectare to make the habitat less attractive for porcupines.
- Attach sheet-metal collars or sleeves (87.5 cm in length) around the lower bole of susceptible trees to prevent porcupines from climbing the stem.
- Manage fisher, the principal predator of porcupines, as a biological control technique to reduce porcupine abundance. Sullivan recommended that this approach be used in a cooperative effort with Wildlife Branch personnel and local trappers.

From studies in southeastern Alaska, there is evidence that the red squirrel (*Tamiasciurus hudsonicus*) is a major consumer of Sitka spruce seed. However, squirrels were not considered to be a deterrent to spruce regeneration because in good seed years, ample seed is available. Furthermore, red squirrels are more abundant in older spruce forests than in cutover lands where the greatest interest in regeneration lies (Meehan 1974).

As described in the next subsection, Sitka spruce is a prominent species in coastal riparian ecosystems. Thus, any wildlife management guidelines proposed for riparian zone management in coastal British Columbia will be of concern to Sitka spruce managers. For Roosevelt elk, Columbian black-tailed deer, and Sitka black-tailed deer, the special features of coastal riparian ecosystems are highlighted in the definitions of special habitats in Appendix 2 of Nyberg and Janz (1990). In the past five years, attention has been focused on these important ecosystems through the preparation of recommendations such as those in the *Riparian Management Area Guidebook* (B.C. Ministry of Forests and B.C. Ministry of Environment, Lands and Parks 1995g). Besides the Carnation Creek studies highlighted in the next subsection, many review articles provide additional information on land, forest, water, wildlife, and fishery relationships in streamside habitats of the Pacific Northwest, coastal British Columbia, and southeastern Alaska where Sitka spruce is a riparian component (Swanson et al. 1982; Raedeke 1988; Bunnell and Dupuis 1993; McLennan 1993; Morgan and Lashmar 1993; White 1993; Swanson 1994).

Water-Related Roles of Sitka Spruce Ecosystems

Hetherington (1987) was among the early researchers who described the hydrologic relationships in British Columbia's west coast ecosystems where Sitka spruce may occur. Recently, Lawford et al. (1996) assembled a comprehensive collection of reviews on the climate, hydrology, ecology, and conservation of rain forests and associated ecosystems of the west coast of the Americas. This collection included a review by Hetherington (1996) of the effects of logging on aquatic ecosystems, using Vancouver Island's Carnation Creek Experimental Watershed as a case study. Some of the results of the Carnation Creek research are described in this subsection. More information is available in the 15-year synthesis edited by Chamberlin (1988) and from Hartman and Scrivener (1990); a 23-year synthesis is now in press, based on the 1994 Carnation Creek–Queen Charlotte Islands Fishery Forestry Workshop.

The water-related roles of west coast watersheds that may contain Sitka spruce involve a variety of fluvial and ecological conditions or processes, including the dynamics and successional fate of coarse woody debris, riparian ecosystem dynamics, stream flows in winter and summer, water quality, and quality of fish habitat. The role of riparian areas in maintaining British Columbia's overall forest biodiversity is well documented (Stevens et al. 1995). While not dealing with Sitka spruce specifically, these authors outlined 11 overall objectives in riparian zone management. We consider Sitka spruce to have a significant role in 3 of these objectives: (1) it contributes to the amount and distribution of coarse woody debris in riparian and aquatic ecosystems; (2) it helps provide ecological and wildlife linkages as riparian habitats become more disturbed and more widely separated; and (3) it plays a role in the associated social and economic values of riparian ecosystems.

British Columbia's *Riparian Management Area Guidebook* (B.C. Ministry of Forests and B.C. Ministry of Environment, Lands and Parks 1995g) recommends the retention of wildlife trees on coastal floodplains. Riparian areas are characterized by large trees, snags, and downed wood, as well as many deciduous trees and shrubs, all of which contribute to structural diversity and increase niche availability for wildlife (Bunnell and Dupuis 1993). Fallen or transported trees in streams influence channel hydraulics and create conditions suitable for fish-rearing habitat (B.C. Ministry of Forests 1992b).

Sitka spruce was a component of the pre-harvest forest, and also a species used for post-harvest planting, in the well-documented Carnation Creek study area on western Vancouver Island. Syntheses of the measured or inferred impacts of forestry practices in the Carnation Creek coastal stream ecosystem (Hartman 1982; Chamberlin 1988; Hartman and Scrivener 1990) do not focus specifically on the harvest and regeneration of Sitka spruce in relation to drainage basin processes, hydrology, fluvial geomorphology, water chemistry and physics, stream biology, and fish production. However, some inferences can be made about Sitka spruce's role in these physical and biological processes. In the 10 km^2 Carnation Creek watershed, Sitka spruce was identified as a component in three of seven vegetation associations mapped (Oswald 1982; Hartman and Scrivener 1990). These three associations were in the parts of the watershed at lowest elevation, closest to the channel of Carnation Creek. The key point for Sitka spruce management is that, of the forest harvest activities documented at Carnation Creek (road construction, yarding, hauling, prescribed burning, scarification, and herbicide use), the most important impacts were caused by falling and yarding in the stream-side zone, which is the area in the watershed where Sitka spruce is most abundant.

The Carnation Creek overview by Hartman and Scrivener (1990) drew upon the work of many other researchers to make the point that the rate at which vegetation recolonizes an area after logging and burning has an important influence upon several watershed processes, including soil erodibility, slope stability, sediment load in streams, evapotranspiration rate, water yield of streams, soil and stream insolation relationships; and stream-side insect fauna that serve as fish food. For the Carnation Creek study area, the specific role of Sitka spruce can only be inferred for the pre- and post-logging vegetation conditions documented by Oswald (1982), King and Oswald (1982), and Smith et al. (1988). The inference can be based on where Sitka spruce is most likely to occur naturally, or to be encouraged silviculturally, within the watershed.

Clearcutting resulted in the original 80-90% of vegetation cover in the Carnation Creek watershed being reduced to virtually no tree cover, with about 24% cover by the shrub/herb and forb/moss layers. Only about 5% total cover by all vegetation layers remained on sites that were clearcut and then burned (Smith et al. 1988). However, three years after disturbance, all three layers of vegetation had become re-established, and total cover by these layers was 36% on logged sites and 27% on logged sites that had also been burned. Of relevance to ecosystems amenable to Sitka spruce, and also to managers interested in encouraging Sitka spruce as a component of future forest crops, is the observation that the three-year recovery of total vegetative cover was greater in valley bottoms than on the slopes of the watershed (King and Oswald 1982). On the slopes, coniferous and red alder regeneration were smaller, and their distribution more clumped, than in the valley bottom (Hartman and Scrivener 1990). Finally, Sitka spruce, which grows best in the lowland and floodplain portions of coastal watersheds, is a natural and ecologically suited component of valley bottom ecosystems that can recover relatively quickly after disturbance, thus shortening the time that logging or burning adversely affect British Columbia's west coast watershed processes.

A dominant riparian role for Sitka spruce does not come quickly after disturbance. It is Sitka spruce's favoured ecosystems that respond quickly to revegetation and recover from disturbance, not Sitka spruce itself. For example, in Ritherdon Creek, a watershed immediately north of and similar to Carnation Creek, a red alder canopy had closed over the stream 16 years after disturbance. In the period from 15 to 25 years after disturbance, such riparian vegetation begins to shift from broadleaf to coniferous species. Carnation Creek researchers speculated that the rapid development of red alder cover in this watershed's riparian zone would be followed by a slow replacement of alder by Sitka spruce, western hemlock, and western redcedar along the stream channel, with conifers reaching pre-harvest levels of canopy coverage about 80 years after the valley bottom was logged (see Figure 99 of Hartman and Scrivener 1990).

As near-coastal watersheds, such as Carnation Creek, are returned to productive coniferous forest after harvesting, regeneration is typically achieved by a combination of natural restocking by hemlock and cedar and planting on lower slopes and valley bottoms. In the case of Carnation Creek, planted cutblocks received various mixtures of Sitka spruce, western redcedar, Douglas-fir, and amabilis fir, depending on elevation, orientation to the sun, and frequency of sea fog. The zone of frequent sea fog is more relevant for Sitka spruce than for other conifers when trying to match the choice of plantation species to site characteristics. Where sea fog is frequent, temperatures tend to be cool enough to deter white pine weevil populations, an important factor for successful Sitka spruce plantations (McMullen 1976a). For this reason, Sitka spruce was planted in the Carnation Creek watershed only in the zone of relatively frequent sea fog (Hartman and Scrivener 1990).

In the old-growth Sitka spruce–western hemlock forests of the Gregory Creek watershed of western Graham Island, British Columbia, Wilford (1982) documented the important role of large organic woody debris for sediment storage on slopes. Downed trees often form a series of cross-slope obstructions. This is especially the case when storm winds topple old trees perpendicular to the slope. Sediments and small organic debris from upslope mass movements are deposited behind these obstructions, forming a series of terraces that temporarily delay the delivery of sediment to stream channels.

Wilford (1982), Maser et al. (1988), and Chatwin et al. (1994) have all identified the watershed role of large logs lying across the slopes of spruce-hemlock forests at study sites on the Queen Charlotte Islands, in southeastern Alaska, and on the Olympic Peninsula. There have been no concerted efforts to remove large organic debris from hill slopes as there have been in 'stream improvement' programs. However, continued harvesting of old-growth forests will eventually deplete the source of large woody debris. As the old large woody debris rots and is not replaced, both forest and fisheries resources may suffer. Highly productive lower-slope terraces are more likely to be gullied and site productivity may drop as runoff is concentrated in channels instead of spreading over lower slopes. In coarse-textured parent materials, the creation of gullies can lead to a lowering of the water table, further reducing site productivity. Wilford (1982) urged greater recognition of the role of large woody debris in sediment storage, and advised foresters and fisheries biologists to consider managing and protecting large woody debris throughout an entire watershed, not just in and along the edges of streams.

Logging does not always increase stream flow in all forest ecosystems. In an old-growth forest in Oregon, Harr (1980) recorded that stream flow decreased slightly when 25% of two small drainages in a watershed were clearcut. Harr hypothesized that loss of fog drip when forest stands were clearcut resulted in diminished stream flow, and he subsequently determined over a 40-week experiment that net precipitation in the old-growth forest was 1,739 mm, 29% more than in 11-year-old clearcuts, which annually was 25% greater than in clearcuts.

From the preceding examples, it is evident that riparian vegetation is critical to the structure, nutrient dynamics, and health of stream ecosystems because it provides shade, organic detritus, and large woody debris. Woody debris is particularly important because it controls the routing of water and storage of sediment, especially on hill slopes (Wilford 1982); governs pool and riffle formation; and provides substrate for biological activity (Meehan et al. 1977). Standing tree size, distance from a stream, and tree species present determine the future quantity and longevity of coarse woody debris (Robison and Beschta 1990), with coniferous debris lasting considerably longer than broadleaf debris. A study of the occurrence of trees, shrubs, and forest regeneration in riparian environments in coastal Oregon revealed that most of the Sitka spruce seedlings occurred adjacent to or beneath hardwood overstoreys, and riparian conifers were most common on streams with steep gradients. As an indicator of disturbance history, it is likely that areas along coastal Oregon streams where red alder is common have been logged during the last 50 years. In contrast, areas where alder is rare have probably been undisturbed for a century or more (Minore and Weatherly 1994).

4
Sitka Spruce in British Columbia's Future

The original motivation for this book was the need for more intensive Sitka spruce silviculture on the Queen Charlotte Islands resulting from the reduction in the working forest land base that accompanied the creation of Gwaii Haanas/South Moresby National Park Reserve. Thus, in addition to a comprehensive review of Sitka spruce's ecological characteristics, considerable attention has been given to silvicultural and crop-planning options for this species.

Not surprisingly, however, there are public interests in Sitka spruce ecosystems that go far beyond silviculture, crop planning, and the setting of allowable annual cuts. Sitka spruce's role in management for biodiversity, non-timber land uses, integrated resource management, protected old-forest ecosystems, and biologically sustainable forests are all part of today's concern with ecosystems that contain Sitka spruce. Based on details presented in Chapters 1 to 3, this final chapter highlights these diverse interests, rounding out the picture of Sitka spruce in western coastal ecosystems.

Except for the longstanding relationships that several First Nations groups of coastal British Columbia have had with west coast forests, Sitka spruce is probably not as prominent in the psyche of this province's residents as more broad-ranging species such as Douglas-fir, western hemlock, western redcedar, and interior spruces. This may soon change as more people are made aware of the ecological, cultural, and economic role of this important tree species in North America's temperate rain forest. Several factors are raising Sitka spruce's profile.

- Sitka spruce's natural occurrence in sites that are culturally important to Aboriginal people, typically involving highly productive lowland sites near coastlines, where marine and forest resources can be harvested from one location, ensures a role for this species in future efforts to achieve sustainable coastal zone management.

- Sitka spruce's prominence in coastal estuaries and in riparian, floodplain, and near-coastal ecosystems ensures that it will be an ecological component in the difficult land-use decisions involving such diverse and highly productive areas.
- Sitka spruce's ability to regenerate well on a broad range of seedbeds and site conditions, including a variety of disturbed surfaces, provided there is little competition from other vegetation, guarantees it an important role in forest renewal and site restoration within its natural range.
- Sitka spruce is still valued for the features that made it important earlier this century. It is the largest spruce in North America (Figure 49), with wood that is lightweight but strong, suited to the manufacture of high-value products dependent on large stems of clear wood.
- As described later in this chapter, Sitka spruce figures prominently in matters of old-forest protection, biodiversity, and integrated resource management in highly productive ecosystems.

Figure 49. The magnificent Sitka spruce forests of the past are now uncommon, but the ecosystems that supported those stands are prime locations for future management of this spruce. This stand was photographed near Quatsino Sound, Vancouver Island, British Columbia (British Columbia Archives and Records Services, Photograph NA-06166, Catalogue no. FS02141-0, date and photographer unknown).

Compared with other native tree species from coastal British Columbia, an unusual feature of Sitka spruce is the vast literature on its silvics, growth patterns, and physiology, largely because of its success as a plantation species in Europe. This book recognizes the wealth of information available for Sitka spruce as a result of ecological and silvicultural studies where this tree has been established as an exotic species. However, as the subtitle suggests, the emphasis here is on the natural range of this tree in British Columbia; it will be the task of future researchers and forest managers to assess the degree to which knowledge about Sitka spruce acquired elsewhere applies to ecological conditions and management circumstances within Sitka spruce's natural range in British Columbia and the Pacific Northwest (Figure 50). Similarly, the degree to which information from its natural range can be applied to sites where this spruce is an introduced species must be judged by researchers and practitioners familiar with conditions at those sites.

Future Silviculture of Sitka Spruce in Its Natural Range

The natural tendency is to classify existing information about Sitka spruce into two main groups – information derived from studies of Sitka spruce in its natural range and information from spruce plantations beyond its natural range. This distinction is followed here through the use of geographic

Figure 50. Sitka spruce's excellent early growth potential makes it an important component of new forests in coastal British Columbia. Experimental plantations have been established by Western Forest Products Limited in the Koprino River watershed, northern Vancouver Island, British Columbia. (Photograph by Western Forest Products Limited, Vancouver)

qualifiers in paragraphs that summarize information about Sitka spruce ecology and management.

A further classification is needed, however. Within British Columbia, there are ecological reasons for considering separately Sitka spruce on the Queen Charlotte Islands and Sitka spruce on Vancouver Island and the British Columbia mainland. Foresters on the Queen Charlotte Islands face different opportunities and challenges from foresters in other parts of Sitka spruce's natural range. On the Queen Charlotte Islands, Sitka spruce management is distinguished by the following factors:

- Sitka spruce makes up a higher proportion of tree standing crop on the Queen Charlotte Islands than elsewhere in coastal British Columbia, probably mainly due to these islands' extensive coastal area and length of shoreline exposed to offshore winds.
- Sitka spruce occupies sites of high productivity free of white pine weevil problems, in contrast to severe weevil constraints on much of Vancouver Island and the coastal mainland.
- Foresters managing Sitka spruce on the Queen Charlotte Islands are spared concerns about managing habitat for elk and grizzly bear. Salmon habitat protection or enhancement can be handled at the district level because there are no large salmon-bearing watersheds that occupy different administrative regions, such as those in the major drainage systems of mainland British Columbia. Also, Queen Charlotte Islands foresters do not need to consider the forest management aspects of wolf influences on deer populations when developing forest management plans related to deer.
- Because the Queen Charlotte Islands are the only place in the world without white pine weevil damage and with less severe windthrow hazard than in the United Kingdom, they are the only place where large-diameter, high-quality Sitka spruce can be grown on long rotations.
- The highest average wind speeds in North America occur on the southern part of the Queen Charlotte Islands, and all windward slopes of this archipelago are prone to extremely high winds, a factor that influences not only silvicultural design outside the national park reserve but also the longevity of natural stands containing Sitka spruce.
- The Queen Charlotte Islands are characterized by an abundance of naturally disturbed sites, especially from frequent landslides and widespread deposition of sediments onto river floodplains, which provide suitable seedbeds for successful spruce regeneration. However, innovative harvest methods are needed (Figure 51) to reduce the risk of mass wasting and landslides.
- The exceptionally high deer population on the Queen Charlotte Islands constrains western redcedar and yellow-cedar regeneration and removes

brush species that compete with conifer regeneration, thereby favouring Sitka spruce development. However, where there is extremely severe deer browsing, there is evidence that Sitka spruce can be browsed also.

- The Queen Charlotte Islands have a high frequency of culturally modified trees, especially at lower altitudes, requiring special forest harvesting practices for heritage protection in ecosystems that often contain Sitka spruce.
- On the Queen Charlotte Islands today, there is a forest land base in which areas protected for Aboriginal, cultural, and parks purposes limit the potential area of working forest, a circumstance that encourages intensive silviculture on the land available for sustainable forest management.
- Apart from forest lands set aside for parks and other protected areas, and lands that may be involved in future negotiations related to Aboriginal land claims, there are no other major land uses that compete with timber production on the Queen Charlotte Islands.

Figure 51. Large-capacity Sikorsky Skycrane helicopter used for experimental harvest of Sitka spruce–western hemlock–western redcedar stands on steep landslide-prone slopes of coastal British Columbia. This example of Sitka spruce removal from sensitive terrain was carried out by Husby Forest Products Ltd. and Canadian Air-Crane Ltd. in an attempt to reduce the mass-wasting impact of conventional logging. The experimental harvest was in the area shown in Plate 3, near Rennell Sound, Queen Charlotte Islands, British Columbia. (Photograph by F. Pendl)

Some of these factors are unique to the Queen Charlotte Islands, whereas others are not. All, however, are particularly important for management of Sitka spruce on that archipelago.

In British Columbia today, Sitka spruce silviculture is synonymous with silviculture to control the white pine weevil. Weevil-related topics dominate the Sitka spruce literature of the past decade, and weevil-damaged plantations are a prominent concern of field foresters. Except on the Queen Charlotte Islands and areas north of the latitude of Kitimat, where there are no weevil problems, most of coastal British Columbia's Sitka spruce is so severely damaged by the weevil that planting this species is discouraged (Alfaro et al. 1994). Planting guidelines, suggested a decade ago (Heppner and Wood 1984) and still in effect, allow only 20% Sitka spruce in areas with high or medium weevil hazard. Most foresters expect to have some Sitka spruce naturally regenerate on clearcuts in many parts of the Coastal Western Hemlock (CWH) Zone, but at present no attempt is made to encourage spruce because of its weevil problem.

Details of the weevil–Sitka spruce relationship are provided under 'Insects and Diseases' (page 113) in Chapter 2 and 'Management to Minimize Losses to Insects and Disease' (page 232) in Chapter 3. In 1993 a survey of forest health research needs in British Columbia (Nevill and Winston 1994) ranked the white pine weevil second among insect pests in the entire province, after the mountain pine beetle. The white pine weevil was the province's highest-ranked insect pest solely affecting regenerating stands. Suggested research included better survey techniques for weevil incidence and intensity, better damage appraisal methods, improved hazard ratings, better definition of differences in population dynamics, improved control measures, and impact of various silvicultural management systems on weevil incidence.

Compared with Britain, British Columbia has virtually no tended, mature Sitka spruce stands. The question of appropriate silvicultural systems is therefore mainly speculative. Most of the site series in which Sitka spruce occurs naturally are very resilient with regard to natural regeneration. There are, however, no precedents for 'geriatric' silvicultural manipulations in old-growth stands other than retrospective studies of historical disturbances. For old forests dominated by Sitka spruce, mixed with hemlock and cedar, the only feasible silvicultural system is clearcutting, possibly clearcutting in patches. Attempts at true selection systems, using helicopter extraction (Figure 51), appear likely to lead to dominance of western hemlock regeneration, often with heavy mistletoe infection.

Overstocked stands, ranging from 12,000 to 25,000 stems per hectare, are commonplace in naturally regenerated sites of the Western Hemlock–Sitka Spruce forest type. Crown closure usually occurs quickly, often in less than 15-20 years. The rate of crown closure is dependent on the rate of ingress of natural regeneration, the survival of any advanced growth of western hemlock, the site index, and the density of the regeneration. Trees in these stands undergo intense competition, and individual tree diameter growth is slow under such dense stocking, but natural mortality is high and there is generally a good expression of dominance. Both Sitka spruce and western hemlock respond well to density control and early commercial thinning, particularly up to the age of 40 years. Height growth response is similar for both species. Because pre-commercial thinning involves thinning from below, it favours Sitka spruce and is therefore a useful method for increasing the spruce content of stands. Conversely, where weevil activity is high, spacing can be used to promote other species by removing the spruce. Spacing is usually done only to obtain the desired species composition and density, and is a basic silvicultural obligation in dense stands. The residual density following pre-commercial thinning depends upon the stand management objectives, which are usually a mix of timber and biodiversity objectives.

Sitka spruce is very responsive to fertilization, and there are over 50 years of experience with N and P fertilization in the United Kingdom. In British Columbia fertilizer treatment has been limited to N and P fertilization of chlorotic plantations growing on salal-dominated cutovers. On nonalluvial sites without salal, it seems reasonable to expect a response to N fertilization. This should be tested in thinned and pruned stands. Sitka spruce has very poor shedding of dead branches, and produces clear wood on very long rotations only, unless pruned. In British Columbia there appears to be a good case for pruning Sitka spruce on the best sites, because of today's premium values for clear grades of lumber.

The growth and yield information base for Sitka spruce in British Columbia is incomplete for use in crop planning. The growth and yield data sets for Sitka spruce are much smaller than those available to support Douglas-fir and western hemlock crop planning. Existing site index curves for Sitka spruce are excellent for site quality estimation when they are correctly applied to appropriate stands. However, they are not suitable for very young stands and very old stands. The existing information base meets all crop-planning requirements for stem volume prediction, and is relatively good in relation to the density management aspects of Sitka spruce crop planning. The response of old-growth Sitka spruce–western hemlock stands to attempts to apply partial-cutting regimes needs to be quantified.

It seems unreasonable to expect in British Columbia a duplication of the high gross yields seen in the British Sitka spruce yield tables, obtained by harvesting all mortality through use of repeated thinning. Only on the Queen Charlotte Islands can Sitka spruce silviculture be taken seriously at present, but these islands are remote, with difficult road construction conditions and thus unlikely economic circumstances for repeated commercial thinning.

Current principles of ecosystem management require that the biological diversity, health, and integrity of ecosystems be considered in both stand- and forest-level management (Kimmins 1996). Natural disturbance and successional patterns at both scales should be observed in natural forests and mimicked if possible (Perry 1994). This concept of using the natural dynamics of Sitka spruce–western hemlock old-growth forest as a basis for human intervention is at the heart of the current biogeoclimatic site classification system and manuals used by field foresters and planners.

The current understanding of disturbance ecology suggests a broader approach to forest management than has been practised (Rogers 1996). Preservation of natural disturbance regimes is essential to promote healthy, dynamic ecosystems, but also requires a good knowledge of the dominant processes at work and the critical functions such as disturbance interactions, as suggested by Noss (1990). However, there are still operational uncertainties about how to implement these ecological criteria in old-growth harvesting practices when terrain, access, tree size, and markets force decision-making into a process of modifying clearcut size, location, and timing. Partial cutting or geriatric silviculture is very difficult to apply, and there are few precedents for these alternative approaches, especially because windthrow, disease, and high-grading are very important concerns for the forest manager.

For the manager involved with Sitka spruce in British Columbia, there is not much recorded experience describing the results of particular silvicultural interventions in relation to various successional stages in specific ecosystem units of a given biogeoclimatic subzone. This is a subject for further research.

Biodiversity Values of Ecosystems Where Sitka Spruce Is Prominent

Biodiversity is a topic that has been so thoroughly analyzed in recent years through major monographs and journal articles that little needs to be added here. Readers interested in this topic can find analyses of worldwide application in Wilson and Peter (1988), others that focus on the biodiversity of North America's west coast temperate rain forests (Hansen et al. 1991; Alaback 1996), and analyses with emphasis on British Columbia's biodiversity (B.C. Ministry of Forests 1992a; Fenger and Harcombe 1993; Fenger et al. 1993; Foster 1993; Pojar 1993; Harding and McCullum 1994).

The stand management guidelines for maintaining biological diversity in coastal British Columbia (B.C. Ministry of Forests 1992a) are built around six biodiversity attributes: snags, large green trees, coarse woody debris, tree species diversity, understorey plant community, and vertical and horizontal structure. Using these attributes, methods for maintaining biodiversity in managed stands have been proposed in relation to nine management activities: silvicultural systems, harvest methods, utilization standards, site preparation, regeneration, vegetation management, spacing and thinning, pruning, and pest management.

Any one of these nine management activities does not necessarily have an impact on all six biodiversity attributes. For example, utilization standards interact with only two attributes, retention of large green trees and coarse woody debris; pruning interacts only with vertical and horizontal structure and understorey plant community. In contrast, site preparation can have an impact on all six biodiversity attributes.

The ecological basis for these biodiversity attributes is that high levels of structural complexity have been identified with many natural forests, especially those representing late-successional conditions in the west coast temperate rain forest (Graham and Cromack 1982; B.C. Ministry of Forests 1992b; Parks and Shaw 1996). Much of this structural complexity survives natural disturbances from windstorms, wildfires, or insects, and is incorporated into the recovering ecosystem after a disturbance (Hansen et al. 1991). Even on landslides of the type common on the Queen Charlotte Islands (debris avalanches), succession is in most cases secondary rather than primary, because it typically starts from a biologically altered soil that contains organic matter. Primary succession tends to be limited mainly to the tops of landslides where bare bedrock or truncated subsoil may be exposed.

In west coast forests, most natural disturbances (again excluding major landslides) typically do not consume or remove much woody material, even though trees are often killed. Large quantities of snags, downed logs, significant numbers of standing green trees, and humus layers containing seed banks and live roots are often left behind. Franklin (1993) refers to these biological carryovers as 'biological legacies.' These legacies greatly enrich the regenerating forest ecosystem and make it structurally, compositionally, and functionally much more complex (Maser et al. 1988).

The multidisciplinary efforts over the past five years to prepare guidelines for maintaining biological diversity in coastal forests led to the recent release of the *Biodiversity Guidebook* under the Forest Practices Code of British Columbia Act (B.C. Ministry of Forests and B.C. Ministry of Environment, Lands and Parks 1995a). Of most relevance to Sitka spruce management are the classes of disturbance types defined in this guidebook. Most of the geographic range of Sitka spruce in British Columbia falls within a zone that is mapped as Natural Disturbance Type 1. Biogeoclimatic subzones and variants that make up this disturbance type include those that are singled out in Tables 14, 15, and 16 in Chapter 2.

The *Biodiversity Guidebook* contains recommendations on the following: desired distributions of early, mature, and old seral stages; temporal and spatial distributions of cut and leave areas; old-stage seral representativeness; connectivity of different landscape components and different seral stages; ways to encourage vertical structure by maintaining a variety of canopy layers, and horizontal structure by appropriate spacing of patches; and stand-level approaches for maintaining a plant species composition similar to that resulting from natural succession.

A complicating factor for Sitka spruce management on the Queen Charlotte Islands is that these biodiversity guidelines are based on natural disturbances, mainly windstorms and wildfires, and, to a lesser extent, insects and landslides. In reality, disturbance of vegetation by introduced Sitka black-tailed deer on the Queen Charlotte Islands is more widespread than the factors that define British Columbia's natural disturbance types. By reducing vegetation that competes with Sitka spruce regeneration, disturbance by deer may in the long term lead to an overall increase in the abundance of Sitka spruce on the Queen Charlotte Islands. Management for biodiversity values on these islands will need to recognize the complicating influence of the deer–Sitka spruce relationship superimposed upon natural disturbance regimes from landslides, alluvial deposition, and windthrow.

A further consideration for biodiversity management is that some of the site series for which Sitka spruce is a potential regeneration species, as defined by the Silviculture Interpretations Working Group (1994), no longer contain Sitka spruce on the Queen Charlotte Islands – past harvesting has removed it from several site series for which it is ecologically suited (G. Wiggins, pers. comm., April 1996). The biodiversity values that Sitka spruce can contribute to temperate rain forest ecosystems could be more widely distributed if it were encouraged in all the site series on the coast where the Silviculture Interpretations Working Group has judged it to be a primary, secondary, or tertiary regeneration species (see Tables 14, 15, and 16 in Chapter 2).

The CWH zone is biologically the most diverse of British Columbia's 14 biogeoclimatic zones, and within this zone the CWHvh and CWHwm subzones are the most diverse (Radcliffe et al. 1994; Stevens 1995b). Numbers do not provide the whole picture, but they can act as indicators. Compared with other biogeoclimatic zones, the CWH zone has the highest percentage of the provincial total for numbers of species of mammals, birds, and amphibians, and the sixth highest for reptiles. This zone has 451 wildlife species, distributed as follows: amphibians, 13; reptiles, 6; birds, 327; and mammals, 105 (Stevens 1995b).

Studies of the arthropod diversity in old-growth Sitka spruce canopies in the upper Carmanah Creek valley on the southwest coast of Vancouver Island indicate the special biological inheritance of these forests (Winchester 1993). Initial results show that canopies of old-growth Sitka spruce contain distinct species assemblages of arthropods that form a discrete epiphytic community. An upward view into the biologically diverse world of a west coast rain forest Sitka spruce canopy is shown in Plate 13. In this canopy, about 1,500 arthropod species have been identified to date, 80 of which

(5% of identified species) are new to science. Winchester and co-workers suggested that old-growth forests contain a set of natural controls, including a high incidence of predators and parasitoids, that keep potential arthropod outbreaks in check.

Another feature of the Carmanah Sitka spruce canopy is the presence of moss mats, 4-28 cm deep underlain by a soil layer, on the upper surfaces of large branches in the canopy. Springtails, mites, and ticks dominate these moss mats, and several of these arthropods are new species records. The moss, litter, and soil layers on the large branches of Sitka spruce are complex systems whose biodiversity rivals that of many tropical forest systems. The arboreal soil-arthropod fauna found there is another scientific first, and so far has not been found in any second-growth sites examined (Winchester 1993; Krajick 1995; Winchester and Ring 1996). Winchester and Ring (1996) indicate that there is a unique set of arthropod species that are old-growth dependent. Several of the newly described species exhibit a habitat specificity that restricts their distribution to structural features of the old-growth Sitka spruce forest, both on the forest floor and in the canopy.

Coarse woody debris has several important biodiversity functions in forest ecosystems: provision of structural diversity to old-growth ecosystems; provision of habitats for plants and animals; provision of refugia for organisms after disturbances; influence on hillside and stream geomorphology; influence on forest floor microtopography and microclimate; and influence on the nutrient and organic-matter dynamics of forests ecosystems (Caza 1993). The highest input rates for coarse woody debris in North America are in the coastal coniferous forests of the Pacific Northwest and British Columbia, and are the result of varying combinations of wind, insects, diseases, suppression, competition, and fire (Harmon et al. 1986).

In many mature and old-growth forests, tree seedlings most commonly root in or on fallen logs – the nurse log phenomenon. For example, 88-97% of tree seedlings in an alluvial spruce-hemlock stand in Olympic National Park grew on logs that represented only 6-11% of the ground cover on the site. Fallen logs and snags provide important components of wildlife habitat for forest birds, small mammals, invertebrates, and amphibians; these roles can vary in succession as wood decays from hard to softer materials, providing different opportunities to wildlife (Harmon et al. 1986). Aboveground input of coarse woody debris to a Sitka spruce–western hemlock ecosystem in Washington state (Harcombe et al. 1990; Caza 1993) for three stand ages (37, 85, and 138 years) indicated that rates of input were below 2.0 Mg/(ha · yr) until about age 85, and then increased to 6.0 Mg/(ha · yr) by age 138. Another study in the Olympic Peninsula (Graham and Cromack 1982) revealed that in two Sitka spruce–western hemlock stands over 200 years old, the biomass of downed logs ranged from 93 to 167

Mg/ha, and the biomass of snags ranged from 26 to 40 Mg/ha. At this location, Sitka spruce typically died upright, whereas hemlock was subject to windthrow.

Dead woody material exerts ecological influences on a site for hundreds of years, first as snags and fallen trees and then as decaying wood and organic matter. A succession of decomposition classes for logs and snags was developed by Bartels et al. (1985), illustrating this progression. The long-term source of future woody debris is today's regeneration of conifers, including Sitka spruce in riparian sites. However, historical harvesting has left many coastal riparian areas of the Pacific Northwest dominated by broadleaf tree species and shrubs, and silvicultural treatment may be required to restore the former level of coniferous coarse woody debris input to riparian sites.

The use of native plants by Aboriginal people is often overlooked as a component of forest biodiversity. The list of Aboriginal uses of Sitka spruce mentioned by Taylor and Taylor (1973), Brough (1990), Douglas (1991), and Turner (1995) included the following:

- Aboriginal people in the Bella Coola region used practically every part of Sitka spruce for medicinal purposes. Elsewhere within the geographic range of Sitka spruce in western North America, it has been used medicinally as follows: (1) spruce branch tips mixed with berries of *Ribes* and *Amelanchier* were taken internally to treat gonorrhea; (2) spruce gum was boiled and taken internally as a diuretic; (3) spruce gum was applied to cuts or sores and, when mixed with baked leaves of *Veratrum californicum*, was used as a poultice on the chest for heart trouble; (4) sap from peeled spruce trunks was taken as a laxative; (5) a bed consisting of inner spruce bark, ripe spruce cones, and leaves of *Oplopanax horridus*, all placed over hot stones, was considered good for backache and rheumatism; (6) ripe spruce cones were boiled and the resulting liquid was drunk for various internal pains.
- The Makah Indians of Washington ate young spruce shoots raw, but there is no record of such use by British Columbia First Nations people.
- Several British Columbia tribes used spruce roots and small branches to make baskets, boxes, hats, and other household articles; many exquisite spruce root articles are still made today by the Haida.
- Spruce gum was widely used as chewing gum, as well as for cementing parts of tools and caulking canoes.
- The Haida, Tsimshian, and Alaskan Tlingit used to scrape off the slimy cambium and secondary phloem tissue between the wood and the bark of Sitka spruce and eat it fresh in spring or dry it into cakes for winter use; Turner (1995) indicated that some Tsimshian people still eat spruce cambium today, and some other Native people use it in summer specifically as a laxative.

- Turner (1995) also records that the Nuu-chah-nulth sometimes tied spruce boughs to submerged wooden fences to intercept herring spawn; they peeled off the spawn after it was dry and did not eat the spruce needles.
- Sitka spruce also held a special place in Native mythology and in religious rituals; some west coast Native people carved spruce wood into charms, believing that the tree had magical powers.

The prospects for maintaining British Columbia's forest biodiversity are promising because the province contains some of the largest unlogged watersheds in the world, several of which are located within the natural range of Sitka spruce. An inventory of 354 coastal watersheds (Moore 1991a) determined that 20% were considered pristine (virtually no evidence of past human or industrial activity and any small-scale activities limited to less than 5 ha), but at the time of this survey, only nine watersheds, or 2.5% (in all size classes), had some form of protected status. A more detailed analysis of undeveloped watersheds in British Columbia, classified as unprotected, partially protected, and fully protected, was prepared and mapped at a scale of 1:2,000,000 by the Ministry of Forests, Recreation Branch (1992). Considering only the CWH zone, the main zone of Sitka spruce occurrence, in 1992 there were 172 undeveloped watersheds; 145 (84.3%) were unprotected, 8 (4.7%) were partially protected, and 19 (11.0%) were fully protected.

The discussion of the regeneration dynamics of Sitka spruce in Chapter 2 (see page 81) used the example of the Carmanah Creek floodplain ecosystem to demonstrate that repeated disturbance from stream channel changes, flooding, and windthrow is a natural part of such alluvial ecosystems, and that these physical processes influence the natural renewal of Sitka spruce stands. For long-term management, the important point is that on alluvial sites periodic disturbances create a continuous supply of new forests that, if not scheduled for harvest, will provide certain sites that will develop into future old-growth stands. On non-alluvial sites, disturbances created by landslides and windthrow from occasional catastrophic windstorms may also be natural sources of new regeneration opportunities for Sitka spruce. Human disturbances, including modern forestry practices, are another source of non-alluvial seedbeds that could be the sites of magnificent Sitka spruce stands in the future. Those who focus solely on maintaining today's old-growth forests may overlook the fact that disturbances that remove the present forest canopy and provide mineral soil surfaces suitable for Sitka spruce germination create opportunities for future old-growth spruce forests, which will be enjoyed by those living in the natural range of Sitka spruce three or more centuries from now.

Sitka Spruce as Part of Integrated Resource Management

Integrated resource management (IRM) is a process of identifying, assessing, and comparing all resource values in order to make decisions on resource use that ensure continuing management of timber, forage, soil, wildlife and wildlife habitat, water, recreation, cultural, and heritage resources. One of the best examples of IRM involving Sitka spruce is that of grizzly bear management. The moist and nutrient-rich alluvial floodplains and lower slopes of the CWH zone of mainland coastal British Columbia are among the most productive tree-growing sites in the province. The same sites support some of the highest densities of grizzly bears in Canada. The increased emphasis on reforestation of these sites to full canopy closure of conifers conflicts with the goal of maintaining grizzly bear forage after logging (Hamilton et al. 1991). The objective of IRM is to manage grizzly bear habitat in a condition suitable for maintaining viable grizzly bear populations, while practising sustainable forestry. This requires that some sites have wildlife management priority, while other sites are managed for a broader range of uses. Such an approach means that planning must be site-specific, but it must also be placed in the context of a broader geographic scale. For example, the impact of proposed silvicultural vegetation management on grizzly bear forage must be examined in the context of bear habitat availability in the management area or watershed. The effective co-management of grizzly bears and floodplain forests is an example of IRM at its best. The integrated management of timber and deer in coastal forests of British Columbia and Alaska (Nyberg et al. 1989) is an equally good example, although in this case IRM is somewhat handicapped by the lack of clear policies and objectives on the part of the Ministry of Forests regarding deer production on Crown forest lands.

Sitka spruce occurs in areas of high scenic and recreational value, but its management for such land uses will be limited to a relatively small number of areas that are readily accessible to recreational users of the forest land base. Despite the international publicity accompanying the creation of Gwaii Haanas/South Moresby National Park Reserve, tourism has not yet filled the economic void left by the cessation of forest harvesting on South Moresby Island. The relevance of this for foresters managing Sitka spruce in the portion of the Queen Charlotte Islands allocated to the working forest is that they will not, in the foreseeable future, be dealing with large numbers of recreational users of forest land outside Gwaii Haanas/South Moresby National Park Reserve.

An important point for IRM involving Sitka spruce is that it and its companion species occupy ecosystems where, free of disturbance, trees can survive for 300-800 years, growing into some of the world's tallest, oldest, and most massive trees. The coastal temperate rain forest has the highest standing biomass of any ecosystem on earth, even greater than tropical rain forests (Sierra Club of Western Canada 1992). Vancouver Island, the approximate middle of Sitka spruce's north-south range, possesses some of the most productive examples of temperate rain forest in the world. As Carmanah Walbran Provincial Park will likely demonstrate, the use of such magnificent forests (Figure 49 and Plate 13) by recreational visitors will only increase in future decades.

Sitka Spruce in Sustainable Forest Management

The dependence of an economically sustainable society upon biologically sustainable forests has been made clear by Maser (1994). He makes the discouraging observation that we are not headed towards sustainable forestry because foresters are focusing on plantation management rather than management of the prerequisite foundation – biologically sustainable ecosystems. What does this imply for a manager involved with Sitka spruce?

One implication is that the biologically unique coastal forests where Sitka spruce occurs cannot be sustainably managed using European plantation management methods. Maser argued persuasively that European plantation silviculture is simply incompatible with the goal of maintaining biological diversity and ecosystem sustainability. In reality, ecosystem sustainability is defined by the degree of overlap between what is ecologically possible and what current generations of humans desire for themselves and for future generations (Bormann et al. 1994). In this context, Weetman (1996) also cautioned against the use of European silvicultural practices as models for the silviculture of Canadian forests. He suggests that Canadian silviculture must find its own methods and precedents, particularly for geriatric problems in old natural forests. Even where the same tree species is involved, as is the case with Sitka spruce, European silvicultural methods and precedents fail to satisfy the multi-use objectives of forest management in British Columbia.

Moore (1993) was somewhat more optimistic that British Columbia forest managers could learn much from European forestry practices, particularly with regard to integrating non-timber uses and values into the forest planning process. British Columbia forest managers are fortunate to be working with a land base that already supports a high degree of biodiversity compared with European plantation forests. In contrast, British ecologists are now putting considerable effort into methods for mimicking ecological processes that occur in natural forests, in an attempt to convert British upland spruce forests into more diverse ecosystems (Ratcliffe and Peterken 1995).

The British desire to improve the biodiversity of their Sitka spruce plantations is a difficult challenge; fortunately it is a task that British Columbia foresters have been spared until now. However, the term *ecosystem management* is increasingly used by scientists, managers, policymakers, and the general public in British Columbia and elsewhere in North America. The key elements of this approach are that it is ecosystem-based, it is centred around the concept of sustainability, it attempts to mimic natural disturbance regimes, and it seeks to maintain viable populations of all native species. To implement these goals, planning must be done at a landscape level (50-1,000 km^2) and with long time horizons (more than five years). Furthermore, because of the large scales involved in ecosystem management, interagency coordination and public communication are essential (Galindo-Leal and Bunnell 1995). This gives ecosystem management an immediate link to the goals of integrated resource management.

Both the incentive for this book and the approach used in synthesizing available information were influenced by several special features of the forest region in which Sitka spruce occurs naturally. The majestic Sitka spruce forests of yesterday (Figure 49) are now uncommon. However, the ecosystems that earlier nurtured these impressive old-growth stands of Sitka spruce can be the sites of both intensive silviculture in some locations (Figure 50) and protected old-growth forests in other locations (Plate 13).

Sitka spruce grows in that part of Canada where the nation's most widespread and intensive applications of silviculture occur, mainly because the west coast region has some of the lowest biological risks for growing timber in North America (Weetman 1991). There are several reasons for these relatively low risks: (1) the region produces trees that grow to great ages and sizes; (2) the region's ecological conditions, including relatively infrequent occurrences of hurricane-force winds, allow dead trees to remain standing for a very long time; (3) many old-forest stands remain economically viable to harvest at relatively old ages even if there is some loss of usable wood from stem decay, insect losses, and death of individual trees as stands grow older; (4) except for the white pine weevil problem emphasized earlier, second-growth stands are relatively free of insect and disease problems; and (5) there is generally prompt forest regeneration, although not always with desired species.

Relatively low biological risks compared with other forest regions on this continent, coupled with very high stand yields and high market values for large tree stems, greatly favour intensive silviculture in the part of Canada where Sitka spruce is a significant natural component of old-growth stands, as well as in new forests that have developed on recently disturbed areas. An example of the excellent early growth possible in new forests of Sitka spruce in coastal British Columbia is shown in Figure 50.

The circumstances described here relate to Sitka spruce throughout its natural range, but they are particularly relevant today on the Queen Charlotte Islands, where the creation of a national park reserve has led to a reduction in allowable cut. The creation of a recovery fund for more intensive silviculture over a 10-year period to offset this reduction assumes not only low biological risk but also high technical success in the silviculture of ecosystems that contain Sitka spruce. It is in this context that this book has summarized the ecology of Sitka spruce and reviewed both local and international experience in the management of this species.

Appendices

Appendix 1

Sample page of Sitka spruce managed-stand yields and mean annual increments based on a 1994 version of TIPSY (Version 2.1.2 Gamma), for an assumed index of 24 m at breast-height age 50 and an assumed stand density of 500 trees per hectare. Persons interested in using WinTIPSY for Sitka spruce should consult Mitchell et al. (1995) for a user's guide to WinTIPSY Version 1.3 running under Microsoft Windows®, and Stone et al. (1996) for the economic analysis module for WinTIPSY. This model can be accessed by contacting the British Columbia Ministry of Forests, Research Branch, Forest Productivity and Decision Support Section, Victoria, British Columbia.

PRODUCT	Managed stand yield summary
AGENCY	MOF Research Branch
PROJECT	
SPECIES	Sitka spruce
REGENERATION	Natural
DENSITY	500 trees/ha
TREAT.	Untreated
OAF's 1 & 2	0.00% 0.00% (Operational adjustment factors)
SOURCE	TASS V 2.05.24 93-Jan-07
VERSION	TIPSY V 2.1.2 Gamma
DATE	Jan 18/94; 14:40:22
SITE	24 m @ bh age 50
DELAY	0 years
FILE	ss24-5.out

▶

Cumulative production

Tot. age yr	Gross vol. m³/ha	Merchantable volume (m³/ha) for stem diameters greater than lower diameter limits (cm) listed below: 7.5	12.5	**17.5**	22.5	27.5	32.5	BA m²	Mean DBHg cm	Stems per ha	CC %	250 Prime 12.5+: Merch vol. m³/ha	DBHg cm	LC %	Top ht[1] m	MAI[1] m³/ha/yr
0	0	0	0	**0**	0	0	0	0	0.0	500	0	0	0.0	0	0.0	0.0
10	0	0	0	**0**	0	0	0	0	0.0	487	1	0	0.0	0	1.8	0.0
20	2	1	0	**0**	0	0	0	1	4.5	463	12	0	0.0	0	5.8	0.0
30	23	22	15	**9**	3	0	0	6	13.0	444	41	14	16.7	67	10.9	0.3
40	93	92	80	**75**	58	34	14	17	22.4	424	70	67	25.7	71	16.1	1.9
50	224	223	204	**202**	198	175	124	31	31.3	406	86	165	35.5	69	20.9	4.0
60	371	371	346	**345**	341	330	303	45	37.9	399	92	281	42.8	62	25.1	5.7
70	536	535	505	**504**	501	494	476	59	43.7	391	96	418	49.3	55	28.7	7.2
80	721	719	684	**683**	681	675	663	72	49.0	382	98	579	55.2	50	31.8	8.5
90	892	887	849	**848**	846	841	832	83	53.4	372	99	732	59.9	46	34.5	9.4
100	1051	1043	1002	**1002**	1000	996	988	93	57.0	365	99	879	63.8	44	36.8	10.0
110	1210	1200	1156	**1155**	1154	1151	1143	102	60.1	361	99	1027	67.4	41	38.7	10.5
120	1347	1334	1288	**1288**	1287	1284	1278	110	62.8	354	99	1157	70.1	39	40.4	10.7
130	1466	1451	1403	**1403**	1402	1400	1394	116	65.1	349	100	1271	72.4	38	41.9	10.8[2]
140	1583	1561	1511	**1511**	1511	1509	1505	122	67.3	341	100	1384	74.5	37	43.2	10.8[2]
150	1685	1658	1607	**1607**	1607	1606	1603	126	69.3	334	100	1484	76.3	36	44.3	10.7

[1] Top heights and MAIs shown in these columns are for an assumed merchantability class of +17.5 cm stem diameter (the merchantable volume class shown in bold in the volume, m³/ha, column on the left half of this table).

[2] Max MAI = 10.8 m³/(ha · yr) at ages 130-140 years.

Appendix 2

Sample page of Sitka spruce stand volume and diameter table (close utilization less decay volume), based on VDYP Version Prod 4.5, for an assumed site index of 24 m at breast-height age 50 and for crown closure of 25%. Persons interested in using VDYP should contact the British Columbia Ministry of Forests, Research Branch, Forest Productivity and Decision Support Section, Victoria, British Columbia.

Stand volume and diameter table, AIR method
Sitka spruce 100%

BHA 50 Site index: 24.0

Volumes per hectare | Quadratic stand diameters

Culmination values

Limit: 17.5 | Volume: 439.0 | Age: 70 | MAI: 6.27

Close utilization, less decay volume, for stem size 17.5 cm or greater

Total age, yr	Height, m	Diam. cm	Vol., m³/ha	MAI, m³/ha/yr
10	1.8	0.0	0.0	0.00
20	5.8	21.5	40.6	2.03
30	10.9	25.4	125.8	4.19
40	16.1	29.5	213.9	5.35
50	20.9	33.4	297.0	5.94
60	25.1	37.1	372.3	6.20
70	28.7	40.4	439.0	6.27[1]
80	31.8	43.3	497.6	6.22
90	34.5	46.0	549.0	6.10
100	36.8	48.4	594.1	5.94
110	38.7	50.5	633.8	5.76
120	40.4	52.5	668.9	5.57
130	41.9	54.2	700.2	5.39
140	43.2	55.9	728.1	5.20
150	44.3	57.4	753.3	5.02
160	45.3	58.8	776.1	4.85
170	46.1	60.1	796.9	4.69
180	46.9	61.4	815.9	4.53
190	47.6	62.6	833.4	4.39
200	48.2	63.7	849.6	4.25
210	48.8	64.8	864.7	4.12
220	49.3	65.8	878.7	3.99
230	49.7	66.8	892.0	3.88
240	50.2	67.7	904.4	3.77
250	50.5	68.7	916.1	3.66

▶

◀ *Appendix 2*

Model parameters			
VDYP Version number	Prod 4.5	Measured basal area	0.0
Species 1	SS 100%	Measured BA age	50
Method	AIR	Stocking class	1
Forest Inventory Zone	B	PSYU	0000
Utilization level 1	17.5	Spcl Cruise subcode	(None)
Adjustment factor	1.000	Alternate W2 & B file	(None used)
Site index (BHA 50)	24.0	Start total age	10
Stand height (m)	0.0	Finish total age	250
Stand total age	0	Age increment	10
% Crown closure	25.0%		

[1] Max MAI = 6.27 $m^3/(ha \cdot yr)$ at age 70.

Appendix 3

Scientific and common names of plant and animal species mentioned in the text, excluding species listed in the taxonomic history of Sitka spruce. Species are listed alphabetically by genus name. Guidelines for nomenclature follow Banfield (1974), Taylor and MacBryde (1977), Meidinger (1987), Cannings and Harcombe (1990), Harper et al. (1994), Pojar and MacKinnon (1994), and Farrar (1995).

Tree Species	
Abies spp.	true firs
Abies amabilis	amabilis fir, Pacific silver fir, silver fir
Abies grandis	grand fir
Abies lasiocarpa	subalpine fir
Abies procera	noble fir
Acer glabrum	Douglas maple
Acer macrophyllum	bigleaf maple
Alnus rubra	red alder
Arbutus menziesii	arbutus
Betula papyrifera	paper birch
Chamaecyparis lawsoniana	Port Orford cedar
Chamaecyparis nootkatensis	yellow-cedar, Alaska yellow-cedar
Fagus sp.	beech
Larix kaempferi	Japanese larch
Larix lyallii	subalpine larch
Larix occidentalis	western larch
Libocedrus decurrens	incense-cedar
Malus fusca	Pacific crabapple
Picea engelmannii	Engelmann spruce
Picea excelsa	Norway spruce
Picea sitchensis	Sitka spruce
Pinus aristata	bristlecone pine
Pinus contorta var. *contorta*	shore pine
Pinus contorta var. *latifolia*	lodgepole pine
Pinus lambertiana	sugar pine
Pinus monticola	western white pine
Pinus ponderosa	ponderosa pine
Pinus radiata	radiata pine
Pinus sylvestris	Scots pine
Populus tremuloides	aspen
Populus trichocarpa	black cottonwood
Pseudotsuga menziesii	Douglas-fir
Quercus garryana	Garry oak
Sequoia sempervirens	redwood, coast redwood
Sequoiadendron giganteum	giant sequoia
Taxus brevifolia	Pacific yew
Thuja plicata	western redcedar
Tsuga heterophylla	western hemlock
Tsuga mertensiana	mountain hemlock

►

◄ *Appendix 3*

Shrubs	
Acer circinatum	vine maple
Alnus viridis subsp. *sinuata*	Sitka alder
Amelanchier alnifolia	saskatoon
Cornus sericea	red-osier dogwood
Gaultheria shallon	salal
Lonicera involucrata	black twinberry
Menziesia ferruginea	false-azalea
Myrica californica	sweet gale
Oplopanax horridus	devil's club
Ribes bracteosum	stink currant
Rubus idaeus	red raspberry
Rubus parviflorus	thimbleberry
Rubus pedatus	five-leaved bramble
Rubus spectabilis	salmonberry
Rubus ursinus	trailing blackberry
Salix hookeriana	Hooker's willow
Salix sitchensis	Sitka willow
Sambucus racemosa var. *arborescens*	red elderberry
Shepherdia canadensis	soopolallie
Viburnum edule	highbush-cranberry
Vaccinium alaskaense	Alaskan blueberry
Vaccinium ovalifolium	oval-leaved blueberry
Vaccinium ovatum	evergreen huckleberry
Vaccinium parvifolium	red huckleberry
Herbs, Ferns, and Selaginella	
Angelica spp.	angelica species
Aruncus dioicus	goat's beard
Athyrium filix-femina	lady fern
Blechnum spicant	deer fern
Calamagrostis nootkaensis	Nootka reedgrass
Carex deweyana	Dewey's sedge
Carex obnupta	slough sedge
Conioselinum pacificum	Pacific hemlock-parsley
Coptis aspleniifolia	fern-leaved goldthread
Coptis trifolia	three-leaved goldthread
Cornus canadensis	bunchberry
Cornus unalaschkensis	cordilleran bunchberry
Disporum hookeri	Hooker's fairybells
Dryopteris expansa (D. assimilis)	spiny wood fern
Equisetum spp.	horsetail
Festuca rubra	red fescue
Galium triflorum	sweet-scented bedstraw
Gymnocarpium dryopteris	oak fern
Heracleum sphondylium	cow parsnip

►

◄ *Appendix 3*

Juncus mertensianus	Nevada rush
Lupinus littoralis	seashore lupine
Lysichiton americanum	skunk cabbage
Maianthemum dilatatum	false lily-of-the-valley
Moneses uniflora	single delight
Montia sibirica (Claytonia sibirica)	Siberian miner's-lettuce
Oenanthe sarmentosa	Pacific water-parsley
Oxalis oregana	redwood sorrel
Petasites palmatus	palmate coltsfoot
Polystichum lonchitis	mountain holly fern
Polystichum munitum	sword fern
Potentilla anserina	silverweed
Pteridium aquilinum	bracken fern
Selaginella spp.	selaginella
Streptopus roseus	rosy twistedstalk
Tiarella trifoliata	three-leaved foamflower
Tiarella unifoliata	one-leaved foamflower
Tolmiea menziesii	piggy-back plant
Trifolium wormskjoldii	springbank clover
Trillium ovatum	western trillium
Trisetum cernuum	nodding trisetum
Urtica dioica	stinging nettle
Veratrum californicum	false hellebore
Veratrum viride	Indian hellebore
Viola glabella	stream violet
Mosses and Liverworts	
Conocephalum conicum	snake liverwort
Frullania nisquallensis	hanging millipede liverwort
Hylocomium splendens	step moss
Hypnum circinale	coiled-leaf moss
Kindbergia oregana (Stokesiella oregana)	Oregon beaked moss
Leucolepis menziesii	palm tree moss
Mnium spp.	red-mouthed mnium
Pellia neesiana	ring pellia
Plagiomnium insigne	coastal leafy moss
Plagiothecium undulatum	flat moss
Pleurozium schreberi	red-stemmed feathermoss
Pogonatum spp.	grey haircap moss
Polypodium scouleri	leathery polypody
Porella navicularis	tree-ruffle liverwort
Pseudoisothecium stoloniferum	cat-tail moss
Ptilidium californicum	ptilidium
Radula bolanderi	flat-leaved liverwort
Rhizomnium glabrescens	large leafy moss
Rhytidiadelphus loreus	lanky moss

►

◄ *Appendix 3*

Rhytidiadelphus triquetrus	electrified cat's-tail moss
Rhytidiopsis robusta	pipecleaner moss
Scapania bolanderi	yellow-ladle liverwort
Sphagnum girgensohnii	common green sphagnum
Sphagnum squarrosum	squarrose peat moss
Birds	
Accipiter gentilis laingi	Queen Charlotte Goshawk
Brachyramphus marmoratus	Marbled Murrelet
Dryocopus pileatus	Pileated Woodpecker
Glaucidium gnoma	Northern Pygmy-Owl
Haliaeetus leucocephalus	Bald Eagle
Larus spp.	Gull
Phalacrocorax spp.	Cormorant
Regulus satrapa	Golden-crowned Kinglet
Sphyrapicus ruber	Red-breasted Sapsucker
Synthliboramphus antiquus	Ancient Murrelet
Strix occidentalis caurina	Northern Spotted Owl
Troglodytes troglodytes	Winter Wren
Fish	
Oncorhynchus gorbuscha	pink salmon
Oncorhynchus keta	chum salmon
Oncorhynchus kisutch	coho salmon
Salmo clarki	cutthroat trout
Salmo gairdneri	steelhead trout
Salvelinus malma	Dolly Varden char
Mammals	
Cervus elaphus roosevelti	Roosevelt elk
Canis lupus	gray wolf
Erethizon dorsatum	American porcupine
Martes pennanti	fisher
Odocoileus hemionus columbianus	Columbian black-tailed deer
Odocoileus hemionus sitkensis	Sitka black-tailed deer
Tamiasciurus hudsonicus	red squirrel
Ursus arctos	coastal grizzly bear

Appendix 4

Abbreviations used for tree species and for biogeoclimatic zones, subzones, and variants. Names and abbreviations are based on Banner et al. (1993) for the Prince Rupert Forest Region, on Green and Klinka (1994) for the Vancouver Forest Region, and on the Silviculture Interpretations Working Group (1994).

Codes for Tree Species

Abbreviation	Tree species
Act	cottonwood
Ba	amabilis fir
Bg	grand fir
Bl	subalpine fir
Cw	western redcedar
Dr	red alder
Ep	paper birch
Fd	Douglas-fir
Hm	mountain hemlock
Hw	western hemlock
Pl	shore pine, lodgepole pine
Ss	Sitka spruce
Sw	white spruce
Sx[1]	Roche spruce (*Picea glauca* × *sitchensis* × *engelmannii;* synonymous with *P.* ×*lutzii*)
Sxs[1]	Roche spruce (*Picea sitchensis* × *glauca*)
Yc	yellow-cedar

[1] The definition of Sx and Sxs is from Banner et al. (1993), p. A-18.

Codes for Biogeoclimatic Units

Abbreviation	Biogeoclimatic subzone or variant
PRINCE RUPERT FOREST REGION	
Coastal Western Hemlock (CWH) Zone	
CWHvh2	Central Very Wet Hypermaritime CWH Variant
CWHvm1	Submontane Very Wet Maritime CWH Variant
CWHvm2	Montane Very Wet Maritime CWH Variant
CWHwm	Wet Maritime CWH Subzone
CWHws1	Submontane Wet Submaritime CWH Variant
CWHws2	Montane Wet Submaritime CWH Variant
Interior Cedar-Hemlock (ICH) Zone	
ICHmc1	Nass Moist Cold ICH Variant
ICHmc1a	Nass Moist Cold ICH Variant, Amabilis Fir Phase
ICHmc2	Hazelton Moist Cold ICH Variant
ICHvc	Very Wet Cold ICH Subzone
ICHwc	Wet Cold ICH Subzone

►

◀ *Appendix 4*

Mountain Hemlock (MH) Zone

MHwh1	Windward Wet Hypermaritime MH Variant

Engelmann Spruce–Subalpine Fir (ESSF) Zone

ESSFmc	Moist Cold ESSF Subzone
ESSFwv	Wet Very Cold ESSF Subzone

VANCOUVER FOREST REGION

Coastal Douglas-Fir (CDF) Zone

CDFmm	Moist Maritime CDF Subzone

Coastal Western Hemlock (CWH) Zone

CWHdm	Dry Maritime CWH Subzone
CWHds1	Southern Dry Submaritime CWH Variant
CWHds2	Central Dry Submaritime CWH Variant
CWHmm1	Submontane Moist Maritime CWH Variant
CWHmm2	Montane Moist Maritime CWH Variant
CWHms1	Southern Moist Submaritime CWH Variant
CWHms2	Central Moist Submaritime CWH Variant
CWHvh1	Southern Very Wet Hypermaritime CWH Variant
CWHvh2	Central Very Wet Hypermaritime CWH Variant
CWHvm1	Submontane Very Wet Maritime CWH Variant
CWHvm2	Montane Very Wet Maritime CWH Variant
CWHwh1	Submontane Wet Hypermaritime CWH Variant
CWHwh2	Montane Wet Hypermaritime CWH Variant
CWHws2	Montane Wet Submaritime CWH Variant
CWHxm	Very Dry Maritime CWH Subzone

Interior Douglas-Fir (IDF) Zone

IDFww	Wet Warm IDF Subzone

Mountain Hemlock (MH) Zone

MHwh	Wet Hypermaritime MH Subzone
MHmm1	Windward Moist Maritime MH Variant

Literature Cited

Agee, J.K. 1990. The historical role of fire in Pacific Northwest forests. *In* Natural and prescribed fire in Pacific Northwest forests. J.D. Walstad, S.R. Radosevich, and D.V. Sandberg (editors). Oreg. State Univ. Press, Corvallis, Oreg., pp. 25-38.

–. 1993. Fire ecology in the Pacific Northwest forests. Island Press, Washington, D.C.

Alaback, P.B. 1982a. Dynamics of understory biomass in Sitka spruce–western hemlock forests of southeast Alaska. Ecology 63:1932-1948.

–. 1982b. Forest community structural changes during secondary succession in southeast Alaska. *In* Forest succession and stand development research in the Northwest. J.E. Means (editor). Oreg. State Univ., For. Res. Lab., Corvallis, Oreg., pp. 70-79.

–. 1982c. A comparison of old-growth forest structure in the western hemlock–Sitka spruce forests of southeast Alaska. *In* Fish and wildlife relationships in old-growth forests, Proc. Symp. 12-15 April 1982, Juneau, Alaska. W.R. Meehan, T.R. Merrell, Jr., and T.A. Hanley (editors). Am. Inst. Fish. Res. Biol., Morehead City, N.C., pp. 219-226.

–. 1984. Secondary succession following logging in the Sitka spruce–western hemlock forests of southeast Alaska: implications for wildlife management. U.S. Dep. Agric. For. Serv., Gen. Tech. Rep. PNW-173.

–. 1996. Biodiversity patterns in relation to climate: the coastal temperate rainforests of North America. *In* High-latitude rainforests and associated ecosystems of the west coast of the Americas: climate, hydrology, ecology, and conservation. R.G. Lawford, P.B. Alaback, and E. Fuentes (editors). Springer, New York, N.Y., pp. 105-133.

Alaback, P.B. and F.R. Herman. 1988. Long-term response of understory vegetation to stand density in *Picea-Tsuga* forests. Can. J. For. Res. 18:1522-1530.

Alexandrov, A.H. 1993. Sitka spruce provenance experiment in Bulgaria. *In* Proc. of the IUFRO International Sitka spruce provenance experiment (Sitka Spruce Working Group S2.02.12). IUFRO 1984, Edinburgh, Scotland. C.C. Ying and L.A. McKnight (editors). B.C. Min. For., Victoria, B.C. and Irish For. Board, County Wicklow, Ire., pp. 57-58.

Alfaro, R.I. 1980. Host selection by *Pissodes strobi* Peck: chemical interaction with the host plant. PhD thesis. Simon Fraser Univ., Burnaby, B.C.

–. 1982. Fifty-year-old Sitka spruce plantations with a history of intense weevil attack. J. Entomol. Soc. B.C. 79:62-65.

–. 1989. Probability of damage to Sitka spruce by the Sitka spruce weevil, *Pissodes strobi*. J. Entomol. Soc. B.C. 86:48-54.

–. 1992. Forecasting spruce weevil damage. *In* Proceedings of a spruce weevil symposium, March 12, 1992, Terrace, B.C. T. Ebata (editor). B.C. Min. For., Prince Rupert For. Reg., B.C. Spec. Rep.

–. 1994. The white pine weevil in British Columbia. *In* The white pine weevil: biology, damage and management. R.I. Alfaro, G.K. Kiss, and R.G. Fraser (editors). Can. For. Serv. and B.C. Min. For., Victoria, B.C. FRDA Rep. 226. pp. 7-22.

–. 1995. An induced defense reaction in white spruce to attack by the white pine weevil, *Pissodes strobi*. Can. J. For. Res. 25:1725-1730.

–. 1996. Role of genetic resistance in managing ecosystems susceptible to white pine weevil. For. Chron. 72:374-380.

Alfaro, R.I. and J.H. Borden. 1982. Host selection by the white pine weevil, *Pissodes strobi* Peck. Feeding bioassays using host and non-host plants. Can. J. For. Res. 12:64-70.

–. 1985. Factors determining the feeding of the white pine weevil (Coleoptera: Curculionidae) on its coastal British Columbia host, Sitka spruce. Proc. Entomol. Soc. Ont. 116(Suppl.):63-66.

Alfaro, R.I., J.H. Borden, R.G. Fraser, and A. Yanchuk. 1995. The white pine weevil in British Columbia: basis for an integrated pest management system. For. Chron. 71:66-73.

Alfaro, R.I., J.H. Borden, L.J. Harris, W.W. Nijholt, and L.H. McMullen. 1984. Pine oil, a feeding deterrent for the white pine weevil, *Pissodes strobi*. Can. Entomol. 116:41-44.

Alfaro, R.I., M. Hulme, and C. Ying. 1993. Variation in attack by Sitka spruce weevil, *Pissodes strobi* (Peck), within a resistant provenance of Sitka spruce. J. Entomol. Soc. B.C. 90:24-30.

Alfaro, R.I., G.K. Kiss, and R.G. Fraser (editors). 1994. The white pine weevil: biology, damage and management. Proc. symp. 19-21 Jan. 1994, Richmond, B.C. Can. For. Serv. and B.C. Min. For., Victoria, B.C. FRDA Rep. 226.

Alfaro, R.I. and S.A.Y. Omule. 1990. The effect of spacing on Sitka spruce weevil damage to Sitka spruce. Can. J. For. Res. 20:179-184.

Alfaro, R.I., H.D. Pierce, Jr., J.H. Borden, and A.C. Oehlschlager. 1980. Role of volatile and non-volatile components of Sitka spruce bark as feeding stimulants for *Pissodes strobi* Peck (Coleoptera: Curculionidae). Can. J. Zool. 58:626-632.

–. 1981. Insect feeding and oviposition deterrents from western redcedar foliage. J. Chem. Ecol. 7:39-48.

Alfaro, R.I. and C.C. Ying. 1990. Levels of Sitka spruce weevil, *Pissodes strobi* (Peck), damage among Sitka spruce provenances and families near Sayward, British Columbia. Can. Entomol. 122:607-615.

Allen, E.A. 1994. Damage appraisal in pests of young stands. For. Can. and B.C. Min. For., Victoria, B.C. FRDA Work. Pap. WP-1.5-002.

American Forests. 1996. 1996-97 National register of big trees. Am. For. 102:20-45.

Anderson, H.E. 1954. Clearcutting as a silvicultural system in southeastern Alaska. Proc. Alaska Science Conf. 5:1-4.

–. 1956. The problem of brush control on cutover areas in southeastern Alaska. Alaska For. Res. Cent. Tech. Note 33.

Arney, J.D. 1984. A modelling strategy for the growth projection of managed stands. Can. J. For. Res. 15:511-518.

B.C. Forest Service. 1966. Butt-taper tables for coastal tree species. B.C. Min. For., Inven. Br., Victoria, B.C. For. Surv. Notes 7.

–. 1968. Basic taper curves for the commercial tree species of British Columbia – coast. B.C. Min. For., Inven. Br., Victoria, B.C.

B.C. Ministry of Environment, Lands and Parks. 1995. British Columbia grizzly bear conservation strategy. B.C. Min. Environ., Lands and Parks, Victoria, B.C.

B.C. Ministry of Environment and Parks. 1987. Guide to ecological reserves in British Columbia. Parks and Outdoor Recreation Div., Ecological Reserves Program, Victoria, B.C.

B.C. Ministry of Forests. 1990a. Coastal harvest planning guidelines. B.C. Min. For., Vancouver For. Reg., Vancouver, B.C. Draft.

–. 1990b. Forest estate models. B.C. Min. For., Silv. Br., Victoria, B.C.

–. 1990c. Interim forest landscape management guidelines for the Vancouver Forest Region (July 20, 1990). B.C. Min. For., Vancouver For. Reg., Burnaby, B.C.

–. 1990d. A joint five year strategic plan 1991/92 – 1995/96: research program. B.C. Min. For., Res. Br., Victoria, B.C.

–. 1991. Towards a silviculture strategy (II). B.C. Min. For., Silv. Br., Victoria, B.C.

–. 1992a. Biodiversity guidelines: coastal stand-level biodiversity and landscape-level biodiversity. For. Plan. Can. 9(1):33-41.

–. 1992b. An old growth strategy for British Columbia. Integrated Resour. Policy Br., Victoria, B.C.

–. 1995. Growth intercept method for silviculture surveys. Silv. Prac. Br., Victoria, B.C.
B.C. Ministry of Forests and B.C. Ministry of Environment, Lands and Parks. 1995a. Forest Practices Code of British Columbia. Biodiversity guidebook. Victoria, B.C.
–. 1995b. Forest Practices Code of British Columbia. Coastal watershed assessment procedure guidebook. Victoria, B.C.
–. 1995c. Forest Practices Code of British Columbia. Defoliator management guidebook. Victoria, B.C.
–. 1995d. Forest Practices Code of British Columbia. Gully assessment procedures. 3rd ed. Victoria, B.C.
–. 1995e. Forest Practices Code of British Columbia. Hazard assessment keys for evaluating site sensitivity to soil degrading processes guidebook. Victoria, B.C.
–. 1995f. Forest Practices Code of British Columbia. Mapping and assessing terrain stability. Victoria, B.C.
–. 1995g. Forest Practices Code of British Columbia. Riparian management area guidebook. Victoria, B.C.
–. 1995h. Forest Practices Code of British Columbia. Root disease management guidebook. Victoria, B.C.
–. 1995i. Forest Practices Code of British Columbia. Seed and vegetative material guidebook. Victoria, B.C.
–. 1995j. Forest Practices Code of British Columbia. Site preparation guidebook. Victoria, B.C.
–. 1995k. Forest Practices Code of British Columbia. Silviculture prescription guidebook. Victoria, B.C.
–. 1995l. Forest Practices Code of British Columbia. Soil conservation. Victoria, B.C.
B.C. Ministry of Forests, B.C. Ministry of Environment, Lands and Parks, Department of Fisheries and Oceans, and Council of Forest Industries. 1993. British Columbia coastal fisheries/forestry guidelines. 3rd ed. Victoria, B.C.
B.C. Parks. 1992. Carmanah Pacific Provincial Park interim management statement. B.C. Min. Environ., Lands and Parks, Victoria, B.C.
–. 1994. The British Columbia Ecological Reserves Program. *In* Biodiversity in British Columbia: our changing environment. L.E. Harding and E. McCullum (editors). Can. Wildl. Serv., Pac. & Yukon Reg., Vancouver, B.C., pp. 375-392.
–. 1996. Carmanah Walbran Provincial Park. B.C. Min. Environ., Lands and Parks, Victoria, B.C. Foldout.
B.C. Wildlife Branch. 1991. Managing wildlife to 2001: a discussion paper. British Columbia's environment: planning for the future. B.C. Environ., Victoria, B.C.
Backhouse, F. (compiler). 1993. Wildlife tree management in British Columbia. Wildlife Tree Committee of British Columbia, B.C. Environ., B.C. Min. For., For. Can., and Workers' Compensation Board, Victoria, B.C.
Balisky, A.C., P.O. Salonius, C. Walli, and D. Brinkman. 1993. Seedling roots and the forest floor. Can. Silv. Mag. 1(2):8-14.
Banfield, A.W.F. 1974. The mammals of Canada. Univ. Toronto Press, Toronto, Ont.
Banner, A., R.N. Green, A. Inselberg, K. Klinka, D.S. McLennan, D.V. Meidinger, F.C. Nuszdorfer, and J. Pojar. 1990. Site classification for coastal British Columbia. B.C. Min. For., Victoria, B.C. Pamphlet.
Banner, A., W. Mackenzie, S. Haeussler, S. Thomson, J. Pojar, and R. Trowbridge. 1993. A field guide to site identification and interpretation for the Prince Rupert Forest Region. B.C. Min. For., Victoria, B.C. Land Manage. Handb. 26.
Banner, A., J. Pojar, and R. Trowbridge. 1986. Representative wetland types of the northern part of the Pacific Oceanic Wetland Region. B.C. Min. For., Victoria, B.C. Res. Rep. RR85008-PR.
Baranyay, J.A. 1972. *Rhizina* root rot of conifers. Can. For. Serv., Pac. For. Res. Cent., Victoria, B.C. For. Pest Leafl. 56.
Baranyay, J.A. and R.B. Smith. 1972. Dwarf mistletoes in British Columbia and recommendations for their control. Can. For. Serv., Pac. For. Res. Cent., Victoria, B.C. Inf. Rep. BC-X-72.

Barker, J.E. 1983. Site index relationships for Sitka spruce, western hemlock, western redcedar and red alder. Western Forest Products Limited, Vancouver, B.C. Moresby Tree Farm Licence #24, Queen Charlotte Islands. File Rep.

Barker, J.E. and J.W. Goudie. 1987. Site index curves for Sitka spruce. B.C. Min. For., Res. Br., Victoria, B.C. Unpubl. Rep.

Bartlett, J.C. 1977. Species trial at Sewell Inlet. B.C. Min. For., Res. Br., Victoria, B.C. Working Plan EP 814.

Bartels, R., J.D. Bell, R.L. Knight, and G. Schaefer. 1985. Dead and down woody material. *In* Management of wildife and fish habitats in forests of Oregon and Washington. E.R. Brown (editor). U.S. Dep. Inter., Fish & Wildl. Serv., Washington, D.C. Pub. R6-F&WL-192-1985. pp. 171-186.

Basabe, F.A., R.L. Edmonds, and T.V. Larson. 1987. Ozone levels and fog chemistry in forested areas in western Washington. *In* Proc. 35th Annual West. Intern. For. Disease Work Conf., 18-21 August 1987, Nanaimo, B.C. G.A. DeNitto (compiler). pp.129-133.

Basabe, F.A., R.L. Edmonds, W.L. Chang, and T.V. Larson. 1989. Fog and cloudwater chemistry in western Washington. *In* Effects of air pollution on western forests. R.K. Olson and A.S. Lefohn (editors). APCA Trans. 16, June 1989, Anaheim, Calif. Air & Waste Manage. Assoc., Pittsburgh, Penn., pp. 33-39.

Battigelli, J.P., S.M. Berch, and V.G. Marshall. 1994. Soil fauna communities in two distinct but adjacent forest types on northern Vancouver Island, British Columbia. Can. J. For. Res. 24:1557-1566.

Baumgartner, D.M., R.L. Mahoney, J. Evans, J. Caslick, D.W. Breuer (editors). 1987. Animal damage management in Pacific Northwest forests. Wash. State Univ., Coop. Exten., Pullman, Wash.

Beale, J. 1989. Management guidelines for laminated root rot in the Vancouver Forest Region. B.C. Min. For., Victoria, B.C. FRDA Memo 108.

Beaudry, P.G. and D.L. Hogan. 1990. Flood hazard classification for silviculture. *In* Hydrologic and geomorphic considerations for silvicultural investments on the lower Skeena River floodplain. P.G. Beaudry, D.L. Hogan, and J.W. Schwab. For. Can. and B.C. Min. For., Victoria, B.C. FRDA Rep. 122. pp. 25-35.

Beaudry, P.G., D.L. Hogan, and J.W. Schwab. 1990. Hydrologic and geomorphic considerations for silvicultural investments on the lower Skeena River floodplain. For. Can and B.C. Min. For., Victoria, B.C. FRDA Rep. 122.

Beese, W.J. 1989. An ecological assessment of the proposed Carmanah Creek spruce reserve. MacMillan Bloedel Limited Management Plan Carmanah Valley. MacMillan Bloedel Ltd., Nanaimo, B.C. Folio II.

Benedikz, T. 1976. Progress report on the international ten provenance experiment with Sitka spruce in Iceland. *In* IUFRO. Sitka Spruce *Picea sitchensis* (Bong.) Carr. International Ten Provenance Experiment. Nursery stage results. Dep. Lands, For. Wildl. Serv., Dublin, Ire., pp. 124-139.

Bergan, J. 1970. Afforestation in north Norway. Tidss. for Skogbruk 78(2):251-261.

Berntsen, C.M. 1955. Seedling distribution on a spruce-hemlock clearcut. U.S. Dep. Agric. For. Serv., Pac. N.W. For. Range Exp. Sta., Res. Note 119.

Betts, H.S. 1945. American woods: Sitka spruce. U.S. Dep. Agric. For. Serv., Washington, D.C.

Bier, J.E., R.E. Foster, and P.J. Salisbury. 1946. Studies in forest pathology. IV. Decay of Sitka spruce on the Queen Charlotte Islands. Can. Dep. Agric. Publ. 804. Tech. Bull. 56.

Biggs, W.G. and M.E. Walmsley. 1988 rev. Research requirements to determine the effects of silvicultural herbicides on wildlife habitat: a problem analysis. Can. For. Serv. and B.C. Min. For., Victoria, B.C. FRDA Rep. 40.

Binford, L.C., B.G. Elliott, and S.W. Singer. 1975. Discovery of a nest and the downy young of the Marbled Murrelet. Wilson Bull. 87:303-319.

Binns, W.O., G.J. Mayhead, and J.M. MacKenzie. 1976. Nutrient deficiencies of conifers in British forests: an illustrated guide. For. Comm., Farnham, U.K. Leafl. 76.

Biring, B.S., P.G. Comeau, and J.O. Boateng. 1996. Effectiveness of forest vegetation control methods in British Columbia. B.C. Min. For., Victoria, B.C. FRDA Handb. 11.

Birot, Y. 1976. Juvenile and provisional results of international ten provenance experiments in France. *In* IUFRO. Sitka Spruce *Picea sitchensis* (Bong.) Carr. International Ten Provenance Experiment. Nursery stage results. Dep. Lands, For. Wildl. Serv., Dublin, Ire., pp. 60-66.

Blood, D.A. and G.G. Anweiler. 1982. Forest nesting habitat of ancient murrelets in the Queen Charlotte Islands. *In* Fish and wildlife relationships in old-growth forests, Proc. Symp. 12-15 April 1982, Juneau, Alaska. W.R. Meehan, T.R. Merrell, Jr., and T.A. Hanley (editors). Am. Inst. Fish. Res. Biol., Morehead City, N.C., pp. 297-302.

Bodsworth, F. 1970. The illustrated natural history of Canada: the Pacific coast. NSL Nat. Sci. Can. Limited, Toronto, Ont.

Bones, J.T. 1968. Volume tables and equations for old-growth western hemlock and Sitka spruce in southeast Alaska. U.S. Dep. Agric. For. Serv., Res. Note PNW-91.

Bongard, A.H.G. 1832. Vegetation de Sitcha. Description of *Pinus Sitchensis* s. sp. Acad. Imp. Sci. St. Petersburg, Russia.

Bormann, B.T. 1989. Podzolization and windthrow: natural fluctuations in long-term productivity and implications for management. *In* Maintaining the long-term productivity of Pacific Northwest forest ecosystems. D.A. Perry, R. Meurisse, B. Thomas, R. Miller, J. Boyle, J. Menas, C.R. Perry, and R.F. Powers (editors). Timber Press, Portland, Oreg., p. 245.

Bormann, B.T., M.H. Brookes, E.D. Ford, A.R. Kiester, C.D. Oliver, and J.F. Weigand. 1994. A framework for sustainable-ecosystem management. U.S. Dep. Agric. For. Serv., Gen. Tech. Rep. PNW-GTR-319.

Bormann, F.H. 1985. Air pollution and forests: an ecosystem perspective. BioScience 35:434-441.

Boyce, J.S. 1929. Deterioration of wind-thrown timber on the Olympic Peninsula. U.S. Dep. Agric. Washington, D.C. Tech. Bull. 104.

Brazier, J.D. 1970. The effect of spacing on the wood density and wood yields of Sitka spruce. Forestry 43(Suppl.):22-28.

Brink, V.C. 1954. Survival of plants under flood in the lower Fraser River Valley, British Columbia. Ecology 35:94-95.

Britton, G.M., D.V. Meidinger, and A. Banner. 1996. The development of an ecological classification data management and analysis system for British Columbia. *In* Global to local: ecological land classification, Aug. 14-17, 1994, Thunder Bay, Ont. R.A. Sims, I.G.W. Corns, and K. Klinka (editors). Kluwer Academic Publ., Boston, Mass., pp. 365-372.

Brix, H. 1972. Growth response of Sitka spruce and white spruce seedlings to temperature and light intensity. Dep. Environ., Can. For. Serv., Pac. For. Res. Cent., Victoria, B.C. Inf. Rep. BC-X-74.

Brockley, R.P. 1988. The effects of fertilization on the early growth of planted seedlings: a problem analysis. For. Can. and B.C. Min. For. and Lands, Victoria, B.C. FRDA Rep. 11.

Brooks, J.E. and J.H. Borden. 1992. Development of a resistance index for Sitka spruce against the white pine weevil, *Pissodes strobi*. For. Can. and B.C. Min. For., Victoria, B.C. FRDA Rep. 180.

Brooks, J.E., J.H. Borden, and H.D. Pierce, Jr. 1987. Foliar and cortical monoterpenes in Sitka spruce: potential indicators of resistance to the white pine weevil, *Pissodes strobi* Peck (Coleoptera: Curculionidae). Can. J. For. Res. 17:740-745.

Brough, S.G. 1990. Wild trees of British Columbia. Pacific Educational Press, Vancouver, B.C.

Brown, E.R. (editor). 1985. Management of wildlife and fish habitats in forests of Oregon and Washington. U.S. Dep. Inter., Fish & Wildl. Serv., Washington, D.C. Pub. R6-F&WL-192-1985.

Brown, K.R., W.A. Thompson, E.L. Camm, B.J. Hawkins, and R.D. Guy. 1996a. Effects of N addition rates on the productivity of *Picea sitchensis, Thuja plicata,* and *Tsuga heterophylla* seedlings. II. Photosynthesis, 13C discrimination and N partitioning in foliage. Trees 10:198-205.

Brown, K.R., W.A. Thompson, and G.F. Weetman. 1996b. Effects of N addition rates on the productivity of *Picea sitchensis, Thuja plicata,* and *Tsuga heterophylla* seedlings. I. Growth rates, biomass allocation and macroelement nutrition. Trees 10:189-197.

Brown, T.G. 1987. Characterization of salmonid over-wintering habitat within seasonally flooded land on Carnation Creek floodplain. B.C. Min. For., Victoria, B.C. Land Manage. Rep. 44.

Brunt, K.R. 1987. Man-made forests and elk in coastal British Columbia. For. Chron. 63:155-158.

–. 1990. Ecology of Roosevelt elk. *In* Deer and elk habitats in coastal forests of southern British Columbia. J.B. Nyberg and D.W. Janz (technical editors). B.C. Min. For. and B.C. Min. Environ., Victoria, B.C. Spec. Rep. Ser. 5. pp. 65-98.

Bryant, A.A., J.-P.L. Savard, and R.T. McLaughlin. 1993. Avian communities in old-growth and managed forests of western Vancouver Island, British Columbia. Can. Wildl. Serv., Pac. & Yukon Reg. Tech. Rep. Ser. 167.

Bunnell, F.L. 1990. Ecology of black-tailed deer. *In* Deer and elk habitats in coastal forests of southern British Columbia. J.B. Nyberg and D.W. Janz (technical editors). B.C. Min. For. and B.C. Min. Environ., Victoria, B.C. Spec. Rep. Ser. 5. pp. 31-63.

–. 1995. Forest-dwelling vertebrate faunas and natural fire regimes in British Columbia: patterns and implications for conservation. Conserv. Biol. 9:636-644.

Bunnell, F.L. and L.A. Dupuis. 1993. Riparian habitats in British Columbia: their nature and role. *In* Riparian habitat management and research. K.H. Morgan and M.A. Lashmar (editors). Proc. of a workshop 4-5 May 1993, Kamloops, B.C. Environ. Can., Fraser River Action Plan, North Vancouver, B.C. Spec. Publ. pp. 7-21.

Bunnell, F.L. and G.W. Jones. 1982. Black-tailed deer and old-growth forests – a synthesis. *In* Fish and wildlife relationships in old-growth forests, Proc. Symp. 12-15 April 1982, Juneau, Alaska. W.R. Meehan, T.R. Merrell, Jr., and T.A. Hanley (editors). Am. Inst. Fish. Res. Biol., Morehead City, N.C., pp. 411-420.

Bunnell, F.L. and L.L. Kremsater. 1990. Sustaining wildlife in managed forests. Northw. Environ. J. 6:243-269.

Burgar, R. 1964. The effect of seed size on germination, survival and initial growth in white spruce. For. Chron. 40:93-97.

Burley, J. 1965a. Genetic variation in *Picea sitchensis* (Bong.) Carr.: a literature review. Commonw. For. Rev. 44:47-59.

–. 1965b. Genetic variation in *Picea sitchensis* (Bong.) Carr. PhD thesis. Yale Univ., New Haven, Conn.

–. 1965c. Karyotype analysis of Sitka spruce, *Picea sitchensis* (Bong.) Carr. Silvae Genet. 14:127-132.

–. 1965d. Variation in seed characteristics of Sitka spruce. Advan. Front. Plant Sci. 10:11-24.

–. 1966a. Genetic variation in seedling development of Sitka spruce, *Picea sitchensis* (Bong.) Carr. Forestry 39:68-94.

–. 1966b. Provenance variation in growth of seedling apices of Sitka spruce. For. Sci. 12:170-175.

–. 1966c. Variation in colour of Sitka spruce seedlings. Quart. J. For. 60(1):51-54.

–. 1976. Combined analysis of nursery stage of the IUFRO international ten provenance trial of Sitka spruce (*Picea sitchensis* (Bong.) Carr.). *In* IUFRO. Sitka Spruce *Picea sitchensis* (Bong.) Carr. International Ten Provenance Experiment. Nursery stage results. Dep. Lands, For. Wildl. Serv., Dublin, Ire., pp. 238-256.

Bustard, D.R. and D.W. Narver. 1975. Aspects of the winter ecology of juvenile coho salmon (*Oncorhynchus kisutch*) and steelhead trout (*Salmo gairdneri*). J. Fish. Res. Board Can. 32(5):667-680.

Callan, B.E. 1993 rev. Rhizina root rot of conifers. For. Can., Pac. For. Cent., Victoria, B.C. Pest Leafl. 56.

Campbell, R.W., N.K. Dawe, I. McTaggart-Cowan, J.M. Cooper, G.W. Kaiser, and M.C.E. McNall. 1990. The birds of British Columbia. Volume 1. Nonpasserines: introduction and loons through waterfowl. Environ. Can., Can. Wildl. Serv., and Roy. B.C. Museum, Victoria, B.C.

Canadian Forest Service. 1996. The state of Canada's forests 1995-1996: sustaining forests at home and abroad. Nat. Resour. Can., Ottawa, Ont.

Cannell, M.G.R. and J.E. Jackson (editors). 1985. Attributes of trees as crop plants. Inst. Terr. Ecol., Nat. Environ. Res. Coun., Abbots Ripton, U.K.

Cannell, M.G.R., P. Rotherby, and E.D. Ford. 1984. Competition within stands of *Picea sitchensis* and *Pinus contorta*. Ann. Bot. 53(3):349-362.
Cannings, R.A. and A.P. Harcombe (editors). 1990. The vertebrates of British Columbia: scientific and English names. Roy. B.C. Museum, Victoria, B.C. Heritage Rec. No. 20; Wildl. Br. Rep. R24.
Carey, A.B. 1989. Wildlife associated with old-growth forests in the Pacific Northwest. Nat. Areas J. 9(3):151-162.
Carr, W.W. 1985. Watershed rehabilitation options for disturbed slopes on the Queen Charlotte Islands. B.C. Min. For., Victoria, B.C. Land Manage. Rep. 36.
Carr, W.W. and I. Wright. 1992. A soil erosion and sediment control planning system for managed forest land: a case study at Shomar Creek, Queen Charlotte Islands. B.C. Min. For., Victoria, B.C. Land Manage. Rep. 69.
Carrasco, M.M. 1954. Primeros resultados de reforestacion con coniferas exoticas en las zonas Pre-Cordilleranas de la Provincia de Valdivia [Preliminary results of afforestation with exotic conifers in the Andean foothills of Valdivia Province]. Bosques y Maderas (spec. issue of Chile Maderero) 1954:12-14.
Carriere, E.A. 1855. Traité générale des coniferes. 1st ed. and 2nd ed. Paris, France.
Carter, C.I. 1989. The 1989 outbreak of the green spruce aphid, *Elatobium abietinum*. For. Comm., Farnham, U.K. Res. Inf. Note 161.
Cary, N.L. 1922. Sitka spruce: its uses, growth, and management. U.S. Dep. Agric. For. Serv., Bull. 1060.
Caza, C.L. 1993. Woody debris in the forests of British Columbia: a review of the literature and current research. B.C. Min. For., Victoria, B.C. Land Manage. Rep. 78.
Chaisurisri, K., D.G.W Edwards, and Y.A. El-Kassaby. 1992. Genetic control of seed size and germination in Sitka spruce. Silvae Genet. 41:348-355.
–. 1993. Accelerated aging of Sitka spruce seeds. Silvae Genet. 42:303-308.
–. 1994a. Effects of seed size on seedling attributes in Sitka spruce. New For. 8:81-87.
Chaisurisri, K. and Y.A. El-Kassaby. 1993a. Estimation of clonal contribution to cone and seed crops in a Sitka spruce seed orchard. Annal. Sci. For. 50(5):461-467.
–. 1993b. Genetic control of isoenzymes in Sitka spruce. J. Heredity 84:206-211.
–. 1994a. Genetic diversity in a seed production population vs. natural populations of Sitka spruce. Biodiv. Conserv. 3(6):512-523.
–. 1994b. Variation in the mating system of Sitka spruce (*Picea sitchensis*): evidence for partial assortative mating. Am. J. Bot. 81:1410-1415.
Chaisurisri, K., J.B. Mitton, and Y.A. El-Kassaby. 1994b. Variation in the mating system of Sitka spruce (*Picea sitchensis*): evidence for partial assortative mating. Am. J. Bot. 81:1410-1415.
Chamberlin, T.W. (editor). 1988. Proceedings of the workshop: applying 15 years of Carnation Creek results. Pac. Biol. Sta., Nanaimo, B.C.
Chang, S.X., C.M. Preston, and G.F. Weetman. 1995. Soil microbial biomass and microbial and mineralizable N in a clear-cut chronosequence on northern Vancouver Island, British Columbia. Can. J. For. Res. 25:1595-1607.
Chatwin, S.C., D.E. Howes, J.W. Schwab, and D.N. Swanston. 1994. A guide for management of landslide-prone terrain in the Pacific Northwest. 2nd ed. B.C. Min. For., Victoria, B.C. Land Manage. Handb. 18.
Chatwin, S.C. and R.B. Smith. 1992. Reducing soil erosion associated with forestry operations through integrated research: an example from coastal British Columbia, Canada. *In* Erosion, debris flows and environment in mountain regions. Proc. internat. symp., 5-9 July 1992, Chengdu, China. D.E. Walling, T.R. Davies, and B. Hasholt (editors). IAHS Publ. 209. pp. 377-385.
Cheng, W.C. 1947. Les forêts du Setchouan et du Si-kang oriental. Trav. Lab. for. Toulouse 1(2):1-233.
Church, M. 1983. Patterns of instability in a wandering gravel bed channel. Spec. Publ. Int. Assoc. Sediment 6:169-180.
Clark, B.J. 1965. Variation in cone and seed characteristics of Sitka spruce in British Columbia. BScF thesis. Univ. B.C., Vancouver, B.C.

Cleary, B.D., R.D. Greaves, and R.K. Hermann. 1978. Regenerating Oregon's forests: a guide for the regeneration forester. Oreg. State Univ., Exten. Serv., Corvallis, Oreg.

Clement, C.J.E. 1984. Biogeoclimatic units and ecosystem associations of the Kimsquit drainage. Min. Environ. and Min. For. Cooperative Project, Victoria, B.C. Wildl. Working Rep. WR-5; Wildl. Habitat Res. Rep. WHR-13.

Coates, K.D., M.-J. Douglas, J.W. Schwab, and W.A. Bergerud. 1993. Grass and legume seeding on a scarified coastal alluvial site in northwestern British Columbia: response of native non-crop vegetation and planted Sitka spruce (*Picea sitchensis* (Bong.) Carr.) seedlings. New For. 7:193-211.

Coates, K.D, S. Haeussler, S. Lindeburgh, R. Pojar, and A.J. Stock. 1994. Ecology and silviculture of interior spruce in British Columbia. For. Can. and B.C. Min. For., Victoria, B.C. FRDA Rep. 220.

Coates, K.D., J. Pollack, and J.E. Barker. 1985. Effect of deer browsing on the early growth of three conifer species in the Queen Charlotte Islands. B.C. Min. For., Victoria, B.C. Res. Rep. RR85002-PR.

Cochrane, L.A., and E.D. Ford. 1978. Growth of a Sitka spruce plantation analysis and stochastic description of the development of the branching structure. J. Appl. Ecol. 15(1):227-244.

Comeau, P.G., S.B. Watts, C.L. Caza, J. Karakatsoulis, and S.M. Thomson. 1990. Ecology, responses and utilization of species which compete with conifers in British Columbia: a computerized bibliography. Min. For., Res. Br., Victoria, B.C.

Conard, S.G. 1984. Forest vegetation management in British Columbia: problem analysis. Min. For., Victoria, B.C. Res. Rep. RR84001-HQ.

Condrashoff, S.F. 1968. Biology of *Steremnius carinatus* (Coleoptera: Curculionidae), a reforestation pest in coastal British Columbia. Can. Entomol. 100:386-394.

Conkle, M.T. 1972. Analyzing genetic diversity in conifers: isoenzyme resolution by starch gel electrophoresis. U.S. Dep. Agric. For. Serv., Res. Note PSW-264.

Consortium of Thrower/Blackwell/Oikos. 1995. A SIBEC Project report: working tables estimated average site index for major forested ecosystems in B.C. Volume 1. Site-series statistics. Prep. for B.C. Min. For., Res. and Silv. Br., Victoria, B.C.

Cooper, W.S. 1931. The layering habit in Sitka spruce and the two western hemlocks. Bot. Gaz. 91:441-451.

Copes, D.L. and R.C. Beckwith. 1977. Isoenzyme identification of *Picea glauca, P. sitchensis,* and *P. lutzii* populations. Bot. Gaz. 138(4):512-521.

Cordes, L.D. 1972. An ecological study of the Sitka spruce forest on the west coast of Vancouver Island. PhD thesis. Univ. B.C., Vancouver, B.C.

Cordes, L.D. and G.A. MacKenzie. 1972. A vegetation classification for Phase I of Pacific Rim National Park. *In* Pacific Rim: an ecological approach to a new Canadian national park. J.G. Nelson and L.D. Cordes (editors). Univ. West. Ont., Dep. Geogr., London, Ont., pp. 37-59.

Council of Forest Industries of British Columbia. 1990. British Columbia forest industry statistical tables. The Council, Vancouver, B.C.

Coupé R., C.A. Ray, A. Comeau, M.V. Ketcheson, and R.M. Annas. 1982. A guide to some common plants of the Skeena area, British Columbia. B.C. Min. For., Victoria, B.C. Land Manage. Handb. 4.

Courtin, P. 1992. The relationship between the ecological site quality and the site index and stem form of red alder in southwestern B.C. MF thesis, Univ. B.C., Vancouver, B.C.

Coutts, M.P. 1983a. Root anchorage and tree stability. Plant Soil 71:171-188.

–. 1983b. Development of the structural root system of Sitka spruce. Forestry 56:1-16.

–. 1986. Components of tree stability in Sitka spruce on peaty gley soil Forestry 59:173-197.

Coutts, M.P. and J. Grace (editors). 1995. Trees and wind. Cambridge Univ. Press, Cambridge, U.K.

Coutts, M.P. and J.J. Philipson. 1987. Structure and physiology of Sitka spruce roots. *In* Symposium, Sitka spruce. D.M. Henderson and R. Faulkner (editors). Proc. Roy. Soc. Edinburgh 93B:131-144.

Coutts, M.P., C. Walker, and A.C. Burnand. 1990. Effects of establishment method on root form of lodgepole pine and Sitka spruce and on the production of adventitious roots. Forestry 63:143-159.

Daniel, T.W. and J. Schmidt. 1972. Lethal and nonlethal effects of the organic horizons of forested soils on the germination of seeds from several associated conifer species of the Rocky Mountains. Can. J. For. Res. 2(3):179-184.

Daubenmire, R. 1968. Some geographic variations in *Picea sitchensis* and their ecologic interpretation. Can. J. Bot. 46:787-798.

Davies, E.J.M. 1967. Silviculture of the spruces in West Scotland. Forestry 40(1):37-46.

–. 1972. History and background of Sitka spruce in the United Kingdom to the present day. Scott. For. 26(1):61-68.

Day, W.R. 1957. Sitka spruce in British Columbia: a study in forest relationships. For. Comm., London, U.K. Bull. 28.

Deal, R.L. and W.A. Farr. 1994. Composition and development of conifer regeneration in thinned and unthinned natural stands of western hemlock and Sitka spruce in southeast Alaska. Can. J. For. Res. 24:976-984.

Deal, R.L., C.D. Oliver, and B.T. Bormann. 1991. Reconstruction of hemlock-spruce stands in coastal southeast Alaska. Can. J. For. Res. 21:643-654.

Deans, J.D. 1979. Fluctuations of the soil environment and fine root growth in a young Sitka spruce plantation. Plant Soil 52:195-208.

–. 1981. Dynamics of coarse root production in a young plantation of *Picea sitchensis*. Forestry 54:139-155.

–. 1983. Distribution of thick roots in a Sitka spruce plantation 16 years after planting. Scott. For. 37:17-31.

Deans, J.D., W.L. Mason, M.G.R. Cannell, A.L. Sharpoe, and L.J. Sheppard. 1989. Growing regimes for bare-root stock of Sitka spruce, Douglas fir and Scots pine. I. Morphology at the end of the nursery phase. For. 62(Suppl.):53-60.

Demaerschalk, J.P. and A. Kozak. 1977. The whole-bole system: a conditional dual-equation system for precise prediction of tree profiles. Can. J. For. Res. 7:488-497.

DeMontigny, L.E., C.M. Preston, P.G. Hatcher, and I. Kogel-Knaber. 1993. Comparisons of humus horizons from two ecosystem phases on northern Vancouver Island using 13C CPMAS NMR spectroscopy and CuO oxidation. Can. J. Soil Sci. 73:9-25.

DeYoe, D.R. and K. Cromack. 1983. Mycorrhizae – a hidden benefactor to forest trees. Oreg. State Univ., Exten. Serv., Corvallis, Oreg. Publ. EM8247.

Dighton, J. 1987. Ecology and management of ectomycorrhizal fungi in the U.K. *In* Mycorrhizae in the next decade: practical applications and research priorities. D.M. Sylvia, L.L. Hung, and J.H. Graham (editors). Univ. Fla., Inst. Food Agric. Sciences, Gainesville, Fla., pp. 75-77.

Dighton, J., J.M. Poskitt, and D.M. Howard. 1986. Changes in the occurrence of basidiomycete fruit-bodies during forest stand development with specific reference to mycorrhizal species. Trans. Brit. Mycol. Soc. 87(1):163-171.

Dighton, J., E.D. Thomas, and P.M. Latter. 1987. Interactions between tree roots, mycorrhizas, a saprotrophic fungus and the decomposition of organic substrates in a microcosm. Biol. Fert. Soils 4(3):145-150.

Dinwoodie, J.M. 1963. Variation in the tracheid length in *Picea sitchensis* (Bong.) Carr. G.B. Dep. Sci. & Ind. Res., For. Prod. Res., London, U.K. Spec. Rep. 16.

Donald, D.G.M. and D.G. Simpson. 1985. Shallow conditioning and late fertilizer application effects on the quality of conifer nursery stock in British Columbia. B.C. Min. For., Victoria, B.C. Res. Note 99.

Douglas, G.W. 1975. Spruce (*Picea*) hybridization in west-central British Columbia. B.C. Min. For., Smithers, B.C. Unpubl. rep.

Douglas, S. 1991. Trees and shrubs of the Queen Charlotte Islands: an illustrated guide. Islands Ecol. Res., Queen Charlotte City, B.C.

Drew, T.J. and J.W. Flewelling. 1979. Stand density management: an alternative approach and its application to Douglas-fir plantations. For. Sci. 25:518-532.

Duffield, J.W. 1956. Damage to western Washington forests from November 1955 cold wave. U.S. Dep. Agric. For. Serv., Res. Note 129.

Dunbar, D. and I. Blackburn. 1995. Management options for the northern spotted owl in B.C. BioLine 13(1):13-16.
Duncan, R.W. 1982. Common pitch moths in pine in British Columbia. Can. For. Serv., Pac. For. Res. Cent., Victoria, B.C. For. Pest Leafl. 69.
–. 1996. Common woolly aphids and adelgids of conifers. Can. For. Serv., Pac. For. Cent., Victoria, B.C. For. Pest Leafl. 19.
Dutch, J. 1995. The effect of whole-tree harvesting on early growth of Sitka spruce on an upland restocking site. For. Comm., Farnham, U.K. Res. Inf. Note 261.
Eck, K.C. 1982. Forest characteristics and associated deer habitat values, Prince William Sound islands. *In* Fish and wildlife relationships in old-growth forests, Proc. Symp. 12-15 April 1982, Juneau, Alaska. W.R. Meehan, T.R. Merrell, Jr., and T.A. Hanley (editors). Am. Inst. Fish. Res. Biol., Morehead City, N.C., pp. 235-245.
Edwards, D.G.W. and Y.A. El-Kassaby. 1993. *Ex situ* conservation of forest biodiversity in British Columbia. *In* Proc. of the forest ecosystem dynamics workshop, February 10-11, 1993. V. Marshall (compiler). For. Can. and B.C. Min. For., Victoria, B.C. FRDA Rep. 210. pp.65-67.
Edwards, P.N. and J.M. Christie. 1981. Yield models for forest management. For. Comm., Farnham, U.K. Bull. 48.
Eilers, H.P. 1975. Plants, plant communities, net production and tide levels: the ecological biogeography of the Nehalem Salt Marshes, Tillamook County, Oregon. PhD thesis. Oreg. State Univ., Corvallis, Oreg.
Eklundh, C. 1943. Artkorsningar inom sl. *Abies, Pseudotsuga, Picea, Larix, Pinus* och *Chamaecyparis* tillhoerande fam. Pinaceae [Species crosses within the genera *Abies, Pseudotsuga, Picea, Larix, Pinus,* and *Chamaecyparis,* belonging to the family Pinaceae]. Svensk. Papp. Tidn. 46:55-61.
El-Kassaby, Y.A. 1992. Domestication and genetic diversity – should we be concerned? For. Chron. 68:687-700.
El-Kassaby, Y.A., K. Chaisurisri, D.G.W. Edwards, and D.W. Taylor. 1993. Genetic control of germination parameters of Douglas-fir, Sitka spruce, western redcedar, and yellow-cedar and its impact on container nursery production. *In* Dormancy and barriers to germination, Proc. of Int. Symp. of IUFRO Project Group P2.04-00 (Seed Problems), April 23-26, 1991, Victoria, B.C. D.G.W. Edwards (editor). For. Can., Pac. For. Cent., Victoria, B.C., pp. 37-42.
Embry, R.S. 1963. Prescribed burning for seedbed improvement. Science in Alaska, Proc. Alaska Sci. Conf. 13:88.
Environment Canada. 1980. Temperature and precipitation, 1951-1980, British Columbia. Atmos. Environ. Serv., Downsview, Ont.
Eremko, R.D., D.G.W. Edwards, and D. Wallinger. 1989. A guide to collecting cones of British Columbia conifers. For. Can. and B.C. Min. For., Victoria, B.C. FRDA Rep. 55.
Eyuboglu, A.K. 1986. Trabzon-macka yoresinde denenen sitka ladini (*Picea sitchensis* Bong. Carr.) Orijin denemesinin sonuclari [Results of Sitka spruce (*Picea sitchensis* Bong. Carr.) provenance trial at Trabzon-Macka district]. Ormancilik Arastirma Enstitusu Yayinlari, Ankara, Turkey. Teknik bulten serisi no. 175.
Fabricius, O. 1926. Douglas og Sitkagran [Douglas-fir and Sitka spruce]. Dansk. Skovfor. Tidsskr. 11:405-541.
Farnden, C. 1992. Cost and efficacy variation of forest vegetation management treatments in British Columbia. Prep. for B.C. Environ., Pesticide Manage. Prog., Victoria, B.C.
–. 1996. Stand density management diagrams for lodgepole pine, white spruce, and interior Douglas-fir. Can. For. Serv., Pac. For. Cent., Victoria, B.C. Inf. Rep. BC-X-360.
Farr, W.A. 1984. Site index and height growth curves for unmanaged even-aged stands of western hemlock and Sitka spruce in southeast Alaska. U.S. Dep. Agric. For. Serv., Res. Pap. PNW-326.
Farr, W.A., D.J. DeMars, and J.E. Dealy. 1989. Height and crown width related to diameter for open-grown western hemlock and Sitka spruce. Can. J. For. Res. 19:1203-1207.
Farr, W.A. and E.W. Ford. 1988. Site quality, forest growth, and soil-site studies in the forests of coastal Alaska. *In* Proc. of the Alaska forest soil productivity workshop. C. Slaughter and T. Gasbarro (editors). U.S. Dep. Agric. For. Serv., Gen. Tech. Rep. PNW-GTR-219. pp. 49-54.

Farr, W.A. and A.S. Harris. 1979. Site index of Sitka spruce along the Pacific Coast related to latitude and temperatures. For. Sci. 25:145-153.
–. 1983. Site index research in the western hemlock–Sitka spruce forest type of coastal Alaska. *In* IUFRO Symposium of forest site and continuous productivity, 22-28 August 1982, Seattle, Wash. R. Ballard and S.P. Gessel (technical editors). U.S. Dep. Agric. For. Serv., Gen. Tech. Rep. PNW-163. pp. 87-93.
Farr, W.A., V.J. La Bau, and T.H. Laurent. 1976. Estimation of decay in old growth western hemlock and Sitka spruce in southeast Alaska, USA. U.S. Dep. Agric. For. Serv., Res. Pap. PNW-204.
Farrar, J.L. 1995. Trees in Canada. Can. For. Serv. and Fitzhenry and Whiteside, Markham, Ont.
Feller, M.C. 1982. The ecological effects of slashburning with particular reference to British Columbia: a literature review. B.C. Min. For., Victoria, B.C. Land Manage. Rep. 3.
Fenger, M.A. and A.P. Harcombe. 1993. Biodiversity, old-growth forests, and wildlife in British Columbia. For. Plan. Can. 9(4):36-38.
Fenger, M.A., E.H. Miller, J.F. Johnson, and E.J.R. Williams (editors). 1993. Our living legacy. Proceedings of a symposium on biological diversity. Roy. B.C. Museum, Victoria, B.C.
Ferguson, D.E., A.R. Stage, and R.J. Boyd. 1986. Predicting regeneration in the grand fir–cedar-hemlock ecosystem of the northern Rocky Mountains. For. Sci. Monogr. 26.
Fight, R.D., J.M. Cahill, and D. Thomas. 1992. DFPRUNE users guide. U.S. Dep. Agric. For. Serv., Gen. Tech. Rep. PNW-GTR 300.
Finck, K.E., P. Humphreys, and G.V. Hawkins. 1989. Field guide to pests of managed forests in British Columbia. For. Can. and B.C. Min. For., Victoria, B.C. Joint Publ. 16.
Finck, K.E., G.M. Shrimpton, and D.W. Summers. 1990. Insect pests in reforestation. *In* Regenerating British Columbia's forests. D.P. Lavender, R. Parish, C.M. Johnson, G. Montgomery, A. Vyse, R.A. Willis, and D. Winston (editors). Univ. B.C. Press, Vancouver, B.C., pp. 279-301.
Fletcher, A.M. 1976. Seed collection in north-west America with particular reference to a Sitka spruce seed collection for provenance studies. *In* IUFRO. Sitka Spruce *Picea sitchensis* (Bong.) Carr. International Ten Provenance Experiment. Nursery stage results. Dep. Lands, For. Wildl. Serv., Dublin, Ire., pp. 2-20.
Fonda, R.W. 1974. Forest succession in relation to river terrace development in Olympic National Park, Washington. Ecology 55(5):927-942.
Ford, E.D. 1975. Competition and stand structure in some even-aged plant monocultures. J. Ecol. 63:311-333.
–. 1982. High productivity in a pole stage Sitka spruce, *Picea sitchensis,* stand and its relation to canopy structure. Forestry 55(1):1-18.
–. 1985. Branching, crown structure and the control of timber production. *In* Attributes of trees as crop plants. M.G.R. Cannell and J.E. Jackson (editors). Inst. Terr. Ecol., Nat. Environ. Res. Coun., Abbots Ripton, U.K., pp. 228-252.
Ford, E.D. and J.D. Deans. 1977. Growth of a Sitka spruce plantation: spatial distribution and seasonal fluctuations of lengths, weights and carbohydrate concentrations of fine roots. Plant Soil 47(2):463-486.
Ford, E.D., W.A. Farr, and C. Lu-Ping. 1988. Preliminary analysis of four soil variables and their relation to site index of Sitka spruce in southeast Alaska. *In* Proc. of the Alaska forest soil productivity workshop. U.S. Dep. Agric. For. Serv., Gen. Tech. Rep. PNW-GTR-219. pp. 84-90.
Forest Club of the University of British Columbia. 1959. Forestry handbook for British Columbia, 2nd ed. Univ. B.C., Fac. For., Vancouver, B.C.
Forest Industry Trader. 1996. Vancouver log market prices. For. Indust. Trader 6(127):10.
Forest Renewal B.C. Newsletter. 1996. Value added: an investment in our future. For. Renewal B.C. Newsl. 1:2-3.
Forrest, G.I. 1975. Polyphenol variation in Sitka spruce. Can. J. For. Res. 5(1):26-37.
–. 1980. Geographic variation in the monoterpene composition of Sitka spruce cortical oleoresin. Can. J. For. Res. 10(4):458-463.

Foster, B. 1993. The importance of British Columbia to global biodiversity. *In* Our living legacy: proceedings of a symposium on biological diversity. M.A. Fenger, E.H. Miller, J.F. Johnson, and E.J.R. Williams (editors). Roy. B.C. Museum, Victoria, B.C., pp. 65-81.

Foster, R.E. and A.T. Foster. 1951. Studies on forest pathology. VIII. Decay of western hemlock on the Queen Charlotte Islands. Can. J. Bot. 29:479-552.

Fourt, D.F. 1968. Sitka spruce, shelter and moisture. For. Comm., Farnham, U.K. Res. Devel. Pap. 72.

Fowells, H.A. 1965. Silvics of forest trees of the United States. U.S. Dep. Agric. For. Serv., Washington, D.C.

Fowler, D.P. 1983. The hybrid black spruce × Sitka spruce, *Picea mariana* × *Picea sitchensis,* implications to phylogeny of the genus *Picea.* Can. J. For. Res. 13(1):108-115.

Franklin, J.F. 1982. Ecosystem studies in the Hoh River drainage, Olympic National park. *In* Ecological research in national parks of the Pacific Northwest. Proc. of 2nd conf. on scientific research in national parks, Nov. 1979, San Francisco, Calif. E.E. Starkey, J.F. Franklin, and J.W. Mathews (technical coordinators). Oreg. State Univ. For. Res. Lab., Corvallis, Oreg.

–. 1992. Scientific basis for new perspectives in forests and streams. *In* Watershed management: balancing sustainability and environmental change. R.J. Naiman (editor). Springer-Verlag, New York, N.Y., pp. 25-72.

–. 1993. Lessons from old growth. J. For. 91:10-13.

Franklin, J.F., K. Cromack, W. Denison, A. McKee, C. Maser, J. Sedell, F. Swanson, and G. Juday. 1981. Ecological characteristics of old-growth Douglas-fir forests. U.S. Dep. Agric. For. Serv., Gen. Tech. Rep. PNW-118.

Franklin, J.F. and C.T. Dyrness. 1973. Natural vegetation of Oregon and Washington. U.S. Dep. Agric. For. Serv., Gen. Tech. Rep. PNW-8.

Franklin, J.F., F.C. Hall, C.T. Dyrness, and C. Maser. 1972. Federal research natural areas in Oregon and Washington; a guidebook for scientists and educators. U.S. Dep. Agric. For. Serv., Pac. N.W. For. Range Exp. Sta., Portland, Oreg.

Franklin, J.F., W.H. Moir, M.A. Hemstrom, S.A. Greene, and B.G. Smith. 1988. The forest communities of Mount Rainier National Park. U.S. Dep. Inter., Nat. Park Serv., Washington, D.C. Sci. Monog. Ser. 19.

Franklin, J.F. and A.A. Pechanec. 1968. Comparison of vegetation in adjacent alder, conifer, and mixed alder-conifer communities. I. Understory vegetation and stand structure. *In* Biology of alder. Northwest Sci. Assoc. 40th Annu. Meet. Symp. Proc. J.M. Trappe, J.F. Franklin, R.F. Tarrant, and G.M. Hansen (editors). U.S. Dep. Agric. For. Serv., Pac. N.W. For. Range Exp. Sta., Portland, Oreg., pp. 7-43.

Franklin, J.F., F.J. Swanson, and J.R. Sedell. 1982. Relationships within the valley floor ecosystems in western Olympic National Park: a summary. *In* Ecological research in national parks of the Pacific Northwest. Proc. of 2nd conf. on scientific research in national parks, Nov. 1979, San Francisco, Calif. E.E. Starkey, J.F. Franklin, and J.W. Mathews (technical coordinators). Oreg. State Univ. For. Res. Lab., Corvallis, Oreg., pp. 43-45.

Franklin, J.F. and R.H. Waring. 1979. Distinctive features of the northwestern coniferous forest: development, structure and function. *In* Forests: fresh perspectives from ecosystem analysis. Proc. 40th Annual Biology Colloquium. R.H. Waring (editor). Oreg. State Univ. Press., Corvallis, Oreg., pp. 59-85.

Fraser, B.C.E. and L.D. Cordes. 1967. Sap pressure of active and dormant plants of three Pacific Northwest conifers. *In* Progress Report, National Research Council Grant No. A-92. V.J. Krajina (editor). Univ. B.C., Dep. Bot., Vancouver, B.C., pp. 4-12.

Fraser, R.G. and D.G. Heppner. 1993. Control of white pine weevil, *Pissodes strobi,* on Sitka spruce using implants containing systemic insecticide. For. Chron. 69:600-603.

Furniss, R.L. and V.M. Carolin. 1977. Western forest insects. U.S. Dep. Agric. For. Serv., Washington, D.C. Misc. Publ. 1339.

Galindo-Leal, C. and F.L. Bunnell. 1995. Ecosystem management: implications and opportunities of a new paradigm. For. Chron. 71:601-606.

Geburek, T. and D. Krusche. 1985. Hybrid growth of Sitka and Serbian spruce in comparison with the parent species, Lower Saxony, Schleswig-Holstein, Germany. F.R. Allgem. Forst u. Jagdzeitung 156(3):47-54.

Germain, A.Y. 1985. Fertilization of stagnated Sitka spruce plantations on northern Vancouver Island. MF thesis. Univ. B.C., Vancouver, B.C.

Gimbarzevsky, P. 1988. Mass wasting on the Queen Charlotte Islands: a regional inventory. Min. For. Lands, Victoria, B.C. Land Manage. Rep. 29.

Ginns, J.H. 1974. Rhizina root rot: severity and distribution in British Columbia. Can. J. For. Res. 4:143-146.

Glover, M.M., J.R. Sutherland, C.L. Leadem, and G.M. Shrimpton. 1987. Efficacy and phytotoxicity of fungicides for control of *Botrytis* gray mould on container-grown conifer seedlings. Can. For. Serv. and B.C. Min. For., Victoria, B.C. FRDA Rep. 12.

Godman, R.M. 1949. What kind of trees make the best growth in southeast Alaska. U.S. Dep. Agric. For. Serv., Alaska For. Res. Cent. Tech. Note 2.

Godman, R.M. and R.A. Gregory. 1953. Seasonal distribution of radial and leader growth in the Sitka spruce–western hemlock forests of southeast Alaska. J. For. 53(11):827-833.

Gonor, J.J., J.R. Sedell, and P.A. Benner. 1988. What we know about large trees in estuaries, in the sea, and on coastal beaches. *In* From the forest to the sea: a story of fallen trees. C. Maser, R.F. Tarrant, J.M. Trappe, and J.F. Franklin (technical editors). U.S. Dep. Agric. For. Serv., Gen. Tech. Rep. PNW-GTR-229. pp. 83-112.

Goudie, J. 1991. Guide for assessing permanent sample plot data in FPDS database. B.C. Min. For., Res. Br., Victoria, B.C.

Goulet, F. 1995. Frost heaving of forest tree seedlings: a review. New For. 9:67-94.

Graham, R.L. and K. Cromack, Jr. 1982. Mass, nutrient content, and decay rate of dead boles in rain forests of Olympic National Park. Can. J. For. Res. 12:511-521.

Green, R.N., P.J. Courtin, K. Klinka, R.J. Slaco, and C.A. Ray. 1984. Site diagnosis, tree species selection, and slashburning guidelines for the Vancouver Forest Region. B.C. Min. For., Victoria, B.C. Land Manage. Handb. 8.

Green, R.N. and K. Klinka. 1994. Field guide for site series identification and interpretation for the Vancouver Forest Region. B.C. Min. For., Victoria, B.C. Land Manage. Handb. 28.

Green, R.N., P.L. Marshall, and K. Klinka. 1989. Estimating site index of Douglas-fir (*Pseudotsuga menziesii* (Mirb.) Franco) from ecological variables in southwestern British Columbia. For. Sci. 35:50-63.

Greene, S.E., P.A. Harmon, M.E. Harmon, and G. Spycher. 1992. Patterns of growth, mortality and biomass change in a coastal *Picea sitchensis–Tsuga heterophylla* forest. J. Veg. Sci. 3:697-706.

Gregory, R.A. 1956. The effect of clearcutting and soil disturbance on temperature near the soil surface in southeast Alaska. U.S. Dep. Agric. For. Serv., Alaska For. Res. Cent., Sta. Pap. 7.

Griggs, R.F. 1934. The edge of the forest in Alaska and the reasons for its position. Ecology 15:80-96.

Grossnickle, S.C., B.C.S. Sutton, R.S. Folk, and R.J. Gawley. 1996a. Relationship between nuclear DNA markers and physiological parameters in Sitka × interior spruce populations. Tree Physiol. 16:547-555.

Grossnickle, S.C., B.C.S. Sutton, and R.W. Holcomb. 1996b. Genetics and seedling biology of Sitka spruce hybrids. B.C. Min. For., Victoria, B.C. FRDA Memo 232.

H.A. Simons Ltd., Strategic Services Division. 1992. Wood production strategies and the implications for silviculture in B.C. Prep. for B.C. Min. For., Victoria, B.C.

Hadfield, J.S. 1985. Laminated root rot: a guide to reducing and preventing losses in Oregon and Washington forests. U.S. Dep. Agric. For. Serv., For. Pest Manage., PNW-FPM 61.

Haeussler, S. 1991. Prescribed fire for forest vegetation management. B.C. Min. For., Victoria, B.C. FRDA Memo 198.

Haeussler, S., K.D. Coates, and J. Mather. 1990. Autecology of common plants in British Columbia: a literature review. For. Can. and B.C. Min. For., Victoria, B.C. FRDA Rep. 158.

Hall, J.P. 1990. Development of a land race of Sitka spruce in Newfoundland. *In* Joint Meeting of Western Forest Genetics Association and IUFRO Working Parties S2.02-05, 06, 12, and 14. 20-24 Aug. 1990, Olympia, Wash. Pap. pp. 2/118-127.

Hamer, T.E. and E.B. Cummins. 1991. Relationships between forest characteristics and use of inland sites by Marbled Murrelets in northwestern Washington. Wash. Dep. Wildl., Olympia, Wash. Unpubl. rep. Cited in Rodway et al. 1993.

Hamilton, A.N. 1987. Classification of coastal grizzly bear habitat for forestry interpretations and the role of food in habitat use by coastal grizzly bears. MSc thesis. Univ. B.C., Vancouver, B.C.

Hamilton, A.N. and W.R. Archibald. 1986. Grizzly bear habitat in the Kimsquit River valley, coastal British Columbia: evaluation. *In* Proc. Grizzly Bear Habitat Symp. G.P. Contreras and K.E. Evans (editors). U.S. Dep. Agric. For. Serv., Gen. Tech. Rep. INT-207. pp. 50-57.

Hamilton, A.N. and J. Berry. 1992. Interim guidelines for integrating coastal grizzly bear habitat and silviculture in the Vancouver Forest Region. B.C. Environ., Wildl. Br., and B.C. Min. For., Silv. Sec., Vancouver, B.C. Intern. rep.

Hamilton, A.N., C.A. Bryden, and C.J. Clement. 1991. Impacts of glyphosate application on grizzly bear forage production in the Coastal Western Hemlock Zone. For. Can. and B.C. Min. For., Victoria, B.C. FRDA Rep. 165.

Hamilton, G.J. 1969. The dependence of volume increment of individual trees on dominance, crown dimensions and competition. Forestry 42(2):133-144.

–. 1981. The effects of high intensity thinning on yield. Forestry 54(1):1-15.

Hamilton, G.J. and J.M. Christie. 1971. Forest management tables (metric). For. Comm., London, U.K. Booklet 34.

Hanley, T.A. 1984. Relationships between Sitka black-tailed deer and their habitat. U.S. Dep. Agric. For. Serv., Gen. Tech. Rep. PNW-168.

–. 1993. Balancing economic development, biological conservation, and human culture: the Sitka black-tailed deer *Odocoileus hemionus sitkensis* as an ecological indicator. Biol. Cons. 66:61-67.

Hanley, T.A., R.G. Cates, B. Van Horne, J.D. McKendrick. 1987. Forest stand-age-related differences in apparent nutritional quality of forage for deer in southeastern Alaska. Proc. symp. plant-herbivore interactions, 7-9 August 1985, Snowbird, Utah. U.S. Dep. Agric. For. Serv., Gen. Tech. Rep. INT-222.

Hanley, T.A., C.T. Robbins, and D.E. Spalinger. 1989. Forest habitats and the nutritional ecology of Sitka black-tailed deer: a research synthesis with implications for forest management. U.S. Dep. Agric. For. Serv., Gen. Tech. Rep. PNW-GTR-230.

–. 1991. The influence of the forest environment on nutritional ecology of black-tailed deer in Alaska. *In* Global trends in wildlife management, Proc. of 18th Congress of the International Union of Game Biologists, 23-29 Aug., Krakow, Poland. B. Bobek, K. Perzanowski, and W. Regelin (editors). Swiat Press, Krakow, Poland, pp. 357-361.

Hanley, T.A. and J.J. Rogers. 1989. Estimating carrying capacity with simultaneous nutritional constraints. U.S. Dep. Agric. For. Serv., Res. Note PNW-RN-485.

Hanley, T.A. and C.L. Rose. 1987. Influence of overstory on snow depth and density in hemlock-spruce stands: implications for management of deer habitat in southeastern Alaska. U.S. Dep. Agric. For. Serv., Res. Rep. PNW-459.

Hanover, J.W. and R.C. Wilkinson. 1970. Chemical evidence for introgressive hybridization in *Picea*. Silvae Genet. 19:17-22.

Hansen, A.J., T.A. Spies, F.J. Swanson, and J.L. Ohmann. 1991. Conserving biodiversity in managed forests. BioScience 41:382-392.

Hansen, C. 1892. Pinetum Danicum. J. Roy. Hort. Soc. XIV. Cited in Karlberg 1961.

Harcombe, P.A. 1986. Stand development in a 130-year-old spruce-hemlock forest based on age structure and 50 years of mortality data. For. Ecol. Manage. 14:41-58.

Harcombe, P.A., M.E. Harmon, and S.E. Greene. 1990. Changes in biomass and production over 53 years in a coastal *Picea sitchensis–Tsuga heterophylla* forest approaching maturity. Can. J. For. Res. 20:1602-1610.

Hard, J. 1992. Success of spruce beetle attacks in pruned and unpruned boles of Lutz spruce in south-central Alaska. For. Ecol. Manage. 47:51-70.

Harding, L.E. and E. McCullum (editors). 1994. Biodiversity in British Columbia: our changing environment. Can. Wildl. Serv., Pac. & Yukon Reg., Vancouver, B.C.

Harlow, W.M. and E.S. Harrar. 1958. Textbook of dendrology. 4th ed. McGraw-Hill, New York, N.Y.

Harmon, M.E. 1985. Logs as sites of tree regeneration in *Picea sitchensis–Tsuga heterophylla* forests of coastal Washington and Oregon. PhD thesis. Oreg. State Univ., Corvallis, Oreg.
Harmon, M.E. and J.F. Franklin. 1989. Tree seedlings on logs in *Picea-Tsuga* forests of Oregon and Washington. Ecology 70:48-59.
Harmon, M.E., J.F. Franklin, F.J. Swanson, P. Sollins, S.V. Gregory, J.D. Lattin, N.H. Anderson, S.P. Cline, N.G. Aumen, J.R. Sedell, G.W. Lienkaemper, K. Cromack, Jr., and K.W. Cummins. 1986. Ecology of coarse woody debris in temperate ecosystems. Advan. Ecol. Res. 15:133-302.
Harper, W., S. Cannings, D. Fraser, and W.T. Munro. 1994. Provincial lists of species at risk. *In* Biodiversity in British Columbia: our changing environment. L.E. Harding and E. McCullum (editors). Can. Wildl. Serv., Pac. & Yukon Reg., Vancouver, B.C., pp. 16-28.
Harr, R.D. 1980. Streamflow after patch logging in a small drainage within Bull Run Municipal Watershed, Oregon. U.S. Dep. Agric. For. Serv., Res. Pap. PNW-268.
Harris, A.S. 1964. Sitka spruce – Alaska's new state tree. Am. For. 70(8):32-35.
–. 1966. Effects of slash burning on conifer regeneration in southeast Alaska. U.S. Dep. Agric. For. Serv., Res. Note NOR-18.
–. 1969. Ripening and dispersal of a bumper western hemlock–Sitka spruce seed crop in southeast Alaska. U.S. Dep. Agric. For. Serv., Res. Note PNW-105.
–. 1984 rev. Sitka spruce: an American wood. U.S. Dep. Agric. For. Serv., Washington, D.C. FS-265.
–. 1989. Wind in the forests of southeast Alaska and guides for reducing damage. U.S. Dep. Agric. For. Serv., Gen. Tech. Rep. GTR-PNW-244.
–. 1990. *Picea sitchensis* (Bong.) Carr. Sitka spruce. *In* Silvics of North America. Volume 1, Conifers. R.M. Burns and B.H. Honkala (technical coordinators). U.S. Dep. Agric. For. Serv., Washington, D.C. Agric. Handb. 654. pp.260-267.
Harris, A.S. and W.A. Farr. 1974. The forest ecosystem of southeast Alaska. 7. Forest ecology and timber management. U.S. Dep. Agric. For. Serv., Gen. Tech. Rep. PNW-25.
Harris, A.S. and D.L. Johnson. 1983. Western hemlock–Sitka spruce. *In* Silvicultural systems for the major forest types of the United States. R.M. Burns (technical compiler). U.S. Dep. Agric., Washington, D.C. Agric. Handb. 445. pp. 5-8.
Harris, L.J., J.H. Borden, H.D. Pierce, Jr., and A.C. Oehlschlager. 1983. Cortical resin monoterpenes in Sitka spruce and resistance to the white pine weevil, *Pissodes strobi* (Coleoptera: Curculionidae). Can. J. For. Res. 13(2):350-352.
Hartman, G.F. (editor). 1982. Proceedings of the Carnation Creek Workshop, a 10 year review. Pac. Biol. Sta., Nanaimo, B.C.
Hartman, G.F. and T.G. Brown. 1988. Forestry-fisheries planning considerations on coastal floodplains. For. Chron. 64:47-51
Hartman, G.F., L.B. Holtby, and J.C. Scrivener. 1982. Some effects of natural and logging-related winter stream temperature changes on the early life history of the coho salmon (*Oncorhynchus kisutch*) in Carnation Creek, British Columbia. *In* Fish and wildlife relationships in old-growth forests, Proc. Symp. 12-15 April 1982, Juneau, Alaska. W.R. Meehan, T.R. Merrell, Jr., and T.A. Hanley (editors). Am. Inst. Fish. Res. Biol., Morehead City, N.C., pp. 141-149.
Hartman, G.F. and J.C. Scrivener. 1990. Impacts of forestry practices on a coastal stream ecosystem, Carnation Creek, British Columbia. Can. Bull. Fish. Aquatic Sci. 223:1-148.
Hartman, L. 1974. Sitka spruce weevil trial on Vancouver Island. B.C. Min. For., Res. Br., Victoria, B.C. Establishment Rep. EP 702.03.
Hawkes, B.C., M.C. Feller, and D. Meehan. 1990. Site preparation: fire. *In* Regenerating British Columbia's forests. D.P. Lavender, R. Parish, C.M. Johnson, G. Montgomery, A. Vyse, R.A. Willis, and D. Winston (editors). Univ. B.C. Press, Vancouver, B.C., pp. 131-149.
Hedlin, A.F. 1974. Cone and seed insects of British Columbia. Can. For. Serv., Pac. For. Res. Cent., Victoria, B.C. Inf. Rep. BC-X-90.
Hedlin, A.F., H.O. Yates III, D.C. Tovar, B.H. Ebel, T.W. Koerber, and E.P. Merkel. 1980. Cone and seed insects of North American conifers. U.S. Dep. Agric. For. Serv. and Can. For. Serv. Joint Report.

Hegg, K.M. 1967. A photo identification guide for the land and forest types of interior Alaska. U.S. Dep. Agric. For. Serv., North. For. Res. Sta., Juneau, Alaska. Pap. NOR-3.

Hegyi, F., J.J. Jelinek, J. Viszlai, and D.B. Carpenter. 1981 rev. Site index equations and curves for the major tree species in British Columbia. B.C. Min. For., Inven. Br., Victoria, B.C. For. Inven. Rep. 1.

Heilman, P.E. 1990. Forest management challenged in the Pacific Northwest: increasingly cynical public questions sustainability of current practices. J. For. 88(11):15-23.

Henderson, J.A., D.H. Peter, R.D. Lesher, and D.C. Shaw. 1989. Forested plant associations of the Olympic National Forest. U.S. Dep. Agric. For. Serv., N.W. Reg., Portland, Oreg. R6 Ecol. Tech. Pap. 001-88.

Henderson, R., E.D. Ford, E. Renshaw, and J.D. Deans. 1983. Morphology of the structural root system of Sitka spruce. I. Analysis and quantitative description. Forestry 56:121-135.

Henmanns, D.W. 1963. Pruning conifers for the production of quality timber. For. Comm., Farnham, U.K. Bull. 35.

Hennon, P.E. 1995. Are heart rot fungi major factors of disturbance in gap-dynamic forests? Northw. Sci. 69:284-293.

Heppner, D.G. and P.M. Wood. 1984. Vancouver Forest Region Sitka spruce weevil survey results (1982-1983) with recommendations for planting Sitka spruce. B.C. Min. For., Victoria, B.C. Int. Rep. PM-V-5.

Heritage Forests Society and Sierra Club of Canada. 1988. A proposal to add the Carmanah Creek drainage with its exceptional Sitka spruce forests to Pacific Rim National Park. Heritage Forests Society and Sierra Club of Western Canada, Vancouver, B.C.

Herman, F.R. 1964. Epicormic branching of Sitka spruce. U.S. Dep. Agric. For. Serv., Res. Pap. PNW-18.

Hermann, R.K. 1987. North American tree species in Europe; transplanted species offer good growth potential on suitable sites. J. For. 85(12):27-32.

Hetherington, E.D. 1987. The importance of forests in the hydrological regime. *In* Canadian aquatic resources. M.C. Healey and R.R. Wallace (editors). Can. Bull. Fish. Aquatic Sci. 215. pp. 179-211.

–. 1996. Logging effects on the aquatic ecosystem: a case study in the Carnation Creek experimental watershed on Canada's west coast. *In* High-latitude rainforests and associated ecosystems of the west coast of the Americas: climate, hydrology, ecology, and conservation. R.G. Lawford, P.B. Alaback, and E. Fuentes (editors). Springer, New York, N.Y., pp. 342-352.

Hibbs, D.E. and D.S. DeBell. 1994. Management of young red alder. *In* The biology and management of red alder. D.E. Hibbs, D.S. DeBell, and R.F. Tarrant (editors). Oreg. State Univ. Press, Corvallis, Oreg., pp. 202-215.

Hill, M.O. and E.W. Jones. 1978. Vegetation changes resulting from afforestation of rough grazings in Caed Forest, South Wales. J. Ecol. 66:433-456.

Hines, W.W. 1971. Plant communities in the old-growth forests of north coastal Oregon. MSc thesis. Oreg. State Univ., Corvallis, Oreg.

Hodges, J.D. 1967. Patterns of photosynthesis under natural environmental conditions. Ecology 48:234-242.

Hogan, D.L. and J.W. Schwab. 1990. Floodplain stability and island longevity. Hydrologic and geomorphic considerations for silvicultural investments on the lower Skeena River floodplain. *In* P.G. Beaudry, D.L. Hogan, and J.W. Schwab. For. Can. and B.C. Min. For., Victoria, B.C. FRDA Rep. 122. pp. 13-24.

Holden, J.M., G.W. Thomas, and R.M. Jackson. 1983. Effect of ectomycorrhizal inocula on the growth of Sitka spruce seedlings in different soils. Plant Soil 71:313-317.

Holms, J.C. 1967. Sitka spruce weevil in British Columbia. Dep. For. Rural Devel., For. Res. Lab., Victoria, B.C. For. Pest Leafl.

Holsten, E.H., P.E. Hennon, and R.A. Werner. 1985 rev. Insects and diseases of Alaskan forests. U.S. Dep. Agric. For. Serv., For. Pest Manage., State and Private Forestry, Juneau, Alaska. Alaska Reg. Rep. 181.

Holsten, E.H. and R.A. Werner. 1993. Effectiveness of polyethylene sheathing in controlling spruce beetles (Coleoptera: Scolytidae) in infested stacks of spruce firewood in Alaska. U.S. Dep. Agric. For. Serv., Res. Pap. PNW-RP-466.

Hosie, R.C. 1990. Native trees of Canada. 8th ed. Fitzhenry and Whiteside, Markham, Ont.
Houseknecht, S., S. Haeussler, A. Kokoshke, J. Pojar, D. Holmes, B.M. Geisler, D. Yole, and C. Clement. 1987. A field guide for identification and interpretation of the Interior Cedar-Hemlock Zone, Northwestern Transitional Subzone (ICHg), in the Prince Rupert Forest Region. rev. ed. B.C. Min. For., Victoria, B.C. Land Manage. Handb. 12.
Hulme, M.A. 1995. Resistance by translocated Sitka spruce to damage by *Pissodes strobi* (Coleoptera: Curculionidae) related to tree phenology. J. Econ. Entomol. 88:1525-1530.
Hultén, E. 1937. Bokforlags Aktiebolaget Thule., Stockholm, Sweden.
Hunt, R.S. and L. Unger. 1994. Tomentosus root disease. Can. For. Serv., Pac. For. Cent., Victoria, B.C. For. Pest Leafl. 77.
Illingworth, K. 1976. Provenance trials in British Columbia, Canada. *In* Proc. Div. II (Forest plants and forest protection), XVI IUFRO World Congress, 20 June to 2 July 1976, As, Norway. IUFRO.
Illingworth, K. 1978. Sitka spruce trials three years after planting in British Columbia. *In* Proc. Joint Meeting of Working Parties, 1978, Vancouver, B.C. Vol. 2. Lodgepole pine, Sitka spruce and *Abies* provenances. IUFRO. B.C. Min. For., Victoria, B.C., pp. 311-326.
Ilmurzynski, E. et al. 1968. Investigations on the growth and development of certain N. American tree species in [Polish] nurseries and plantations. Prace Inst. Bad. Lesn. No. 364:84.
Ingelby, K., J. Wilson, P.A. Mason, R.C. Munro, C. Walker, and W.L. Mason. 1994. Effects of mycorrhizal inoculation and fertilizer regime on emergence of Sitka spruce seedlings in bare-root nursery seedbeds. Can. J. For. Res. 24:618-623.
International Union of Forestry Research Organizations (IUFRO). 1976. IUFRO Sitka spruce *Picea sitchensis* (Bong.) Carr. International Ten Provenance Experiment. Nursery stage results. Dep. Lands, For. Wildl. Serv., Dublin, Ire.
–. 1984. Minutes of the Sitka Spruce Working Group S.2.02 – 12th meeting, 2-7 September 1984. Edinburgh, Scotland. Unpubl. rep.
Isaac, L.A. 1940. 'Water sprouts' on Sitka spruce. U.S. Dep. Agric. For. Serv., Res. Note 31:6-7.
Jack, W.H. 1971. The influence of tree spacing on Sitka spruce growth. Irish For. 28:13-33.
James, G.A. and R.A. Gregory. 1959. Natural stocking of a mile square clearcutting in southeast Alaska. U.S. Dep. Agric. For. Serv., Alaska For. Res. Cent., Sta. Pap.
Jarvis, N.J. and C.E. Mullins. 1987. Modelling the effects of drought on the growth of Sitka spruce in Scotland. Forestry 60:13-30.
John, A. and W. Mason. 1987. Vegetative propagation of Sitka spruce. Proc. Roy. Soc. Edinburgh 93B:197-203.
Johnson, L.P.V. 1939. A descriptive list of natural and artificial interspecific hybrids in North American forest-tree genera. Can. J. Res. 17(Sec. C):411-444.
Jones, G.W. and F.L. Bunnell. 1982. Response of black-tailed deer to winters of different severity on northern Vancouver Island. *In* Fish and wildlife relationships in old-growth forests, Proc. Symp. 12-15 April 1982, Juneau, Alaska. W.R. Meehan, T.R. Merrell, Jr., and T.A. Hanley (editors). Am. Inst. Fish. Res. Biol., Morehead City, N.C., pp. 385-390.
Jones, S.K., U. Bergsten, and P.G. Gosling. 1993. A comparison of 5°C and 15°C as dormancy breakage treatments for Sitka spruce seeds (*Picea sitchensis*). *In* Dormancy and barriers to germination, Proc. of Int. Symp. of IUFRO Project Group P2.04-00 (Seed Problems), 23-26 April 1991, Victoria, B.C. D.G.W. Edwards (editor). For. Can., Pac. For. Cent., Victoria, B.C., pp. 51-55.
Karakatsoulis, J., J.P. Kimmins, and R.E. Bigley. 1989. Comparison of the effects of chemical (glyphosate) and manual conifer release on conifer seedling physiology and growth on Vedder Mountain, British Columbia. Appendix 1. *In* Proceedings of the Carnation Creek Herbicide Workshop. P.E. Reynolds (editor). For. Can. and B.C. Min. For., Victoria, B.C. FRDA Rep. 63. pp. 168-188.
Karlberg, S. 1961. Development and yield of Douglas fir [*Pseudotsuga taxifolia* (Poir) Britt.] and Sitka spruce [*Picea sitchensis* (Bong.) Carr.] in southern Scandinavia and on the Pacific coast. Kungl. Skogshoegsk. Skr. 34:1-141.
Kayahara, G.J. and A.F. Pearson. 1996. Relationships between site index, and soil moisture and nutrient regimes for western hemlock and Sitka spruce. B.C. Min. For., Victoria, B.C. Work. Pap. 17/1996.

Keenan, R.J. 1993. Structure and function of western redcedar and western hemlock forests on northern Vancouver Island. PhD thesis, Univ. B.C., Vancouver, B.C.

Keenan, R.J., J.P. Kimmins, and J. Pastor. 1995. Modeling carbon and nitrogen dynamics in western red cedar and western hemlock forests. *In* Carbon forms and functions in forest soils. W.W. McFee and J.M. Kelly (editors). Soil Sci. Soc. Am., Madison, Wis., pp. 547-568.

Keenan, R.J., C.E. Prescott, and J.P. Kimmins. 1993. Mass and nutrient content of woody debris and forest floor in western red cedar and western hemlock forests on northern Vancouver Island. Can. J. For. Res. 23:1052-1059.

Kellogg, L.D. and E.D. Olsen. 1988. Economic evaluation of thinning alternatives in a western hemlock–Sitka spruce forest. West. J. Appl. For. 3:14-17.

Kellogg, L.D., E.D. Olsen, and M.A. Hargrave. 1986. Skyline thinning a western hemlock–Sitka spruce stand: harvesting costs and stand damage. Oreg. State Univ., For. Res. Lab., Corvallis, Oreg. Res. Bull. 53.

Kessler, W.B. 1982. Management potential of second-growth forest for wildlife objectives in southeast Alaska. *In* Fish and wildlife relationships in old-growth forests, Proc. Symp. 12-15 April 1982, Juneau, Alaska. W.R. Meehan, T.R. Merrell, Jr., and T.A. Hanley (editors). Am. Inst. Fish. Res. Biol., Morehead City, N.C., pp. 381-384.

Ketcheson, M.V., T.F. Braumandl, D. Meidinger, G. Utzig, D.A. Demarchi, and B.M. Wikeem. 1991. Interior Cedar-Hemlock Zone. *In* Ecosystems of British Columbia. D. Meidinger and J. Pojar (editors). B.C. Min. For., Victoria, B.C. Spec. Rep. Ser. 6. pp. 167-181.

Khalil, M.A.K. 1977. Provenance experiments with Sitka spruce in Newfoundland. For. Chron. 53(3):150-154.

Kilpatrick, D.J., J.M. Sanderson, and P.S. Savill. 1981. The influence of five early respacing treatments on the growth of Sitka spruce. Forestry 54(1):17-29.

Kimmins, J.P. 1996. The health and integrity of forest ecosystems: are they threatened by forestry? Ecosys. Health 2:5-18.

King, J.N. 1994. Delivering durable resistant Sitka spruce for plantations. *In* The white pine weevil: biology, damage and management. R.I. Alfaro, G.K. Kiss, and R.G. Fraser (editors). Can. For. Serv. and B.C. Min. For., Victoria, B.C. FRDA Rep. 226. pp. 134-149.

King, J.N. and Y.A. El-Kassaby. 1990. Caveats for early selection. *In* Joint Meeting of Western Forest Genetics Association and IUFRO Working Parties S2.02-05, 06, 12, and 14. 20-24 Aug. 1990, Olympia, Wash., pp. 4/81-84.

King, R.K. and E.T. Oswald. 1982. Revegetation of the Carnation Creek watershed. *In* Proceedings of the Carnation Creek Workshop, a ten-year review. G.F. Hartman (editor). Pac. Biol. Sta., Nanaimo, B.C., pp. 110-128.

Kirk, R. 1966. The Olympic rain forest. Univ. Wash. Press, Seattle, Wash.

Kirk, R. and J.F. Franklin. 1992. The Olympic rain forest: an ecological web. Univ. Wash. Press, Seattle, Wash.

Kiss, G.K. 1989. Engelmann × Sitka spruce hybrids in central British Columbia. Can. J. For. Res. 19:1190-1193.

Kjersgard, O. 1976. IUFRO Sitka spruce ten provenance experiment in Denmark. *In* IUFRO. Sitka Spruce *Picea sitchensis* (Bong.) Carr. International Ten Provenance Experiment. Nursery stage results. Dep. Lands, For. Wildl. Serv., Dublin, Ire., pp. 52-58.

Kleinschmit, J. and A. Sauer. 1976. IUFRO Sitka spruce provenance experiment in Germany – results of nursery performance. *In* IUFRO. Sitka Spruce *Picea sitchensis* (Bong.) Carr. International Ten Provenance Experiment. Nursery stage results. Dep. Lands, For. Wildl. Serv., Dublin, Ire., pp.68-89.

Klenner, W. and L. Kremsater. 1993. Forest management and biodiversity. B.C. Min. For. and Univ. B.C., Vancouver, B.C. Unpubl. workshop notes.

Klinka, K. and R.E. Carter. 1990. Relationship between site index and synoptic environmental factors in immature coastal Douglas-fir. For. Sci. 36:815-830.

Klinka, K., R.E. Carter, M. Feller, and Q. Wang. 1989. Relations between site index, salal plant communities, and sites in coastal Douglas-fir ecosystems. Northw. Sci. 63:19-28.

Klinka, K., M.C. Feller, R.N. Green, D.V. Meidinger, J. Pojar, and J. Worrall. 1990. Ecological principles: applications. *In* Regenerating British Columbia's forests. D.P. Lavender, R. Parish, C.M. Johnson, G. Montgomery, A. Vyse, R.A. Willis, and D. Winston (editors). Univ. B.C. Press, Vancouver, B.C., pp. 64-72.

Klinka, K., R.N. Green, P.J. Courtin, and F.C. Nuszdorfer. 1984. Site diagnosis, tree species selection, and slashburning guidelines for the Vancouver Forest Region. B.C. Min. For., Victoria, B.C. Land Manage. Rep. 25.

Klinka, K., F.C. Nuszdorfer, and L. Skoda. 1979. Biogeoclimatic units of central and southern Vancouver Island. B.C. Min. For., Victoria, B.C.

Klinka, K., J. Pojar, and D.V. Meidinger. 1991. Revision of biogeoclimatic units of coastal British Columbia. Northw. Sci. 65:32-47.

Klinka, K., W.C. van der Horst, F.C. Nuszdorfer, and R.G. Harding. 1980. An ecosystem approach to a subunit plan – Koprino River watershed study. B.C. Min. For., Victoria, B.C. Land Manage. Rep. 5.

Kolomiets, N.G. and D.A. Bogdanova. 1979. Ecology of the European spruce beetle *Dendroctonus micans* in western Siberia, USSR. Sov. J. Ecol. 10(2):136-140.

Koot, H.P. 1992. Spruce aphid. For. Can., Pac. For. Res. Cent., Victoria, B.C. For. Pest Leafl. FPL 16.

Kovats, M. 1993. A comparison of British, Swedish, and British Columbian growth and yield prediction for lodgepole pine. For. Chron. 69:450-457.

Kozak, A. 1988. A variable-exponent taper equation. Can. J. For. Res. 18:1363-1368.

Krag, R.K., E.A. Sauder, and G.V. Wellburn. 1986. A forest engineering analysis of landslides in logged areas on the Queen Charlotte Islands, British Columbia. B.C. Min. For., Victoria, B.C. Land Manage. Rep. 43.

Krajick K. 1995. The secret life of backyard trees. Discover 16(11):92-101.

Krajina, V.J. 1969. Ecology of forest trees in British Columbia. Ecol. West. North Am. 2(1):1-147.

Krajina, V.J., K. Klinka, and J. Worrall. 1982. Distribution and ecological characteristics of trees and shrubs of British Columbia. Fac. For., Univ. B.C., Vancouver, B.C.

Kriebel, H.B. 1954. Bark thickness as a factor in resistance to white pine weevil injury. J. For. 52(11):842-845.

Kriek, W. 1976. Early results of the Sitka spruce ten provenance experiment at Wageningen, the Netherlands. *In* IUFRO. Sitka Spruce *Picea sitchensis* (Bong.) Carr. International Ten Provenance Experiment. Nursery stage results. Dep. Lands, For. Wildl. Serv., Dublin, Ire., pp. 173-204.

Krueger, K.W. and R.H. Ruth. 1969. Comparative photosynthesis of red alder, Douglas-fir, Sitka spruce and western hemlock seedlings. Can. J. Bot. 47:519-527.

Krygier, J.T. and R.H. Ruth. 1961. Effect of herbicides on salmonberry and on Sitka spruce and western hemlock seedlings. Weeds 9(3):416-422.

Kumi, J.W. 1987. The use of screening trials to diagnose Sitka spruce nutritional problems. MacMillan Bloedel Ltd., Woodlands Services Div., Nanaimo, B.C., and B.C. Min. For. and Lands, Victoria, B.C. Intern. contract rep. Proj. NV 99017.

Laing, E.V. 1951. Botanical studies of variation in certain conifer species. Report on forest research for the year ended March 1950. For. Comm., Farnham, U.K.

Lambert, A.B. 1832. *Pinus menziesii* Dougl. *In* Description of the Genus *Pinus*. D. Don (editor). London, England. 3rd ed. Vol. 2, pp. 144-145.

Lande, R. 1995. Mutation and conservation. Conserv. Biol. 9:872-891.

Langner, W. 1952. Die Forschungsstaette fuer Forstgenetik und Forstpflanzenzuechtungg in Schmalenbeck. Zeit. fuer Weltforst. 15:15-18.

Larsen, C.S. 1948. Arboretet i Horsholm og fortbotanisk have i Charlottenlund 1948. Foereningens foer dendrologi och parkvard arsbook Lunstgarden 1947-48.

Larsen, N.J. 1945. Sitkagran × Hvidgran [Sitka spruce × white spruce]. Dansk Skovforen. Tidsskr. 30:450-451.

Larson, F.R. 1992. Downed woody material in southeast Alaska forest stands. U.S. Dep. Agric. For. Serv., Res. Pap. PNW-RP-452.

Last, F.T., P.A. Mason, J. Wilson, and J.W. Deacon. 1983. Fine roots and sheathing mycorrhizas: their formation, function and dynamics. Plant Soil 71(1-3):9-21.

Last, F.T., J. Wilson, and P.A. Mason. 1990. Numbers of mycorrhizas and seedling growth of *Picea sitchensis* – what is the relationship? Agric. Ecosys. Environ. 28:293-298.

Laurent, T.H. 1966. Dwarf mistletoe on Sitka spruce – a new host record. Plant Dis. Rep. 50(12):921.

Lavender, D.P., R. Parish, C.M. Johnson, G. Montgomery, A. Vyse, R.A. Willis and D. Winston (editors). 1990. Regenerating British Columbia's forests. Univ. B.C. Press, Vancouver, B.C.

Lawford, R.G., P.B. Alaback, and E. Fuentes (editors). 1996. High-latitude rainforests and associated ecosystems of the west coast of the Americas: climate, hydrology, ecology, and conservation. Springer, New York, N.Y.

Lee, S.J. 1990. Potential gains from genetically improved Sitka spruce. For. Comm., Farnham, U.K. Res. Inf. Note 190.

–. 1992. Likely increases in volume and revenue from planting genetically improved Sitka spruce. *In* Super Sitka for the 90s. D.A. Rook (editor). For. Comm., Farnham, U.K. Bull. 103. pp. 61-74.

–. 1994. Sitka spruce genetic gain trials. For. Comm., Farnham, U.K. Res. Inf. Note 245.

Lehto, T. 1992a. Effect of drought on *Picea sitchensis* seedlings inoculated with mycorrhizal fungi. Scand. J. For. Res. 7:177-182.

–. 1992b. Mycorrhizas and drought resistance of *Picea sitchensis* (Bong.) Carr. I. In conditions of nutrient deficiency. New Phytol. 122:661-668.

–. 1992c. Mycorrhizas and drought resistance of *Picea sitchensis* (Bong.) Carr. II. In conditions of adequate nutrition. New Phytol. 122:669-673.

Lester, D.T. 1993. Utilizing genetic resources of conifers in British Columbia. B.C. Min. For., Victoria, B.C. Res. Rep. 93001-HQ.

Lester, D.T. and A.D. Yanchuk. 1996. A survey of the protected status of conifers in British Columbia: *in situ* gene conservation. B.C. Min. For., Victoria, B.C. Res. Rep. 04.

Lewis, T. 1982. Ecosystems of the Port McNeill Block (Block 4) of Tree Farm Licence 25. Western Forest Products Limited, Port McNeill, B.C. Unpubl. rep.

Lindley, J. 1833. *Abies menziesii. In* A natural system of botany. Penn. Cycl. 1:32.

Lindley, J. and G. Gordon. 1850. *Abies sitchensis*. J. Hort. Soc. London. 1850:212.

Lines, R. 1967. Standardization of methods for provenance research and testing. *In* IUFRO Congress. Munich, Germany. Vol. 3, Sec. 22:672-714.

–. 1968. A tour of Icelandic forests, 20-30 June 1967. For. Comm., Farnham, U.K. Res. Develop. Pap. 66.

–. 1987. Choice of seed origins for the main forest species in Britain. For. Comm., Farnham, U.K. Bull. 66.

Lines, R. and A.F. Mitchell. 1966. Differences in phenology of Sitka spruce provenances. *In* Report on forest research. For. Comm., London, U.K., pp. 173-184.

Little, E.L., Jr. 1944. Note on the nomenclature in Pinaceae. Am. J. Bot. 31:587-596.

–. 1953. A natural hybrid spruce in Alaska. J. For. 51:745-747.

Longman, K.A. 1978. Control of flowering for forest tree improvement and seed production. Scient. Hort. 30:1-10.

–. 1985. Variability in flower initiation in forest trees. *In* Attributes of trees as crop plants. M.G.R. Cannell and J.E. Jackson (editors). Inst. Terr. Ecol., Monks Wood, Hunts., U.K., pp. 398-408.

Lotan, J.E., M.E. Alexander, S.F. Arno, R.E. French, O.G. Langdon, R.M. Loomis, R.A. Norum, R.C. Rothermel, W.C. Schmidt, and J.W. Van Wagtendonk. 1981. Effects of fire on flora. U.S. Dep. Agric. For. Serv., Gen. Tech. Rep. WO-16. pp. 2-11.

Loucks, D.M., H.C. Black, M.L. Roush, and S.R. Radosevich. 1990. Assessment and management of animal damage in Pacific Northwest forests: an annotated bibliography. U.S. Dep. Agric. For. Serv., Gen. Tech. Rep. PNW-GTR-262.

Loudon, J.C. 1838. Arboretum et fructicetum Britannicum. London, U.K.

Low, A.J. 1975. Production and use of tubed seedlings. For. Comm. Bull 53:1-46.

–. 1987. Sitka spruce silviculture in Scottish forests. *In* Symposium, Sitka spruce. D.M. Henderson and R. Faulkner (editors). Proc. Roy. Soc. Edinburgh 93B:93-106.

Lynch, T.J. 1980. Thinning and spacing research in Sitka spruce and lodgepole pine. Irish For. 37(2, Suppl.):45-67.

–. 1988. A thinning experiment in Avoca Forest: results over 23 years. Irish For. 45(1):55-66.

McClellan, M.H., B.T. Bormann, and K. Cromack, Jr. 1990. Cellulose decomposition in southeast Alaskan forests: effects of pit and mound microrelief and burial depth. Can. J. For. Res. 20:1242-1246.

MacDonald, J. 1931. Sitka spruce in Great Britain, its growth, production, and thinning. Forestry 5:100-107.
MacDonald, J.A.B. 1979. Norway or Sitka spruce? Forestry 40:129-138.
MacHutchon, A.G., S. Himmer, and C.A. Bryden. 1993. Khutzeymateen Valley grizzly bear study: final report. B.C. Min. For. and B.C. Min. Environ., Lands and Parks, Victoria, B.C. Wildl. Rep. R-25; Wildl. Hab. Res. Rep. 31.
McIntosh, R. 1981. Fertilizer treatment of Sitka spruce in the establishment phase in upland Britain. Scott. For. 35:3-13.
–. 1983. Nitrogen deficiency in establishment phase Sitka spruce in upland Britain. Scott. For. 37:185-193.
–. 1995. The history and multi-purpose management of Kielder Forest. For. Ecol. Manage. 79:1-12.
McKay, H.M. 1994a. Frost hardiness and cold storage tolerance of the root system of *Picea sitchensis, Pseudotsuga menziesii, Larix kaempferi* and *Pinus sylvestris* bare root seedlings. Scan. J. For. Res. 9:203-213.
–. 1994b. The quality of Sitka spruce at the time of planting. For. Comm., Farnham, U.K. Res. Inf. Note 243.
McKay, H.M. and R. Howes. 1996. Recommended lifting dates for direct planting and cold storage of Sitka spruce in Britain. For. Comm., Farnham, U.K. Res. Inf. Note 281.
McKee, A., G. LaRoi, and J.F. Franklin. 1982. Structure, composition, and reproductive behaviour of terrace forests, South Fork Hoh River, Olympic National Park. *In* Ecological research in national parks of the Pacific Northwest. Proc. of 2nd conf. on scientific research in national parks, Nov. 1979, San Francisco, Calif. E.E. Starkey, J.F. Franklin, and J.W. Mathews (technical coordinators). Oreg. State Univ. For. Res. Lab., Corvallis, Oreg., pp. 22-29.
McLean, J.A. 1989. Effect of red alder overstory on the occurrence of *Pissodes strobi* (Peck) during the establishment of a Sitka spruce plot. *In* Proc. IUFRO Working Group Meeting on Insects Affecting Reforestation S2.07-03, 3-9 July 1988, Vancouver, B.C. R.I. Alfaro and S. Glover (editors). For. Can., Pac. For. Cent., Victoria, B.C., pp. 167-176.
MacKinnon, A. and M. Eng. 1995. Old forests inventory for British Columbia. Cordillera 2(1):20-33.
MacKinnon, A., D.V. Meidinger, and K. Klinka. 1992. Use of the biogeoclimatic ecosystem classification system in British Columbia. For. Chron. 68:100-120.
MacLean, D.A. 1996. Silvicultural approaches to integrated insect management: the Green Plan Silvicultural Insect Management network. For. Chron. 72:367-369.
McLennan, D.S. 1993. Silvicultural options on alluvial floodplains in coastal British Columbia. *In* Riparian habitat management and research. K.H. Morgan and M.A. Lashmar (editors). Proc. of a workshop 4-5 May 1993, Kamloops, B.C. Environ. Can., Fraser River Action Plan, North Vancouver, B.C. Spec. Publ. pp. 119-133.
MacMillan Bloedel Ltd. 1989. Management Plan Carmanah Valley. MacMillan Bloedel Ltd., Franklin River Division, Port Alberni, B.C. Folio I.
McMullen, L.H. 1976a. Effect of temperature on oviposition and brood development of *Pissodes strobi* (Coleoptera: Curculionidae). Can. Entomol. 108:1167-1172.
–. 1976b. Spruce weevil damage: ecological basis and hazard rating for Vancouver Island. Can. For. Serv., Pac. For. Res. Cent., Victoria, B.C. Inf. Rep. BC-X-141.
McMullen, L.H., A.J. Thomson, and R.V. Quenet. 1987. Sitka spruce weevil (*Pissodes strobi*) population dynamics and control: a simulation model based on field relationships. Can. For. Serv., Pac. For. Cent., Victoria, B.C. Inf. Rep. BC-X-288.
McNay, R.S. and R. Davies. 1985. Interactions between black-tailed deer and intensive forest management: problem analysis. Min. Environ. and Min. For., Victoria, B.C. IWIFR-22.
MacSiurtain, M.P. 1981. Distribution, management, variability and economics of Sitka spruce (*Picea sitchensis* (Bong.) Carr.) in coastal British Columbia. MSc thesis. Univ. B.C., Vancouver, B.C.
Magnesen, S. 1976. The international Sitka spruce ten provenance experiment in West Norway – nursery result. *In* IUFRO. Sitka Spruce *Picea sitchensis* (Bong.) Carr. International Ten Provenance Experiment. Nursery stage results. Dep. Lands, For. Wildl. Serv., Dublin, Ire., pp. 216-236.

Malcolm, D.C. 1987a. Nitrogen supply for spruce on infertile sites (an ecological problem). Univ. B.C., Vancouver, B.C. Leslie L. Schaffer Lectureship in Forest Science, 5 November 1987.

–. 1987b. Some ecologic aspects of Sitka spruce. *In* Symposium on Sitka spruce. D.M. Henderson and R. Faulkner (editors). Proc. Roy. Soc. Edinburgh 93B:85-92.

Malcolm, D.C. and E.A. Caldwell. 1971. Environmental effects on shoot growth in conifers. For. Comm., London, U.K. Rep. For. Res. 141.

Maser, C. 1994. Sustainable forestry: philosophy, science, and economics. St. Lucie Press, Delray Beach, Fla.

Maser, C., R.F. Tarrant, J.M. Trappe, and J.F. Franklin. 1988. From the forest to the sea; a story of fallen trees. U.S. Dep. Agric. For. Serv., Gen. Tech. Rep. PNW-GTR-229.

Mason, P.A., J. Wilson, F.T. Last, and C. Walker. 1983. The concept of succession in relation to the spread of sheathing mycorrhizal fungi on inoculated tree seedlings growing in unsterile soils. Plant Soil 71:247-256.

Mason, W.L. 1984. Vegetative propagation of conifers using stem cuttings. I. Sitka spruce. For. Comm., Farnham, U.K. Res. Inf. Note 90.

Mason, W.L. and J.G.S. Gill. 1986. Vegetative propagation of conifers as a means of intensifying wood production in Britain. Forestry 59(2):155-171.

Mason, W.L. and W.C.G. Harper. 1987. Forest use of improved Sitka spruce cuttings. For. Comm., Farnham, U.K. Res. Inf. Note 119.

Mason, W.L. and C.P. Quine. 1995. Silvicultural possibilities for increasing structural diversity in British spruce forests: the case of Kielder Forest. For. Ecol. Manage. 79: 13-28.

Mason, W.L. and A.L. Sharpe. 1992. The establishment and silviculture of Sitka spruce cuttings. *In* Super Sitka for the 90s. D.A. Rook (editor). For. Comm., Farnham, U.K. Bull. 103. pp. 42-53.

Massie, M.R.C. 1992. Incremental silviculture pruning financial analysis. Prep. for B.C. Min. For., Silv. Br., Victoria, B.C. Unpubl. rep.

Mathiasen, R.L. 1994. Natural infection of new hosts by hemlock dwarf mistletoe. U.S. Dep. Agric. For. Serv., Res. Note RM-RN-530.

Meehan, W.R. 1974. The forest ecosystem of southeast Alaska. 4. Wildlife habitats. U.S. Dep. Agric. For. Serv., Gen. Tech. Rep. PNW-16.

Meehan, W.R., T.R. Merrell, Jr., and T.A. Hanley (editors). 1982. Fish and wildlife relationships in old-growth forests. Proc. Symp. 12-15 April 1982, Juneau, Alaska. Am. Inst. Fish. Res. Biol., Morehead City, N.C.

Meehan, W.R., F.J. Swanson, and J.R. Sedell. 1977. Influences of riparian vegetation on aquatic ecosystems with particular reference to salmonid fishes and their food supply. *In* Importance, preservation and management of riparian habitat, a symposium. R.R. Johnson and D.A. Jones (coordinators). U.S. Dep. Agric. For. Serv., Gen. Tech. Rep. RM-43. pp. 137-145.

Meidinger, D.V. 1987. Recommended vernacular names for common plants of British Columbia. B.C. Min. For. and Lands, Victoria, B.C. Res. Rep. RR87002-HQ.

Meidinger, D.V. and J. Pojar (editors). 1991. Ecosystems of British Columbia. B.C. Min. For., Victoria, B.C. Spec. Rep. Series 6.

Meir, R. 1990. Tree improvement and progeny testing of Sitka spruce in western Washington. *In* Joint Meeting of Western Forest Genetics Association and IUFRO Working Parties S2.02-05, 06, 12, and 14. 20-24 Aug. 1990, Olympia, Wash. pp. 4/103-108.

Mergen, F. and B.A. Thielges. 1967. Intraspecific variation in nuclear volume in four conifers. Evolution 21(4):720-724.

Meslow, E.R.C., C. Maser, and J. Verner. 1981. Old-growth forest as wildlife habitat. Trans. N. Am. Wildl. Conf. 46:329-344.

Messier, C. 1993. Factors limiting early growth of western redcedar, western hemlock and Sitka spruce seedlings on ericaceous-dominated clearcut sites in coastal British Columbia. For. Ecol. Manage. 60:181-206.

Messier, C. and J.P. Kimmins. 1991. Nutritional stress in *Picea sitchensis* plantations in coastal British Columbia: the effects of *Gaultheria shallon* and declining site fertility. Water, Air, Soil Pollut. 54:257-267.

Meyer, W.H. 1937. Yield of even-aged stands of Sitka spruce and western hemlock. U.S. Dep. Agric., Washington, D.C. Tech. Bull. 544.
Millard, P. and M.F. Proe. 1992. Storage and internal cycling of nitrogen in relation to seasonal growth of Sitka spruce. Tree Physiol. 10:33-43.
–. 1993. Nitrogen uptake, partitioning and internal cycling in *Picea sitchensis* (Bong.) Carr. as influenced by nitrogen supply. New Phytol. 125:113-119.
Miller, H.G., C. Alexander, J. Cooper, J. Keenleyside, H. McKay, J.D. Miller, and B.L. Williams. 1986. Maintenance and enhancement of forest productivity through manipulation of the nitrogen cycle. Macaulay Inst. for Soil Research, Aberdeen, Scotland.
Miller, H.G. and J.D. Miller. 1987. Nutritional requirements of Sitka spruce. *In* Symposium, Sitka spruce. D.M. Henderson and R. Faulkner (editors). Proc. Roy. Soc. Edinburgh 93B:75-83.
Miller, K.G. 1985. Windthrow hazard classification. For. Comm., London, U.K. Leafl. 85.
Miller, P.R. and J.R. McBride. 1975. Effects of air pollutants on forests. *In* Responses of plants to air pollution. T.T. Kozlowski and J.B. Mudd (editors). Academic Press, New York, N.Y., pp. 195-235.
Miller, R.E., W. Scott, and J.W. Hazard. 1996. Soil compaction and conifer growth after tractor yarding at three coastal Washington locations. Can. J. For. Res. 26:225-236.
Mills, L.S., R.J. Fredrickson, and B.B. Moorhead. 1993. Characteristics of old-growth forests associated with northern spotted owls in Olympic National Park. J. Wildl. Manage. 57:315-321.
Minore, D. 1979. Comparative autecological characteristics of northwestern tree species: a literature review. U.S. Dep. Agric. For. Serv., Gen. Tech. Rep. PNW-87.
Minore, D. and C.E. Smith. 1971. Occurrence and growth of four northwestern tree species over shallow water tables. U.S. Dep. Agric. For. Serv., Res. Note PNW-160.
Minore, D. and H.G. Weatherly. 1994. Riparian trees, shrubs, and forest regeneration in the coastal mountains of Oregon. New For. 8:249-263.
Mitchell, K.J. and I.R. Cameron. 1985. Managed stand yield tables for coastal Douglas-fir: initial density and precommercial thinning. B.C. Min. For., Victoria, B.C. Land Manage. Rep. 31.
Mitchell, K.J., S.E. Grout, and R.N. Macdonald. 1995. WinTIPSY: user's guide for producing managed stand yield tables with WinTIPSY Version 1.3 under Microsoft Windows. B.C. Min. For., Res. Br., Victoria, B.C.
Mitchell, K.J. and K.R. Polsson. 1988. Site index curves and tables for British Columbia: coastal species. Coastal Forest Productivity Council and B.C. Min. For., Victoria, B.C. FRDA Rep. 37.
Mitchell, R.G., N.E. Johnson, and K.H. Wright. 1974. Susceptibility of 10 spruce species and hybrids to the white pine weevil (Sitka spruce weevil) in the Pacific Northwest. U.S. Dep. Agric. For. Serv., Res. Note PNW-225.
Mitchell, R.G., K.H. Wright, and N.E. Johnson. 1990. Damage by Sitka spruce weevil (*Pissodes strobi*) and growth patterns for 10 spruce species and hybrids over 26 years in the Pacific Northwest. U.S. Dep. Agric. For. Serv., Res. Pap. PNW-RP-434.
Molina, R. and J.M. Trappe. 1982. Patterns of ectomycorrhizal host specificity and potential among Pacific Northwest conifers and fungi. For. Sci. 28:423-458.
Molnar, A.C., J.W.E. Harris, D.A. Ross, and J.H. Ginns. 1968. British Columbia region. Annual report of the Insect and Disease Survey, 1967. Can. Dep. For. Rural Devel., Ottawa, Ont.
Moore, K. 1991a. Coastal watersheds: an inventory of watersheds in the coastal temperate forests of British Columbia. Earthlife Canada Foundation and Ecotrust/Conservation International, Vancouver, B.C.
–. 1991b. Partial cutting and helicopter yarding on environmentally sensitive floodplains in old-growth hemlock/spruce forests. For. Can. and B.C. Min. For., Victoria, B.C. FRDA Rep. 166.
Moore, P. 1993. Can B.C. learn from European forestry practices? For. Plan. Can. 9(4):39-48.
Morgan, K.H. and M.A. Lashmar (editors). 1993. Riparian habitat management and research. Proc. of a workshop 4-5 May 1993, Kamloops, B.C. Environ. Can., Fraser River Action Plan, North Vancouver, B.C. Spec. Publ.

Morgenstern, E.K. 1996. Geographic variation in forest trees. Genetic basis and application of knowledge in silviculture. UBC Press, Vancouver, B.C.

Morrison, D.J. 1981. Armillaria root disease: a guide to disease diagnosis, development and management in British Columbia. Can. For. Serv., Pac. For. Cent., Victoria, B.C. Inf. Rep. BC-X-203.

–. 1989. Factors affecting infection of precommercial thinning stumps by *Heterobasidion annosum* in coastal British Columbia. *In* Proc. symp. on research and management of Annosus root disease (*Heterobasidion annosum*) in western North America. U.S. Dep. Agric. For. Serv., Gen. Tech. Rep. PSW-116. pp. 95-100.

Morrison, D.J., D. Chu, and A.L.S. Johnson. 1985. Species of *Armillaria* in British Columbia. Can. J. Plant Pathol. 7:242-246.

Morrison, D.J. and A.L.S. Johnson. 1970. Seasonal variation of stump infection by *Fomes annosus* in coastal British Columbia. For. Chron. 46:200-202.

–. 1978. Stump colonization and spread of *Fomes annosus* 5 years after thinning. Can. J. For. Res. 8:177-180.

Morrison, D.J., M.D. Larock, and A.J. Waters. 1986. Stump infection by *Fomes annosus* in spaced stands in the Prince Rupert Forest Region of British Columbia. Can. For. Serv., Pac. For. Cent., Victoria, B.C. Inf. Rep. BC-X-285.

Morrison, D.J., G.A. Macaskill, S.C. Gregory, and D.B. Redfern. 1994. Number of *Heterobasidion annosum* vegetative compatibility groups in roots of basidiospore-infected stumps. Plant Pathol. 43:907-912.

Morrison, D.J. and D.B. Redfern. 1994. Long-term development of *Heterobasidion annosum* in basidiospore-infected Sitka spruce stumps. Plant Pathol. 43:897-906.

Murphy, M.L., J.F. Thedinga, K.V. Koski, and G.B. Grette. 1982. A stream ecosystem in an old-growth forest in southeast Alaska. Part V: Seasonal changes in habitat utilization by juvenile salmonids. *In* Fish and wildlife relationships in old-growth forests, Proc. Symp. 12-15 April 1982, Juneau, Alaska. W.R. Meehan, T.R. Merrell, Jr., and T.A. Hanley (editors). Am. Inst. Fish. Res. Biol., Morehead City, N.C., pp. 89-98.

Murphy, M.L. and J.D. Hall. 1981. Varied effects of clear-cut logging on predators and their habitat in small streams of the Cascade Mountains, Oregon. Can. J. Fish. Aquat. Sci. 38:137-145.

Myren, R.T. and R.J. Ellis. 1982. Evapotranspiration in forest succession and long-term effects upon fishery resources: a consideration for management of old-growth forests. *In* Fish and wildlife relationships in old-growth forests, Proc. Symp. 12-15 April 1982, Juneau, Alaska. W.R. Meehan, T.R. Merrell, Jr., and T.A. Hanley (editors). Am. Inst. Fish. Res. Biol., Morehead City, N.C., pp. 183-186.

Nanson, A. 1976. Sitka spruce provenance experiments – first nursery results from Belgium. *In* IUFRO. Sitka Spruce *Picea sitchensis* (Bong.) Carr. International Ten Provenance Experiment. Nursery stage results. Dep. Lands, For. Wildl. Serv., Dublin, Ire., pp. 34-50.

Nelson, D.G. 1990. Chemical control of Sitka spruce natural regeneration. For. Comm., Farnham, U.K. Res. Inf. Note 187.

–. 1991. Management of Sitka spruce natural regeneration. For. Comm., Farnham, U.K. Res. Inf. Note 204.

Nelson, E.E. and R.N. Sturrock. 1993. Susceptibility of western conifers to laminated root rot (*Phellinus weirii*) in Oregon and British Columbia field tests. West. J. Appl. For. 8:67-70.

Nelson, J.G. and L.D. Cordes. 1972. Pacific Rim: an ecological approach to a new Canadian National Park. Univ. Calgary, Calgary, Alta. Studies in Land Use History and Landscape Change, National Park Series 4.

Nevill, R., N. Humphreys, and A. Van Sickle. 1995. Three-year overview of forest health in young managed stands in British Columbia 1992-1994. B.C. Min. For. and Can. For. Serv., Victoria, B.C. FRDA Rep. 236.

Nevill, R. and D. Winston. 1994. Forest health research needs survey. For. Can. and B.C. Min. For., Victoria, B.C. FRDA Rep. 212; BCFRAC Rep. 3.

Newton, M. and P.G. Comeau. 1990. Control of competing vegetation. *In* Regenerating British Columbia's forests. D.P. Lavender, R. Parish, C.M. Johnson, G. Montgomery, A. Vyse, R.A. Willis, and D. Winston (editors). Univ. B.C. Press, Vancouver, B.C., pp. 256-265.

Nicoll, B.C., D.B. Redfern, and H.M. McKay. 1996. Autumn frost damage: clonal variation in Sitka spruce. For. Ecol. Manage. 80:107-112.

Nigh, G.D. 1995. Site index conversion equations for mixed Sitka spruce/western hemlock stands. B.C. Min. For., Victoria, B.C. Exten. Note 2.

–. 1996a. A Sitka spruce height-age model. B.C. Min. For., Res. Br., Victoria, B.C. Unpubl. rep.

–. 1996b. A variable growth intercept model for Sitka spruce. B.C. Min. For., Victoria, B.C. Exten. Note 3.

Norse, E.A. 1990. Ancient forests of the Pacific Northwest. Island Press, Washington, D.C.

Norse, E. A., K.L. Rosenbaum, D.S. Wilcove, B.A. Wilcox, W.H. Romme, D.W. Johnston, and M.L. Stout. 1986. Conserving biological diversity in our national forests. Prep. by Ecological Society of America for the Wilderness Society, Washington, D.C.

Northway, S. 1989. XENO technical report. MacMillan Bloedel Ltd., Woodland Services, Nanaimo, B.C. Unpubl. rep.

Noss, R.F. 1990. Indicators for monitoring biodiversity: a hierarchical approach. Conserv. Biol. 4:355-364.

Nussbaum, A.F. 1996. Site index curves and tables for British Columbia: coastal species. 2nd ed. B.C. Min. For., Victoria, B.C. Land Manage. Handb. Fld. Guide Insert 3.

Nuszdorfer, F.C., K.L. Kassay, and A.M. Scagel. 1985. Biogeoclimatic units of the Vancouver forest region. B.C. Min. For., Victoria, B.C. Map 1:500,000.

Nyberg, J.W. 1990. Interactions of timber management with deer and elk. *In* Deer and elk habitats in coastal forests of southern British Columbia. J.B. Nyberg and D.W. Janz (technical editors). B.C. Min. For. and B.C. Min. Environ., Victoria, B.C. Spec. Rep. Ser. 5. pp. 99-131.

Nyberg, J.B., A.S. Harestad, and F.L. Bunnell. 1987. 'Old growth' by design: managing young forests for old growth wildlife. Trans. N. Am. Wildl. Nat. Resour. Conf. 52:70-81.

Nyberg, J.B. and D.W. Janz (technical editors). 1990. Deer and elk habitats in coastal forests of southern British Columbia. B.C. Min. For. and B.C. Min. Environ., Victoria, B.C. Spec. Rep. Ser. 5.

Nyberg, J.B., R.S. McNay, M.D. Kirchhoff, R.D. Forbes, F.L. Bunnell, and E.L. Richardson. 1989. Integrated management of timber and deer: coastal forests of British Columbia and Alaska. U.S. Dep. Agric. For. Serv., Gen. Tech. Rep. PNW-GTR-226.

Nyberg, J.B., L.D. Peterson, L.A. Stordeur, and R.S. McNay. 1990. Deer use of old-growth and immature forests following snowfalls on southern Vancouver Island. B.C. Min. Environ. and B.C. Min. For., Victoria, B.C. IWIFR-36.

O'Carroll, N. 1978. The nursing of Sitka spruce. I. Japanese larch. Irish For. 35:60-65.

O'Driscoll, J. 1976a. International ten provenance experiment – working plan. *In* IUFRO. Sitka Spruce *Picea sitchensis* (Bong.) Carr. International Ten Provenance Experiment. Nursery stage results. Dep. Lands, For. Wildl. Serv., Dublin, Ire., pp. 22-32.

–. 1976b. Sitka spruce international ten provenance experiment – nursery stage results in Ireland. *In* IUFRO. Sitka Spruce *Picea sitchensis* (Bong.) Carr. International Ten Provenance Experiment. Nursery stage results. Dep. Lands, For. Wildl. Serv., Dublin, Ire., pp. 140-163.

O'Hare, V.L. 1989. Forest pruning bibliography. Univ. Wash., Inst. For. Res., Seattle, Wash. Contr. 67.

Ogilvie, R.T. 1994. Rare and endemic vascular plants of Gwaii Haanas (South Moresby) Park, Queen Charlotte Islands, British Columbia. Can. For. Serv. and B.C. Min. For., Victoria, B.C. FRDA Rep. 214.

Oksbjerg, E. 1953. Om *Picea omorika*. Dansk. Skovfor. Tidsskr. 2:179-192.

Oliver, C.D. and T.M. Hinckley. 1987. Species, stand structures, and silvicultural manipulation patterns for the streamside zone. *In* Streamside management: forestry and fishery interactions. E.O. Salo and T.W. Cundy (editors). Univ. Wash., Coll. For. Resour., Seattle, Wash. Contrib. 57.

Oliver, C.D and B.C. Larson. 1990. Forest stand dynamics. McGraw-Hill, New York, N.Y.

Olson, R.K., D. Binkley, and M. Böhm (editors). 1992. The responses of western forests to air pollution. Springer-Verlag, New York, N.Y.

Omule, S.A.Y. 1987. Comparative height growth to age 28 for seven species in the CWHd subzone. For. Can. and B.C. Min. For. and Lands, Victoria, B.C. FRDA Rep. 5.

–. 1988. Early growth of four species planted at three spacings on Vancouver Island. For. Can. and B.C. Min. For. and Lands, Victoria, B.C. FRDA Rep. 9.

Omule, S.A.Y. and G.J. Krumlik. 1987. Juvenile height growth of four species on four sites in the CWHb1 variant. For. Can. and B.C. Min. For. and Lands, Victoria, B.C. FRDA Rep. 7.

Oswald, E.T. 1982. Preharvest vegetation and soils of Carnation Creek watershed. *In* Proceedings of the Carnation Creek Workshop, a ten-year review. G.F. Hartman (editor). Pac. Biol. Sta., Nanaimo, B.C., pp. 17-35.

Otchere-Boateng, J. and L.J. Herring. 1990. Site preparation: chemical. *In* Regenerating British Columbia's forests. D.P. Lavender, R. Parish, C.M. Johnson, G. Montgomery, A. Vyse, R.A. Willis, and D. Winston (editors). Univ. B.C. Press, Vancouver, B.C., pp. 165-178.

Palmer, M. and T. Nichols. 1981. How to identify and control cutworm damage on conifer seedlings. U.S. Dep. Agric. For. Serv., Pest Leafl. 767-160.

Parks, C.G. and D.C. Shaw. 1996. Death and decay: a vital part of living canopies. Northw. Sci. 70(Spec. Issue):46-51.

Parminter, J. 1983. Fire history and fire ecology in the Prince Rupert Forest Region. *In* Prescribed fire – forest soils symp. proc., 2-3 March 1982, Smithers, B.C. R.L. Trowbridge and A. Macadam (editors). B.C. Min. For., Victoria, B.C. Land Manage. Rep. 16. pp. 1-35.

–. 1992. Historic patterns of wildfire incidence by biogeoclimatic zone, B.C. B.C. Min. For., Victoria, B.C. Unpubl. rep. and map (1994).

Parrish, R. and D. Lester. 1993. Genes, trees and forests. B.C. Min. For., Victoria, B.C.

Pearce, M.L. 1976. International ten provenance experiment. Report of phase I and phase II of nursery experiment Great Britain (south). *In* IUFRO. Sitka Spruce *Picea sitchensis* (Bong.) Carr. International Ten Provenance Experiment. Nursery stage results. Dep. Lands, For. Wildl. Serv., Dublin, Ire., pp. 106-122.

Pearson, A.F. 1992. Relationship between site index of Sitka spruce and measures of ecological site quality in the eastern Queen Charlotte Islands. MSc thesis. Univ. B.C., Vancouver, B.C.

Pederick, L.A. 1979. Provenance variation in Sitka spruce *Picea sitchensis*. Forests Commission, Victoria, Melbourne, Australia. Tech. Pap. 27:33-40.

Perry, D.A. 1985. The competition process in forest stands. *In* Attributes of trees as crop plants. M.G.R. Cannell and J.E. Jackson (editors). Inst. Terr. Ecol., Nat. Environ. Res. Coun., Abbots Ripton, U.K., pp. 481-506.

–. 1994. Ecosystem silviculture: ecological principles, implications for communities. Can. Silv. Mag. 2(1):6-12.

Perry, D.A., R. Meurisse, B. Thomas, R. Miller, J. Boyle, J. Means, C.R. Perry, and R.F. Powers (editors). 1989. Maintaining the long-term productivity of Pacific Northwest forest ecosystems. Timber Press, Portland, Oreg.

Peterson, J., D. Schmoldt, D. Peterson, J. Eilers, R. Fisher, and R. Bachman. 1992. Guidelines for evaluating air pollution impacts on Class 1 wilderness areas in the Pacific Northwest. U.S. Dep. Agric. For. Serv., Gen. Tech. Rep. PNW-GTR-299.

Petty, S.J., P.J. Garson, and R. McIntosh (editors). 1995. Preface. Kielder – the ecology of a man-made spruce forest. For. Ecol. Manage. 79(Spec. Issue):vii-ix.

Phelps, V.H. 1973. Sitka spruce: a literature review with special reference to British Columbia. Environ. Can., Can. For. Serv., Victoria, B.C. Rep. BC-X-83.

Phillips, E.W.J. 1963. Timber improvement by tree selection and breeding. *In* World Consultation Forest Genetics Tree Improvement Proc. Stockholm, Sweden. Sec. 7, pp. 1-5.

Pintaric, K. and F. Mekic. 1993. The growth in height and diameter of Sitka spruce (*Picea sitchensis*) of different provenances. *In* Proc. of the IUFRO international Sitka spruce provenance experiment (Sitka spruce Working Group S2.02.12). IUFRO 1984, Edinburgh, Scotland. C.C. Ying and L.A. McKnight (editors). B.C. Min. For., Victoria, B.C., and Irish For. Board, County Wicklow, Ire., pp. 219-228.

Pirags, D. 1976. International ten provenance experiment of Sitka spruce in Latvian SSR. *In* IUFRO. Sitka Spruce *Picea sitchensis* (Bong.) Carr. International Ten Provenance Experiment. Nursery stage results. Dep. Lands, For. Wildl. Serv., Dublin, Ire., pp. 164-172.

Pojar, J. 1983. Forest ecology. *In* Forestry handbook for British Columbia. 4th ed. S.B. Watts (editor). For. Undergrad. Soc., Fac. For., Univ. B.C., Vancouver, B.C., pp. 221-318.

–. 1993. Terrestrial diversity of British Columbia. *In* Our living legacy: proceedings of a symposium on biological diversity. M.A. Fenger, E.H. Miller, J.F. Johnson, and E.J.R. Williams (editors). Roy. B.C. Museum, Victoria, B.C., pp. 177-190.

Pojar, J. and A. Banner. 1982. Old-growth forests and introduced black-tailed deer on the Queen Charlotte Islands, British Columbia. *In* Fish and wildlife relationships in old-growth forests, Proc. Symp. 12-15 April 1982, Juneau, Alaska. W.R. Meehan, T.R. Merrell, Jr., and T.A. Hanley (editors). Am. Inst. Fish. Res. Biol., Morehead City, N.C., pp. 247-257.

Pojar, J., K. Klinka, and D.A. Demarchi. 1991. Coastal Western Hemlock Zone. *In* Ecosystems of British Columbia. D. Meidinger and J. Pojar (editors). B.C. Min. For., Victoria, B.C. Spec. Rep. Ser. 6. pp. 95-111.

Pojar, J., K. Klinka, and D.V. Meidinger. 1987. Biogeoclimatic ecosystem classification in British Columbia. For. Ecol. Manage. 22:119-154.

Pojar, J., T. Lewis, H. Roemer, and D.J. Wilford. 1980. Relationships between introduced black-tailed deer and the plant life of the Queen Charlotte Islands, British Columbia. B.C. Min. For., Victoria, B.C.

Pojar, J. and A. MacKinnon (compilers and editors). 1994. Plants of coastal British Columbia. Lone Pine Publishing, Vancouver, B.C.

Pojar, J. and D. Meidinger. 1991. British Columbia: the environmental setting. *In* Ecosystems of British Columbia. D. Meidinger and J. Pojar (editors). B.C. Min. For., Victoria, B.C. Spec. Rep. Ser. 6. pp. 39-67.

Pojar, J., F. Nuszdorfer, D. Demarchi, M. Fenger, and B. Fuhr. 1988. Biogeoclimatic and ecoregion units of the Prince Rupert Forest Region. B.C. Min. For., Res. Br., Victoria, B.C. 2 maps (1:500,000).

Pollack, J.C., F. van Thienen, and T. Nash. 1985. A plantation performance assessment guide for the Prince Rupert Forest Region. B.C. Min. For., Victoria, B.C. Land Manage. Rep. 35.

Pollack, J.C., F. van Thienen, and P. LePage. 1990. The influence of initial espacement on the growth of a 27-year-old Sitka spruce plantation. B.C. Min. For., Victoria, B.C. Res. Note 104.

Pollard, D.F.W. and F.T. Portlock. 1990. Certification of tree seed for export from Canada, 1980-1989: lodgepole pine, Sitka spruce, Douglas-fir, alpine fir and grand fir. *In* Joint Meeting of Western Forest Genetics Association and IUFRO Working Parties S2.02-05, 06, 12, and 14. 20-24 Aug. 1990, Olympia, Wash., pp. 2/247-254.

Portlock, F.T. 1996a. Forest tree seed certification in Canada under the OECD scheme and ISTA rules: summary report for 1991-1995. Can. For. Serv., Pac. For. Cent., Victoria, B.C. Inf. Rep BC-X-361.

–. 1996b. List of approved basic material for Canada under the OECD forest tree seed certification scheme, 1970-1995. Can. For. Serv., Pac. For. Cent., Victoria, B.C. Inf. Rep BC-X-359.

Poulin, V.A. 1982. An approach to solving conflicts related to fish and old-growth logging on the Queen Charlotte Islands. *In* Fish and wildlife relationships in old-growth forests, Proc. Symp. 12-15 April 1982, Juneau, Alaska. W.R. Meehan, T.R. Merrell, Jr., and T.A. Hanley (editors). Am. Inst. Fish. Res. Biol., Morehead City, N.C., pp. 167-174.

Prasad, R. 1989. Crop tolerance of three west coast conifer species to glyphosate. *In* Proceedings of the Carnation Creek Herbicide Workshop. P.E. Reynolds (editor). For. Can. and B.C. Min. For., Victoria, B.C. FRDA Rep. 63. pp. 189-196.

Prescott, C.E. and M.A. McDonald. 1994. Effects of carbon and lime additions on mineralization of C and N in humus from cutovers of western red cedar–western hemlock forests on northern Vancouver Island. Can. J. For. Res. 24:2432-2438.

Prescott, C.E., M.A. McDonald, and G.F. Weetman. 1993. Availability of N and P in the forest floors of adjacent stands of western red cedar–western hemlock and western hemlock–amabilis fir on northern Vancouver Island. Can. J. For. Res. 23:605-610.

Prescott, C.E. and G.F. Weetman. 1994. Salal cedar hemlock integrated research program: a synthesis. Fac. For., Univ. B.C., Vancouver, B.C.

Prescott, C.E., G.F. Weetman, L.E. DeMontigny, C.M. Preston, and R.J. Keenan. 1995. Carbon chemistry and nutrient supply in cedar-hemlock and hemlock–amabilis fir forest floors. *In* Carbon forms and functions in forest soils. W.W. McFee and J.M. Kelly (editors). Soil Sci. Soc. Am., Madison, Wis., pp. 377-396.

Price, D.T., T.A. Black, and F.M. Kelliher. 1986. Effects of salal understory removal on photosynthetic rate and stomatal conductance of young Douglas-fir trees. Can. J. For. Res. 16:90-97.

Proe, M.F. and P. Millard. 1994. Relationships between nutrient supply, nitrogen partitioning and growth in young Sitka spruce (*Picea sitchensis* (Bong.) Carr.). Tree Physiol. 14:75-88.

–. 1995a. Effect of P supply upon seasonal growth and internal cycling of P in Sitka spruce (*Picea sitchensis* (Bong.) Carr.). Plant Soil 168-169:313-317.

–. 1995b. Effect of N supply upon the seasonal partitioning of N and P uptake in young Sitka spruce (*Picea sitchensis*). Can. J. For. Res. 25:1704-1709.

Puettmann, K.J. 1994. Growth and yield of red alder. *In* The biology and management of red alder. D.E. Hibbs, D.S. DeBell, and R.F. Tarrant (editors). Oreg. State Univ. Press, Corvallis, Oreg., pp. 229-242.

Puget Sound Research Center Advisory Committee, Committee on Standardization of Growth Computations. 1953. Volume tables for permanent sample plots as recommended by the Puget Sound Research Center Advisory Committee for use in western Washington. U.S. Dep. Agric. For. Serv., Pac. N.W. For. Range Exp. Sta., Portland, Oreg.

Quine, C.P. 1994. An improved understanding of windthrow – moving from hazard towards risk. For. Comm., Edinburgh, U.K. Res. Inf. Note 257.

–. 1995. Assessing the risk of wind damage; practice and pitfalls. *In* Trees and wind. M.P. Coutts and J. Grace (editors). Cambridge Univ. Press, Cambridge, U.K.

Quine, C.P., M. Coutts, B. Gardiner, and G. Pyatt. 1995. Forest and wind: management to minimise damage. For. Comm., Farnham, U.K. Bull. 114.

Quinlan, S.E. and J.H. Hughes. 1990. Location and description of a Marbled Murrelet tree nest site in Alaska. Condor 92:1068-1073.

Radcliffe, G., B. Bancroft, G. Porter, and C. Cadrin. 1994. Biodiversity of the Prince Rupert Forest Region. B.C. Min. For., Victoria, B.C. Land Manage. Rep. 82.

Raedeke, K.J. (editor). 1988. Streamside management: riparian wildlife and forestry interactions. Univ. Wash., Inst. For. Resour., Seattle, Wash. Contrib. 59.

Rafinesque, C.S. 1832. *Abies falcata*. Atlan. J. 1:120.

Ratcliffe, P.R. and G.F. Peterken. 1995. The potential for biodiversity in British upland spruce forests. For. Ecol. Manage. 79:153-160.

Recreation Branch. 1992. An inventory of undeveloped watersheds in British Columbia. B.C. Min. For., Victoria, B.C. Tech. Rep. 1992:2.

Redfern, D.B. 1982. Infection of *Picea sitchensis* and *Pinus contorta* stumps by basidiospores of *Heterobasidion annosum*. Eur. J. For. Path. 12(1):11-25.

–. 1989. Factors affecting infection of Sitka spruce stumps by *Heterobasidion annosum* and the implications for disease development. *In* Proc. of the seventh international conference on root and butt rots, August 1988, Victoria, B.C. IUFRO. D.J. Morrison (editor). For. Can., Pac. For. Cent., Victoria, B.C., pp. 297-307.

–. 1993. The effect of wood moisture on infection of Sitka spruce stumps by basidiospores of *Heterobasidion annosum*. Eur. J. For. Path. 23:218-235.

Redfern, D.B. and M.G.R. Cannell. 1982. Needle damage in Sitka spruce caused by early autumn frosts. Forestry 55:39-45.

Reynolds, G. and G.W. Wallis. 1966. Seasonal variation in spore deposition of *Fomes annosus* in coastal forests of British Columbia. Can. For. Serv. Bi-Month. Res. Notes 22:6-7.

Reynolds, P.E., D.G. Pitt, R. Whitehead, and K. King. 1989. Three-year herbicide efficacy, crop tolerance and crop growth response results for a 1984 glyphosate conifer release trial at Carnation Creek, British Columbia. *In* Proceedings of the Carnation Creek Herbicide Workshop. P.E. Reynolds (editor). For. Can. and B.C. Min. For., Victoria, B.C. FRDA Rep. 63. pp. 141-167.

Reynolds, P.E., J.C. Scrivener, L.B. Holtby, and P.D. Kingsbury. 1993. Review and synthesis of Carnation Creek herbicide research. For. Chron. 69:323-330.

Richardson, J., J.P. Hall, and R.S. van Nostrand. 1984. North Pond: demonstrations of reforestation research results. Can. For. Serv., Nfld. For. Res. Cent., St. John's, Nfld. Inf. Rep. N-X-227.

Ripple, W.J. 1994. Historic spatial patterns of old forests in western Oregon. J. For. 92:45-49.
Ripple, W.J., D.H. Johnson, K.T. Hershey, and E.C. Meslow. 1991. Old-growth and mature forests near spotted owl nests in western Oregon. J. Wildl. Manage. 55:316-318.
Robak, H. 1957. The relation between day length and the end of the annual growth period in some conifers of interest to Norwegian forestry. Medd. Vestlandet. Forstl. Forsoksta. 31:62.
Robison, E.G. and R.L. Beschta. 1990. Identifying trees in riparian areas that can provide coarse woody debris to streams. For. Sci. 36:790-801.
Roche, L. 1969. A genecological study of the genus *Picea* in British Columbia. New Phytol. 68:505-554.
Roche, L. and D.P. Fowler. 1975. Genetics of Sitka spruce. U.S. Dep. Agric. For. Serv., Washington, D.C. Res. Pap. WO-26.
Rodway, M.S., H.M. Regehr, and J.-P.L. Savard. 1993. Activity levels of Marbled Murrelets in different inland habitats in the Queen Charlotte Islands, British Columbia. Can. J. Zool. 71:977-984.
Rodway, M.S., J.-P.L. Savard, and H.M. Regehr. 1991. Habitat use and activity patterns of Marbled Murrelets at inland and at-sea sites in the Queen Charlotte Islands, British Columbia. Can. Wildl. Serv., Pac. & Yukon Reg., Delta, B.C. Tech. Rep. Ser. 122.
Roemer, H.L., J. Pojar, and K.R. Joy. 1988. Protected old-growth forests in coastal British Columbia. Natural Areas J. 8(3):146-159.
Roff, J.W. and H.W. Eades. 1959. Deterioration of logging residue on the British Columbia coast. Can. For. Serv., For. Prod. Lab., Vancouver, B.C. Tech. Note 11.
Rogers, P. 1996. Disturbance ecology and forest management: a review of the literature. U.S. Dep. Agric. For. Serv., Gen. Tech. Rep. INT-GTR-336.
Rollerson, T.P. 1980. Queen Charlotte Woodlands Division windthrow study. MacMillan Bloedel Ltd., Woodlands Services, Nanaimo, B.C.
–. 1989. Evaluation of windthrow potential, Carmanah Creek spruce reserve. MacMillan Bloedel Limited Management Plan Carmanah Valley. MacMillan Bloedel Ltd., Woodlands Services Division, Nanaimo, B.C. Folio II.
Rollinson, T.J.D. 1985. Thinning control. 3rd ed. For. Comm., Edinburgh, U.K. Booklet 54.
–. 1988. Respacing Sitka spruce. Forestry 61(1):1-22.
Rook, D.A. (editor). 1992. Super Sitka for the 90s. For. Comm., London, U.K. Bull. 103.
Rose, C.L. 1982. Response of deer to forest succession on Annette Island, southeastern Alaska. *In* Fish and wildlife relationships in old-growth forests, Proc. Symp. 12-15 April 1982, Juneau, Alaska. W.R. Meehan, T.R. Merrell, Jr., and T.A. Hanley (editors). Am. Inst. Fish. Res. Biol., Morehead City, N.C., pp. 285-290.
Rosskam, C.S. and C. Hyde. 1995. FRDA integrated resource management research 1994 update. FRDA Memo 225. pp. 3 and 6.
Roth, A.L. 1989. Mycorrhizal inoculum potential in forest clearcuts of the Pacific Northwest. *In* Mycorrhizal status of conifer seedlings before and after outplanting on eastern Vancouver Island. A.L. Roth and S.M. Berch. Prep. for B.C. Min. For., Victoria, B.C., pp. 69-81.
Roulund, H. 1971. Experiments with cuttings of *Picea abies, Picea sitchensis* and the hybrid *Picea omorika* × *Picea sitchensis*. For. Tree Improve., Arboretet Horsholm 3:25-27.
Ruggiero, L.F., K.B. Aubry, S.W. Buskirk, L.J. Lyon, and W.J. Zielinski. 1994. The scientific basis for conserving forest carnivores: American marten, fisher, lynx, and wolverine in the western United States. U.S. Dep. Agric. For. Serv., Gen. Tech. Rep. RM-254.
Russell, K. 1978. Mycorrhizae. *In* Regenerating Oregon's forests: a guide for the regeneration forester. B.D. Cleary, R.D. Greaves, and R.K. Hermann (editors). Oreg. State Univ. Exten. Serv., Corvallis, Oreg., pp. 277-278, Appendix G.
Russell, K. 1989. Mycorrhizal inoculum potential in forest clearcuts of the Pacific Northwest. *In* Mycorrhizal status of conifer seedlings before and after outplanting on eastern Vancouver Island. A.L. Roth and S.M. Berch. Prep. for B.C. Min. For., Victoria, B.C., pp. 69-71, Appendix 3.1.
Ruth, D.S., G.E. Miller, and J.R. Sutherland. 1982. A guide to common insect pests and diseases in spruce seed orchards in British Columbia. Can. For. Serv., Pac. For. Res. Cent., Victoria, B.C. Inf. Rep. BC-X-231.

Ruth, R.H. 1954. Cascade Head climatological data 1936 to 1952. U.S. Dep. Agric. Pac. N.W. For. Range Exp. Sta., Portland, Oreg.

–. 1958. Silvical characteristics of Sitka spruce. U.S. Dep. Agric. For. Serv., Pac. N.W. For. Range Exp. Sta., Portland, Oreg. Silv. Series 8.

–. 1964. Silviculture of the coastal Sitka spruce–western hemlock type. *In* Proc. Soc. Am. For. 1964, Denver, Colo., pp. 32-36.

Ruth, R.H. and C.M. Berntsen. 1955. A 4-year record of Sitka spruce and western hemlock seed fall on the Cascade Head Experimental Forest. U.S. Dep. Agric. For. Serv., Res. Pap. PNW-12.

Ruth, R.H. and A.S. Harris. 1979. Management of western hemlock–Sitka spruce forests for timber production. U.S. Dep. Agric. For. Serv., Gen. Tech. Rep. PNW-88.

Ruth, R.H. and R.A. Yoder. 1953. Reducing wind damage in the forests of the Oregon Coast Range. U.S. Dep. Agric. For. Serv., Res. Pap. PNW-7.

Sadoway, K.L. 1988. Effects of intensive forest management on breeding birds of Vancouver Island: problem analysis. B.C. Min. Environ. and Parks and Min. For., Victoria, B.C. IWIFR-25.

Sahota, T.S. 1993. Spruce weevil – enemy of reforestation. Infor. For. (Pac. & Yukon Reg.) March 1993, pp. 4-5.

Sahota, T.S., J.F. Manville, and E. White. 1994a. Interaction between Sitka spruce weevil and its host, *Picea sitchensis* (Bong.) Carr.: a new mechanism for resistance. Can. Entomol. 126:1067-1074.

Sahota, T.S., J.F. Manville, E. White, and A. Ibaraki. 1994b. Towards an understanding of Sitka spruce resistance against *Pissodes strobi*. *In* The white pine weevil: biology, damage and management. R.I. Alfaro, G.K. Kiss, and R.G. Fraser (editors). Can. For. Serv. and B.C. Min. For., Victoria, B.C. FRDA Rep. 226. pp. 7-22.

Sanders, P.R.W. and D.J. Wilford. 1986. Silvicultural alternatives for the management of unstable sites in the Queen Charlotte Islands: a literature review and recommendations. Can. For. Serv. and B.C. Min. For., Victoria, B.C. Land Manage. Rep. 42.

Sauder, E.A. and G.V. Wellburn. 1987. Studies of yarding operations on sensitive terrain, Queen Charlotte Islands, B.C. B.C. Min. For., Victoria, B.C. Land Manage. Rep. 52; FERIC Spec. Rep. SR-43.

–. 1989. Planning logging: two case studies on the Queen Charlotte Islands, B.C. B.C. Min. For., Victoria, B.C. Land Manage. Rep. 59.

Savill, P.S. 1976. The effects of drainage and ploughing of surface water gleys on rooting and windthrow of Sitka spruce in Northern Ireland. Forestry 49(2):133-142.

Schober, R. 1962. Die Sitka-Fichte: ein biologisch-ertragskundliche Untersuchung. Schrift. Forst. Fak. Univer. Göttingen, Mittel. Niedersach. Forst. Versuch. Band 24/25.

Schoen, J.W., M.D. Kirchhoff, and J.H. Hughes. 1988. Wildlife and old-growth forests in southeastern Alaska. Nat. Areas. J. 8:138-145.

Schreiner, E.J. 1937. Improvement of forest trees. U.S. Dep. Agric., Washington, D.C. Yearb. pp. 1242-1279.

Scott, G.R. 1969. Some morphological and physiological differences between normal Sitka spruce and a yellow mutant. BSF thesis, Univ. B.C., Vancouver, B.C.

Seaby, D.A. and D.J. Mowat. 1993. Growth changes in 20-year-old Sitka spruce *Picea sitchensis* after attack by the green spruce aphid *Elatobium abietinum*. Forestry 66:371-379.

Sedell, J.R., P.A. Bisson, F.J. Swanson, and S.V. Gregory. 1988. What we know about large trees that fall into streams and rivers. *In* From the forest to the sea: a story of fallen trees. C. Maser, R.F. Tarrant, J.M. Trappe, and J.F. Franklin (technical editors). U.S. Dep. Agric. For. Serv., Gen. Tech. Rep. PNW-GTR-229. pp. 47-81.

Shaw, C.G., III. 1989. Root disease threat minimal in young stands of western hemlock and Sitka spruce in southeastern Alaska. Plant Dis. 73(7):573-577.

Shaw, C.G., III, R.M. Jackson, and G.W. Thomas. 1984. Effects of fertilization and fungal strain on ectomycorrhizal development of Sitka spruce seedlings. *In* Proc. 6th North Am. conf. on mycorrhizae, 25-29 June 1984, Bend, Oreg. R. Molina (editor). Oreg. State Univ., Coll. For., For. Res. Lab., Corvallis, Oreg., p. 217.

Shaw, C.G., III, R. Molina, and J. Walden. 1982. Development of ectomycorrhizae following inoculation of containerized Sitka and white spruce seedlings. Can. J. For. Res. 12:191-195.

Shaw, C.G., III, R.C. Sidle, and A.S. Harris. 1987. Evaluation of planting sites common to a southeast Alaska clear-cut. III. Effects of microsite type and ectomycorrhizal inoculation on growth and survival of Sitka spruce seedlings. Can. J. For. Res. 17:334-339.

Sheppard, L.J. and M.G.R. Cannell. 1985. Performance and frost hardiness of *Picea sitchensis* × *Picea glauca* hybrids in Scotland. Forestry 58:67-74.

Sidle, R.C. 1991. A conceptual model of changes in root cohesion in response to vegetation management. J. Environ. Qual. 20:43-52.

Sidle, R.C. and C.G. Shaw III. 1987. Evaluation of planting sites common to a southeast Alaska clearcut. 1. Nutrient levels in ectomycorrhizal Sitka spruce seedlings. Can. J. For. Res. 17:340-345.

Sierra Club of Western Canada. 1992. The rainforest story: the ancient temperate rainforests of British Columbia. Sierra Club of Western Canada, Victoria, B.C.

Sigurgeirsson, A., A.E. Szmidt, and J.N. Alden. 1990. A molecular study of interspecific hybridization in the spruce complex of Alaska. *In* Joint Meeting of Western Forest Genetics Association and IUFRO Working Parties S2.02-05, 06, 12, and 14. 20-24 Aug. 1990, Olympia, Wash., pp. 6/14-23.

Silver, G.T. 1968. Studies on the Sitka spruce weevil, *Pissodes sitchensis,* in British Columbia. Can. Entomol. 100:93-100.

Silviculture Interpretations Working Group. 1994. Correlated guidelines for tree species selection (first approximation) and stocking standards (second approximation) for the ecosystems of British Columbia. 2nd rev. ed. For. Can. and B.C. Min. For., Victoria, B.C. FRDA II Rep.

Singer, S.W., N.L. Naslund, S.A. Singer, and C.J. Ralph. 1991. Discovery and observation of two tree nests of the Marbled Murrelet. Condor 93:330-339.

Skinner, E.C. 1959. Cubic volume tables for red alder and Sitka spruce. U.S. Dep. Agric. For. Serv., Res. Note PNW-170.

Smith, J.H.G. 1964. Root spread can be estimated from crown width of Douglas-fir, lodgepole pine and other British Columbia tree species. For. Chron. 40:456-473.

Smith, J.H.G. and J.A. McLean. 1993. Methods are needed to prevent devastation of Sitka spruce plantations by the Sitka spruce weevil *Pissodes strobi* Peck. *In* Proc. of the IUFRO international Sitka spruce provenance experiment (Sitka spruce Working Group S2.02.12). IUFRO 1984, Edinburgh, Scotland. C.C. Ying and L.A. McKnight (editors). B.C. Min. For., Victoria, B.C., and Irish For. Board, County Wicklow, Ire., pp. 81-93.

Smith, N.J. 1993. Estimating site index in TFL 39 and TFL 44. MacMillan Bloedel Ltd., Nanaimo, B.C. Unpubl. rep.

Smith, R.B. 1974. Infection and development of dwarf mistletoes on plantation-grown trees in British Columbia. Can. For. Serv., Pac. For. Res. Cent., Victoria, B.C. Inf. Rep. BC-X-97.

Smith, R.B., P.R. Commandeur, and M.W. Ryan. 1984. Vegetative succession, soil development and forest productivity on landslides, Queen Charlotte Islands, British Columbia, Canada. *In* Effects of forest land use on erosion and slope stability, IUFRO Symp., 7-11 May 1984, Honolulu, Hawaii, pp. 109-116.

–. 1986. Soils, vegetation, and forest growth on landslides and surrounding logged and old-growth areas on the Queen Charlotte Islands. B.C. Min. For., Victoria, B.C. Land Manage. Rep. 41.

Smith, R.B., W. Hays, and R.K. King. 1988. Some implications of vegetative changes induced by forest management. *In* Proceedings of the workshop: applying 15 years of Carnation Creek results. T.W. Chamberlin (editor). Pac. Biol. Sta., Nanaimo, B.C., pp. 93-98.

Smith, S.M. 1990. New models for yield estimation and projection. B.C. Min. For., Inven. Br., Victoria, B.C.

Society of American Foresters. 1980. Forest cover types of the United States and Canada. Soc. Am. Foresters, Washington, D.C.

Sommers, G.L. 1983. An analysis of the effect of the spruce terminal weevil (*Pissodes strobi* (Peck)) on Sitka spruce (*Picea sitchensis* (Bong.) Carr.). BSF thesis, Univ. B.C., Vancouver, B.C.

South Moresby Forest Replacement Account (SMFRA). 1994. Year 6, 1993/94 annual report. B.C. Min. For. and Can. For. Serv., Victoria, B.C.

Spalt, K.W. and W.E. Reifsnyder. 1962. Bark characteristics and fire resistance: a literature review. U.S. Dep. Agric. For. Serv., Occas. Pap. S-193.

Squillace, A.E. 1976. Analysis of monoterpenes of conifers by gas-liquid chromatography. *In* Modern methods in forest genetics. J.B. Milsche (editor). Springer-Verlag, New York, N.Y., pp. 120-158.

Stage, A.R. 1973. Prognosis model for stand development. U.S. Dep. Agric. For. Serv., Res. Pap. INT-137.

Stanek, W. and V.J. Krajina. 1964. Preliminary report on some ecosystems of the western coast on Vancouver Island. *In* Ecology of the forests of the Pacific Northwest. V.J. Krajina (editor). Univ. B.C., Dep. Biol. Bot., Progress Rep., Nat. Res. Coun. Grant No. T-92, Vancouver, B.C., pp. 57-66.

Stathers, R.J. 1989. Summer frost in young forest plantations. For. Can. and B.C. Min. For., Victoria, B.C. FRDA Rep. 73.

Stathers, R.J., T.P. Rollerson, and S.J. Mitchell. 1994. Windthrow handbook for British Columbia forests. B.C. Min. For., Victoria, B.C. Res. Prog. Work. Pap. 9401.

Stephens, F.R., C.R. Gass, and R.F. Billings. 1969. Soils and site index in southeast Alaska. Report number two of the soil-site index administrative study. U.S. Dep. Agric. For. Serv., Alaska Reg., Juneau, Alaska.

Stevens, V. 1995a. Database for wildlife diversity in British Columbia: distribution and habitat use of amphibians, reptiles, birds, and mammals in biogeoclimatic zones. B.C. Min. For., Victoria, B.C. Work. Pap. 5/1995.

–. 1995b. Wildlife diversity in British Columbia: distribution and habitat use of amphibian, reptiles, birds, and mammals in biogeoclimatic zones. B.C. Min. For., Victoria, B.C. Work. Pap. 4/1995.

Stevens, V., F. Backhouse, and A. Eriksson. 1995. Riparian management in British Columbia: an important step towards maintaining biodiversity. B.C. Min. For., Res. Br., and B.C. Min. Environ., Lands and Parks, Wildl. Br., Victoria, B.C. Work. Pap. 13/1995.

Steventon, J.D. 1994. Biodiversity and forest management in the Prince Rupert Forest Region: a discussion paper. *In* Biodiversity of the Prince Rupert Forest Region, and Biodiversity and forest management in the Prince Rupert Forest region: a discussion paper. B.C. Min. For., Victoria, B.C. Land Manage. Rep. 82 (part 2).

Stirling, J. 1996. Deer thriving on seedling diet. Logging and Sawmilling J. 27(1):33-34.

Stoltmann, R. 1993. Guide to the record trees of British Columbia. West. Can. Wilderness Comm., Vancouver, B.C.

Stone, E.L. and M.H. Stone. 1943. 'Dormant' versus 'adventitious' buds. Science 98:62.

Stone, M., S. Grout, and C. Watmough. [1996]. The TIPSY economist: an economic analysis module for WinTIPSY. B.C. Min. For., Victoria, B.C. (in prep.).

Straw, N.A. 1995. Climate change and the impact of green spruce aphid, *Elatobium abietinum* (Walker), in the U.K. Scott. For. 49:134-144.

Sudworth, G.B. 1908. Forest trees of the Pacific slope. U.S. For. Serv., Washington, D.C.

Sullivan, T.P. 1990. Reducing mammal damage to plantations and juvenile stands in young forests of British Columbia. FRDA Memo 153:1-12.

Sullivan, T.P., W.T. Jackson, J. Pojar, and A. Banner. 1986. Impact of feeding damage by the porcupine on western hemlock–Sitka spruce forests of north-coastal British Columbia. Can. J. For. Res. 16:642-647.

Sutherland, J.R. and R.S. Hunt. 1990. Diseases in reforestation. *In* Regenerating British Columbia's forests. D.P. Lavender, R. Parish, C.M. Johnson, G. Montgomery, A. Vyse, R.A. Willis, and D. Winston (editors). Univ. B.C. Press, Vancouver, B.C., pp. 266-278.

Sutherland, J.R. and W. Lock. 1977. Fungicide-drenches ineffective against damping-off of Sitka and white spruces. Can. For. Serv., Bi-Monthly Res. Notes 33(1):6-7.

Sutherland, J.R., T. Miller, and R.S. Quinard. 1987. Cone and seed diseases of North American conifers. Can. For. Serv., Intern. For. Branch, Victoria, B.C. N. Am. For. Comm. Publ. 1.

Sutherland, J.R., G.M. Shrimpton, and R.N. Sturrock. 1989. Diseases and insects in British Columbia forest seedling nurseries. For. Can. and B.C. Min. For., Victoria, B.C. FRDA Rep. 65.

Sutton, B.C.S., D.J. Flanagan, and Y.A. El-Kassaby. 1991. A simple and rapid method for estimating representation of species in spruce seedlots using chloroplast DNA restricted fragment length polymorphism. Silvae Genet. 40:119-123.

Sutton, B.C.S., S.C. Pritchard, J.R. Gawley, and C.H. Newton. 1994. Analysis of Sitka spruce–interior spruce introgression in British Columbia using cytoplasmic and nuclear DNA probes. Can. J. For. Res. 24:278-285.

Swanson, F.J. 1994. Natural disturbance effects on riparian areas. *In* Riparian resources, a symposium on the disturbances, management, economics, and conflicts associated with riparian ecosystems, 18-19 April 1991, Logan, Utah. G.A. Rasmussen and J.P. Dobrowolski (editors). Nat. Resour. Environ. Issues 1:11-14.

Swanson, F.J., S.V. Gregory, J.R. Sedell, and A.G. Campbell. 1982. Land-water interactions: the riparian zone. *In* Analysis of coniferous forest ecosystems in the western United States. R.L. Edmonds (editor). Hutchinson Ross Pub., Stroudsburg, Penn., pp. 267-290.

Swanson, F.J. and G.W. Lienkaemper. 1982. Interactions among fluvial processes, forest vegetation, and aquatic ecosystems, south fork Hoh River, Olympic National Park. *In* Ecological research in national parks of the Pacific Northwest. Proc. of 2nd conf. on scientific research in national parks, Nov. 1979, San Francisco, Calif. E.E. Starkey, J.F. Franklin, and J.W. Mathews (technical coordinators). Oreg. State Univ. For. Res. Lab., Corvallis, Oreg., pp. 30-34.

Swanston, D.N. 1974. The forest ecosystem of southeast Alaska. 5. Soil mass movement. U.S. Dep. Agric. For. Serv., Gen. Tech. Rep. PNW-17.

Tabbush, P.M. 1986. Rough handling, soil temperature, and root development in outplanted Sitka spruce and Douglas-fir. Can. J. For. Res. 16:1385-1388.

–. 1987a. Effect of desiccation on water status and forest performance of bare-rooted Sitka spruce and Douglas-fir transplants. Forestry 60(1):31-43.

–. 1987b. Safe dates for handling and planting Sitka spruce and Douglas fir. For. Comm., Farnham, U.K. Res. Inf. Note 118.

Taylor, A.F.S. and I.J. Alexander. 1990. Demography and population dynamics of ectomycorrhizas of Sitka spruce fertilized with N. Agric. Ecosys. Environ. 28:493-496.

Taylor, A.H. 1990. Disturbance and persistence of Sitka spruce (*Picea sitchensis* (Bong.) Carr.) in coastal forests of the Pacific Northwest, North America. J. Biogeogr. 17:47-58.

Taylor, C.M.A. and P.M. Tabbush. 1990. Nitrogen deficiency in Sitka spruce plantations. For. Comm., Farnham, U.K. Bull. 89.

Taylor, R.F. 1934. Available nitrogen as a factor influencing the occurrence of Sitka spruce and western hemlock seedlings in the forests of southeastern Alaska. PhD thesis. Yale Univ., New Haven, Conn.

Taylor, R.L. and B. MacBryde. 1977. Vascular plants of British Columbia: a descriptive resource inventory. Univ. B.C. Press, Vancouver, B.C. Univ. B.C. Bot. Gardens Tech. Bull. 4.

Taylor, R.L. and S. Taylor. 1973. *Picea sitchensis* (Bongard) Carriere – Sitka spruce. Davidsonia 4(4):41-45.

Thaarup, P. 1945. Bastarden Sitkagran × Hvidgran. Dansk Skovforen. Tidsskr. 30:381-384.

Thedinga, J.F. and K.V. Koski. 1982. A stream ecosystem in old-growth forest in southeast Alaska. Part VI. The production of coho salmon, *Oncorhynchus kisutch,* smolts and adults from Porcupine Creek. *In* Fish and wildlife relationships in old-growth forests, Proc. Symp. 12-15 April 1982, Juneau, Alaska. W.R. Meehan, T.R. Merrell, Jr., and T.A. Hanley (editors). Am. Inst. Fish. Res. Biol., Morehead City, N.C., pp. 99-108.

Thies, W.G., K.M. Russell, and L.C. Weir. 1977. Distribution and damage appraisal of *Rhizina undulata* in western Oregon and Washington. Plant Dis. Rep. 61:859-862

–. 1979. Rhizina root rot of little consequence in Washington and Oregon. J. For. 77(1):22-24.

Thies, W.G. and R.N. Sturrock. 1995. Laminated root rot in western North America. U.S. Dep. Agric. For. Serv., Gen. Tech. Rep. PNW-GTR-349.

Thomas, G.W. and R.M. Jackson. 1983. Growth response of Sitka spruce seedlings to mycorrhizal inoculation. New Phytol. 95:223-229.

Thomas, G.W., D. Rogers, and R.M. Jackson. 1983. Changes in the mycorrhizal status of Sitka spruce following outplanting. Plant Soil 71:319-323.
Thomas, R.C. and H.G. Miller. 1994. The interaction of green spruce aphid and fertilizer applications on the growth of Sitka spruce. Forestry 67:329-342.
Thompson, D.A. 1992. Growth of Sitka spruce and timber quality. *In* Super Sitka for the 90s. D.A. Rook (editor). For. Comm., London, U.K. Bull. 103. pp. 54-60.
Thrower, J.S. and A.F. Nussbaum. 1991. Site index curves and tables for British Columbia: coastal species. B.C. Min. For., Victoria, B.C. Land Manage. Handb. Fld. Guide Insert 3.
Tinus, R.W. 1970. Growing seedlings in a controlled environment. West. For. Conserv. Assoc. West. Reforest. Coord. Comm. Proc. 1970:34-37.
Tomlin, E.S., J.H. Borden, and H.D. Pierce, Jr. 1996. Relationship between cortical resin acids and resistance of Sitka spruce to the white pine weevil. Can. J. Bot. 74:599-606.
Townend, J. and A.L. Dickinson. 1995. A comparison of rooting environments in containers of different sizes. Plant Soil 175:139-146.
Trappe, J.M. 1962. Fungus associates of ectotropic mycorrhizae. Bot. Rev. 38:538-606.
Tripp, D.B. and V.A. Poulin. 1992. The effects of logging and mass wasting on juvenile salmonid populations in streams on the Queen Charlotte Islands. B.C. Min. For., Victoria, B.C. Land Manage. Rep. 80.
Trofymow, J.A. and R. van den Driessche. 1991. Mycorrhizas. *In* Mineral nutrition of conifer seedlings. R. van den Driessche (editor). CRC Press, Boca Raton, Fla., pp. 183-227.
Turner, N.J. 1979. Plants in British Columbia Indian technology. B.C. Min. Prov. Sec. Gov. Serv., Victoria, B.C. Handb. 38.
–. 1995. Food plants of coastal First Peoples. Univ. B.C. Press, Vancouver, B.C. Roy. B.C. Museum. Handb. 34.
Turner, N.J., L.C. Thompson, M.T. Thompson, and A.Z. York. 1990. Thompson ethnobotany: knowledge and usage of plants by the Thompson Indians of British Columbia. Roy. B.C. Museum, Victoria, B.C. Mem. 3.
Turnquist, R.D. and R.I. Alfaro. 1996. Spruce weevil in British Columbia. Can. For. Serv., Pac. For. Cent., Victoria, B.C. For. Pest Leafl. 2.
Ugolini, F.C., B.T. Bormann, and F.H. Bowers. 1989. The effect of treethrow on forest soil development in southeast Alaska. Univ. Wash., Coll. For. Resour., Seattle, Wash. Unpubl. rep.
Vaartaja, O. 1959. Evidence of photoperiodic ecotypes in trees. Ecol. Monogr. 29:91-111.
Vadla, K. 1990a. Tidsforbruk ved stammekvisting av sitkagran (*Picea sitchensis* (Bong.) Carr.) [Consumption of time when pruning Sitka spruce (*Picea sitchensis* (Bong.) Carr.)]. Rapp. Nor. Inst. Skogforsk. 3/90:1-21.
–. 1990b. Overvoksing av kvistsår etter stammekvisting av sitkagran (*Picea sitchensis* (Bong.) Carr.) [Healing after pruning of Sitka spruce (*Picea sitchensis* (Bong.) Carr.)]. Rapp. Nor. Inst. Skogforsk. 13/90:1-19.
van Barneveld, J.W., M. Rafiq, G.F. Harcombe, and R.T. Ogilvie. 1980. An illustrated key to gymnosperms of British Columbia. B.C. Min. Environ. and Min. Prov. Sec. Gov. Serv., Victoria, B.C.
Van Ballenberghe, V. and T.A. Hanley. 1982. Predation on deer in relation to old-growth forest management in southeastern Alaska. *In* Fish and wildlife relationships in old-growth forests, Proc. Symp. 12-15 April 1982, Juneau, Alaska. W.R. Meehan, T.R. Merrell, Jr., and T.A. Hanley (editors). Am. Inst. Fish. Res. Biol., Morehead City, N.C., pp. 291-296.
van den Driessche, R. 1983. Rooting of Sitka spruce cuttings from hedges, and after chilling. Plant Soil 71:495-499.
vandenBrink, M. 1992a. Fall and winter wildlife use of precommercial thinned second growth forests of the Queen Charlotte Islands, B.C. Prep. for Min. Environ., Lands and Parks, Queen Charlotte City, B.C.
–. 1992b. Winter wildlife use of riparian habitats of the Queen Charlotte Lowlands and the Queen Charlotte Ranges ecoregions. Prep. for Min. Environ., Lands and Parks, Queen Charlotte City, B.C.
van der Kamp, B. 1991. Pathogens as agents of diversity in forested landscapes. For. Chron. 67:353-354.

VanderSar, T.J.D. 1978. Resistance of western white pine to feeding and oviposition by *Pissodes strobi* Peck in western Canada. J. Chem. Ecol. 4:641-647.

VanderSar, T.J.D. and J.H Borden. 1977a. Aspects of host selection behaviour of *Pissodes strobi* (Coleoptera: Curculionidae) as revealed in laboratory feeding bioassays. Can. J. Zool. 55(2):405-414.

–. 1977b. Visual orientation of *Pissodes strobi* (Coleoptera: Curculionidae) in relation to host selection behaviour. Can. J. Zool. 55(12):2042-2049.

Van de Sype, H. and B. Roman-Amat. 1990. Genetic variability of Sitka spruce of the IUFRO collection. *In* Joint Meeting of Western Forest Genetics Association and IUFRO Working Parties S2.02-05, 06, 12, and 14. 20-24 Aug. 1990, Olympia, Wash., pp. 2/333-343.

van Hees, W.W.S. 1988. Timber productivity of seven forest ecosystems in southeastern Alaska. U.S. Dep. Agric. For. Serv., Res. Pap. PNW-391.

Van Sickle, G.A. 1992. A review of innovations in disease and insect management and control. For. Chron. 68:742-746.

Van Sickle, G.A. and C.S. Wood. 1996. Pacific and Yukon Region. *In* Forest insect and disease conditions in Canada 1994. J.P. Hall (editor). Can. For. Serv., Ottawa, Ont.

Viereck, L.A. and E.L. Little, Jr. 1972. Alaska trees and shrubs. U.S. Dep. Agric., Washington, D.C. Agric. Handb. 410.

–. 1975. Atlas of United States trees. Volume 2. Alaska trees and common shrubs. U.S. Dep. Agric. For. Serv., Washington, D.C. Misc. Publ. No. 1293.

von Rudloff, E.M. 1978. Variation in leaf oil terpene composition of Sitka spruce. Phytochemistry 17(1):127-130.

Walker, C. 1987. Sitka spruce mycorrhizas. Proc. Roy. Soc. Edinburgh. 93B(1-2):117-129.

Walker, C., P. Biggin, and D.C. Jardine. 1986. Differences in mycorrhizal status among clones of Sitka spruce. For. Ecol. Manage. 14:275-283.

Walstad, J.D., S.R. Radosevich, and D.V. Sandberg (editors). 1990. Natural and prescribed fire in Pacific Northwest forests. Oreg. State Univ. Press, Corvallis, Oreg.

Walters, M.A., R.O. Teskey, and T.M. Hinckley. 1980. Impact of water level changes on woody riparian and wetland communities. U.S. Dep. Inter., Fish & Wildl. Serv., FWS/OBS-78/94.

Wang, G.G. 1995. White spruce site index in relation to soil, understory vegetation and soil nutrients. Can. J. For. Res. 25:29-38.

Wang, Q., G.G. Wang, K.D. Coates, and K. Klinka 1994a. Use of site factors to predict lodgepole pine and interior spruce site index in the Sub-Boreal Spruce Zone. B.C. Min. For., Victoria. Res. Note 114.

–. 1994b. Relationships between ecological site quality and site index of lodgepole pine and white spruce in northern British Columbia. Chinese J. Appl. Ecol. 5(1):1-15.

Waring, R.H. and J.F. Franklin. 1979. Evergreen coniferous forests of the Pacific Northwest. Science 204:1380-1386.

Warkentin, D.L., D.L. Overhulser, R.I. Gara, and T.M. Hinckley. 1992. Relationships between weather patterns, Sitka spruce (*Picea sitchensis*) stress, and possible tip weevil (*Pissodes strobi*) infestation levels. Can. J. For. Res. 22:667-673.

Warrack, G.C. 1957. Natural regeneration following increased logging in the Queen Charlotte Islands. B.C. For. Serv., For. Res. Rev. 1957:25.

Watson, B. and A. Cameron. 1995. Some effects of nursing species on stem form, branching habit and compression wood content of Sitka spruce. Scott. For. 49(3):145-152.

Watts, S.B. (editor). 1983. Forestry handbook for British Columbia. 4th ed. For. Undergrad. Soc., Fac. For., Univ. B.C., Vancouver, B.C.

Weatherell, J. 1957. The use of nurse species in the afforestation of upland heaths. Quart. J. For. 51:298-304.

Weetman, G.F. 1982. Ultimate forest productivity in North America. *In* IUFRO symposium on forest site and continuous productivity. R. Ballard and S.P. Gessel (technical editors). U.S. Dep. Agric. For. Serv., Gen. Tech. Rep. PNW-163. pp. 70-80.

–. 1991. Biological uncertainty in growth and yield benefits of second-growth silviculture actions and old-growth protection and liquidation strategies in British Columbia. Univ. B.C., Vancouver, B.C. For. Econ. Pol. Analysis Res. Unit, Work. Pap. 148.

–. 1996. Are European silvicultural systems and precedents useful for British Columbia silviculture prescriptions? Can. For. Serv. and B.C. Min. For., Victoria, B.C. FRDA Rep. 239.

Weetman, G.F., R. Fournier, J. Barker, E. Schnorbus-Panozzo, and A. Germain. 1989. Foliar analysis and response of fertilized chlorotic Sitka spruce plantations on salal-dominated cedar-hemlock cutovers on Vancouver Island. Can. J. For. Res. 19:1501-1511.

Weetman, G.F., R. Fournier, E. Schnorbus-Panozzo, and J. Barker. 1990. Post-burn nitrogen and phosphorus availability of deep humus soils in coastal British Columbia cedar/hemlock forests and the use of fertilization and salal eradication to restore productivity. Sustained productivity of forest soils. *In* 7th North American Forest Soils Conf. S.P. Gessel, D.S. Lacate, G.F. Weetman, and R.F. Powers (editors). Univ. B.C., Fac. For. Publ., Vancouver, B.C., pp. 451-499.

Weetman, G.F. and A. Vyse. 1990. Natural regeneration. *In* Regenerating British Columbia's forests. D.P. Lavender, R. Parish, C.M. Johnson, G. Montgomery, A. Vyse, R.A. Willis, and D. Winston (editors). Univ. B.C. Press, Vancouver, B.C., pp. 118-129.

Weston, G.C. 1957. Indigenous v. exotic species in New Zealand forestry. *In* Exotic forest trees in New Zealand, Seventh Brit. Commonw. For. Conf., Australia and New Zealand, pp. 37-38.

White, E.E., R.F. Watkins, and D.P. Fowler. 1993. Comparative restriction site maps of chloroplast DNA of *Picea abies, Picea glauca, Picea mariana,* and *Picea sitchensis*. Can. J. For. Res. 23:427-435.

White, R.J. 1993. Old-growth forests and the Northwest's aquatic resources. Northw. Environ. J. 9:17-19.

White, R.L. 1986. Field testing of selected Sitka spruce families and clones for white pine weevil resistance. B.C. Min. For., Res. Br., Victoria, B.C. Establ. Rep. EP 702.06.

Whitford, H.N. and R.D. Craig. 1918. Forests of British Columbia. Can. Comm. Conserv., Ottawa, Ont.

Wiedemann, A.M. 1966. Contributions to the plant ecology of the Oregon coastal sand dunes. PhD thesis. Oreg. State Univ., Corvallis, Oreg.

Wilford, D.J. 1982. The sediment-storage function of large organic debris at the base of unstable slopes. *In* Fish and wildlife relationships in old-growth forests, Proc. Symp. 12-15 April 1982, Juneau, Alaska. W.R. Meehan, T.R. Merrell, Jr., and T.A. Hanley (editors). Am. Inst. Fish. Res. Biol., Morehead City, N.C., pp. 115-119.

Wilford, D.J. and J.W. Schwab. 1982. Soil mass movements in the Rennell Sound area, Queen Charlotte Islands, British Columbia. *In* Can. Hydrol. Symp., National Research Council, Fredericton, N.B., pp. 521-541.

Wilkinson, R.C., J.W. Hanover, J.W. Wright, and R.H. Flake. 1971. Genetic variation in the monoterpene composition of white spruce. For. Sci. 17:83-90.

Wilmes, L.W. 1953. De aanleg van de beplantingen en de boscomlexen in de Noordoostpolder [Tree plantings and the forest in the northeast polder]. Van Zee tot Land 9:2-63.

Wilson, E.O. and F.M. Peter. 1988. Biodiversity. National Academy Press, Washington, D.C.

Wilson, J., P.A. Mason, F.T. Last, K. Ingelby, and R.C. Munro. 1987. Ectomycorrhiza formation and growth of Sitka spruce seedlings on first-rotation forest sites in northern Britain. Can. J. For. Res. 17:957-963.

Winchester, N.N. 1993. Coastal Sitka spruce canopies: conservation of biodiversity. BioLine 11(2):9-14.

Winchester, N.N. and R.A. Ring. 1996. Northern temperate coastal Sitka spruce forests with special emphasis on canopies: studying arthropods in an unexpected frontier. Northw. Sci. 70(Spec. Issue):94-103.

Wood, C.S. 1977 revised ed. Cooley spruce gall aphid. Can. For. Serv., Pac. For. Res. Cent., Victoria, B.C. For. Pest Leafl. 6.

Wood, C.S. and G.A. Van Sickle. 1989. Forest insect and disease conditions: British Columbia and Yukon 1988. Can. For. Serv., Pac. For. Cent., Victoria, B.C. Inf. Rep. BC-X-306.

–. 1993. Forest insect and disease conditions: British Columbia and Yukon – 1992. Can. For. Serv., Pac. For. Cent., Victoria, B.C. Inf. Rep. BC-X-340.

Wood, R.F. 1955. Studies of north-west American forests in relation to silviculture in Great Britain. For. Comm., London, U.K. Bull. 25.

Woods, J. 1988. Nursery trials of Sitka–interior spruce hybrids. FRDA Memo 59.

Woods, J.H., M.U. Stoehr, and J.E. Webber. 1996. Protocols for rating seed orchard seedlots in British Columbia. B.C. Min. For., Victoria, B.C. Res. Rep. 06.

Woodward, A., E.G. Schreiner, D.B. Houston, and B.B. Moorhead. 1994. Ungulate-forest relationships in Olympic National Park: retrospective exclosure studies. Northw. Sci. 68:97-110.

Worrall, J. 1990. Subalpine larch: oldest trees in Canada? For. Chron. 66(5):478-479.

Wright, E. and L.A. Isaac. 1956. Decay following logging injury to western hemlock, Sitka spruce, and true firs. U.S. Dep. Agric. For. Serv., Washington, D.C. Tech. Bull. 1148.

Wright, J.W. 1955. Species crossability in spruce in relation to distribution and taxonomy. For. Sci. 1(4):319-349.

Wright, K.H. 1970 revised ed. Sitka-spruce weevil. U.S. Dep. Agric. For. Serv., For. Pest Leafl. 47.

Yanchuk, A.D. and D.T. Lester. 1996. Setting priorities for conservation of the conifer genetic resources of British Columbia. For. Chron. 72:406-415.

Yeh, F.C. and J.T. Arnott. 1986. Electrophoretic and morphological differentiation of *Picea sitchensis, Picea glauca,* and their hybrids. Can. J. For. Res. 16:791-798.

Yeh, F.C. and Y.A. El-Kassaby. 1980. Enzyme variation in natural populations of Sitka spruce (*Picea sitchensis*). 1. Genetic variation patterns among trees from 10 IUFRO provenances. Can. J. For. Res. 10(3):415-422.

Yeh, F.C. and S. Rasmussen. 1985. Heritability of height growth in 10-year-old Sitka spruce. Can. J. Genet. Cytol. 27(6):726-735.

Ying, C.C. 1990. Adaptive variation in Douglas-fir, Sitka spruce, and true fir: a summary of provenance research in coastal British Columbia. *In* Joint Meeting of Western Forest Genetics Association and IUFRO Working Parties S2.02-05, 06, 12, and 14. 20-24 Aug. 1990, Olympia, Wash., pp. 2/378-379.

–. 1991. Genetic resistance to white pine weevil in Sitka spruce. B.C. Min. For., Victoria, B.C. Res. Note 106.

–. 1995. Long-term provenance tests as source information for gene conservation of forest species. *In* Forest genetic resource conservation and management in Canada. T.C. Nieman, A. Mosseler, and G. Murray (compilers). Can. For. Serv., Petawawa Nat. For. Inst., Chalk River, Ont. Inf. Rep. PI-X-119. pp. 15-20.

Ying, C.C. and T. Ebata. 1994. Provenance variation in weevil attack in Sitka spruce. *In* The white pine weevil: biology, damage and management. R.I. Alfaro, G.K. Kiss, and R.G. Fraser (editors). Can. For. Serv. and B.C. Min. For., Victoria, B.C. FRDA Rep. 226. pp. 98-109.

Yole, D., T. Lewis, A. Inselberg, J. Pojar, and D. Holmes. 1989. A field guide for identification and interpretation of the Engelmann Spruce–Subalpine Fir Zone in the Prince Rupert Forest Region, British Columbia. B.C. Min. For., Victoria, B.C. Land Manage. Handb. 17.

Young, W. 1989. The Green Timbers plantations: a British Columbia forest heritage. For. Chron. 65(3):183-184.

Zasada, J.C., P.W. Owston, and D. Murphy. 1990. Field performance in southeast Alaska of Sitka spruce seedlings produced at two nurseries. U.S. Dep. Agric. For. Serv. Res. Note PNW-RN-494.

Ziemer, R.R. and D.N. Swanston. 1977. Root strength changes after logging in southeast Alaska. U.S. Dep. Agric. For. Serv. Res. Note PNW-306.

Ziller, W.G. 1974. The tree rusts of western Canada. Can. For. Serv., Pac. For. Res. Cent., Victoria, B.C. Publ. 1329.

Index

Index of Biogeoclimatic Zones,
with Subzones and Variants Listed from Driest to Wettest Precipitation Regimes within Each Zone